D0138210

APPLIED PROBABILITY MODELS

APPLIED PROBABILITY MODELS

D. L. (PAUL) MINH

California State University, Fullerton

DUXBURY

THOMSON LEARNING™

Australia • Canada • Mexico • Singapore • Spain • United Kingdom • United States

DUXBURY
THOMSON LEARNING™

Sponsoring Editor: *Curt Hinrichs*
Assistant Editor: *Seema Atwal*
Editorial Associate: *Emily Davidson*
Marketing Team: *Tom Ziolkowski, Laura Hubrich, Ericka Thompson*
Production Coordinator: *Kelsey McGee*
Production Service: *The Book Company*
Manuscript Editor: *Frank Hubert*
Permissions Editor: *Sue Ewing*
Interior Design: *Ellen Pettengell*
Cover Coordinator: *Roy R. Neuhaus*
Cover Design: *Denise Davidson*
Cover Photo: *Digital Vision*
Interior Illustration: *Lotus Arts*
Print Buyer: *Nancy Panziera*
Typesetting: *WestWords*
Cover Printing, Printing and Binding: *Phoenix Color Corp.*

For more information about this or any other Duxbury product, contact:
DUXBURY
511 Forest Lodge Road
Pacific Grove, CA 93950 USA
www.duxbury.com
1-800-423-0563 (Thomson Learning Academic Resource Center)

Printed in United States of America

10 9 8 7 6 5 4 3 2 1

Library of Congress Cataloging-in-Publication Data
Minh, D. L. (Do Le), (date)
Applied probability models / D. L. Minh
p. cm.
Includes index.
ISBN 0-534-38157-X
1. Probabilities I. Title
QA273 .M552 2000 00-040353
519.2--dc21

To my wife, Tina, and my two sons, David and Timothy

CONTENTS IN BRIEF

CONTENTS

CHAPTER 1 Review of Probability Theory 1

CHAPTER 2 Discrete-Time Markov Chains 32

CHAPTER 3 Transient and Limiting Results 59

PREFACE

Applied Probability Models introduces elements of probability models and their applications. I have tried carefully to balance theory with the art of application to real-life phenomena.

AUDIENCE AND PREREQUISITES

The book is geared toward upper-division undergraduate students and graduate students in applied disciplines that include engineering, management science, computer science, economics, finance, biology, and psychology. Students majoring in mathematics will also find the book a welcome resource.

Understanding the material requires only a modest level of college mathematics. Basic knowledge of matrix algebra and calculus is sufficient. Although a course in elementary probability is desirable, it is not necessary.

UNIQUE FEATURES

I took great pains to make the material in *Applied Probability Models* interesting, relevant, and useful to students and professionals in applied fields. To this end, this book has the following features:

- I introduce numerous real-life examples from many disciplines to illustrate the material being discussed. The inside front cover displays a list of these examples and the text pages where they may be found. Within the text, the icon ■ indicates the end of each of these examples. This wide selection of applications should motivate students and professionals to look to their own fields of interest to find applications of the concepts presented in the text. Data from these models can be downloaded in Microsoft® Excel format from our website at *http://www.duxbury.com*.

- A major goal of this text is to train students to think probabilistically. Consequently, I omit complicated proofs of uniqueness, existence, and convergence that tend to distract students from the main thrust of the book. I support most other results with a measured level of proof, neither too rigorous nor too casual, that represents a middle-ground approach. I present proofs in a way that will assist the student to explore the model's structure, visualize its development, discover the results, and provide intuitive interpretations of the results. In many instances, I discuss and prove both intermediate and final results from different points of view. The interrelating of seemingly different models such as absorbing chains and irreducible chains helps students gain a more unified view of the subject. The main tools I use throughout the book are conditional and regenerative arguments.
- The approximately 500 problems in the text are distributed evenly among the chapters. Many have multiple parts. Most of the problems are theoretical in nature, not just hypothetical vignettes that require students to provide numerical answers from existing results. They give a balance with a more leisurely discussion in the exposition and provide a challenge to serious students. Many of these problems require students to prove the same result in different ways, which leads to a better understanding of relationships between techniques.

  Although all problems can be solved without additional resources, some problems can be quite time-consuming because they are very detailed. The solutions of approximately one-third of the problems are available on our website at *http://www.duxbury.com*. An instructor's manual that contains the solutions to all the problems is available free of charge to instructors who adopt the book.

ORGANIZATION AND COVERAGE

The material in this text can be covered in a one-semester course. Although I designed it to be studied in sequence, instructors can omit some material for a one-quarter course. Chapters 1, 7, and 13 are the most likely candidates for omission. In Chapter 6, Sections 6.8 through 6.13 may also be omitted.

Material in early chapters is presented in a leisurely manner, but the pace gradually quickens in the latter part of the book. A brief overview of the chapter material follows.

- Chapter 1 reviews basic probability theory concepts, especially the probability distributions and the conditional arguments.
- Chapters 2 through 6 focus on finite discrete-time chains, especially absorbing and irreducible chains, discussed in Chapters 5 and 6. My approach to absorbing chains, unlike that of most authors, is that it is not only an important stand-alone subject, but is also critical to the understanding of the regenerative nature of irreducible chains.
- Chapter 7 introduces infinite chains. Separating the classifications of Markov chains into two parts, one for finite chains (Chapter 4) and one for infinite chains at this point permits the instructor to introduce the more difficult materials related to infinite chains later or to omit them entirely.
- Chapters 8 through 12 discuss continuous-time chains. Traditionally, the continuous-time Markov chains are introduced before the semi-Markov chains because of the

relative ease of obtaining the differential equations for the Markov models. My approach here is to develop the subjects from a modeling rather than a mathematical perspective. Consequently, the continuous-time Markov chains are introduced in Chapter 11 as a special case of the semi-Markov chains discussed in Chapter 10. In this way, students can gain a better appreciation of the relationship between the two models. In Chapter 12, many queuing models are presented to illustrate continuous-time Markov chains. The student obtains the results of each model in the end-of-chapter problems.

- Chapter 13 discusses techniques for studying non-Markovian processes such as approximation and simulation. The simulation part of this chapter might be introduced along with Chapter 1 for those who prefer to teach the course using simulation as a visual tool.

ACKNOWLEDGMENTS

I am grateful to the staff of Duxbury Press, especially editor Curt Hinrichs, for constant encouragement and assistance. I also gratefully acknowledge the colleagues and students who helped to bring this book into being: Attahiru Alfa, Tran Nam Bịnh, Peter Le, Charles Pearce, Tony Shannon, Verne Stanton, and Lawrence Weill.

I thank the following reviewers for their comments and suggestions:

Christos Alexopoulos, Georgia Institute of Technology
Stergios Fotopoulos, Washington State University
Anant Godbole, Michigan Technological University
Colleen Kelly, University of Rhode Island
Jon Lee, University of Kentucky
Steven Lippman, University of California/Los Angeles
Marcel F. Neuts, University of Arizona
Wolcott Smith, Temple University
James J. Swain, University of Alabama/Huntsville
Yehuda Vardi, Rutgers University

This project was a multi-year effort. It started when my two sons were first learning arithmetic, continued through the time that they knew enough to refer to the subject as "sarcastic processes," and finally appears in print when they can understand the material themselves. I dedicate this book to my family, for their encouragement and understanding during the writing of this book.

D. L. (Paul) Minh

CHAPTER 1

Review of Probability Theory

1.1 EXPERIMENTS AND OUTCOMES

We perform many activities in life because we need to know their *outcomes*. We shower to be clean; we eat to be nourished. Not only do different activities yield different outcomes, but the same activity might also result in different kinds of outcomes, depending on our purpose in performing the experiment. Some persons work for survival, whereas others work just for the fun of it. Some people toss a die into the air to obtain a number, whereas others only want to confirm that it will eventually come down or to see whether it will land on a table or not.

In probability theory, any activity whose outcome is not known in advance with certainty is called an *experiment*, although it is not necessarily performed in any scientific laboratory. If we toss a die to see whether it will eventually come down or not, then we do not have an experiment because nature always provides a positive answer. However, if we toss a die to obtain a number, then we do have an experiment, and the knowledge of probability theory might prove be to very helpful in this case.

Actually, identically repeated activities must have the same outcome. If we use the same die and toss it in exactly the same way—with the same shaking hand, from the same starting position, into the same environment—then we should always get the same face. If science is advanced enough, we should be able to predict this face, and thus there is no need for probability theory. Unfortunately, science is not that omniscient—and never will be. Life is full of practical experiments in which our knowledge has gaps, where we do not know all the influencing factors, and hence we cannot predict the outcomes with certainty or explain their variations. This is where probability theory tries to provide some insight into the experiments' outcomes and variations.

1.2 SAMPLE SPACE AND EVENTS

In an experiment, the set of all possible and distinct outcomes is called a *sample space*. Depending on what we are interested in, the sample space for tossing a die could be:

1. $\{Y, N\}$, if we want to know whether it will land on a table or not.
2. $\{1, 2, 3, 4, 5, 6\}$, if we want to observe the number facing up.
3. $[0, \infty)$,[1] if we want to know the distance from its landing position to a corner of a room. Actually, this sample space may be rather conservative. If the maximum distance is only 10 feet, then it can be narrowed down to $[0, 10]$.

Any subset of a sample space is called an *event*. An event having more than two distinct outcomes is called a *combined* event. An empty set $\emptyset$ is an *impossible* event. The sample space S itself is a *sure* event.

If A and B are two events, then

1. $A \cup B$ is an event that occurs if A *or* B *or both* occur.
2. $A \cap B$ is an event that occurs if *both A and* B occur.
3. A^c is an event that occurs if A *does not* occur.

Furthermore, two events A and B are *mutually exclusive* if they cannot both occur; that is, $A \cap B = \emptyset$.

1.3 PROBABILITIES DEFINED ON EVENTS

Suppose that each event E of a sample space S is now assigned a real value $\Pr\{E\}$. We call $\Pr\{E\}$ a *probability* of event E if the following three *axioms* hold:

1. $0 \le \Pr\{E\} \le 1$
2. $\Pr\{S\} = 1$
3. For any infinite *sequence of mutually exclusive events* $E_1, E_2, E_3, \ldots,$

$$\Pr\{\bigcup_{n=1}^{\infty} E_n\} = \sum_{n=1}^{\infty} \Pr\{E_n\}$$

Consider the following function assigning a value to each outcome of tossing a die:

$$\Pr\{1\} = \frac{1}{10} \qquad \Pr\{2\} = \frac{3}{10} \qquad \Pr\{3\} = \frac{2}{10}$$

$$\Pr\{4\} = \frac{1}{10} \qquad \Pr\{5\} = \frac{2}{10} \qquad \Pr\{6\} = \frac{1}{10}$$

Because these values satisfy the three axioms, they are valid probabilities. Now someone may disagree with this and insist that each outcome should be given an equal proba-

[1] The interval (a, b) excludes both a and b; the interval $[a, b)$ includes a but excludes b; the interval $(a, b]$ excludes a but includes b; the interval $[a, b]$ includes both a and b.

bility; that is, $\Pr\{i\} = 1/6$ for all $i = 1, 2, \ldots, 6$. This disagreement implies that our probabilities are unrealistic but not that they are invalid.

1.4 PROBABILITY MODELS

If probabilities assigned to a particular event can be different, how are they obtained in practice?

1. Sometimes, they are *subjective*, depending on personal interpretation and belief. (I know that pregnancy could happen, but it will not happen to me this time.)
2. Sometimes, they come from the *equal probability* assumption. If a die is fair, then $\Pr\{i\} = 1/6$ for all $i = 1, 2, \ldots, 6$. For a fair coin, $\Pr\{H\} = \Pr\{T\} = 1/2$.
3. Sometimes, they are obtained *empirically*. We repeat the same experiment, observe the *relative frequency* of each outcome, and then use *inductive* reasoning to go from these specific observations to the general case. This is the field of *statistical inference*. In marketing, for example, we survey a group of persons about a proposed product to determine the probability that it will be a hit with the general public.
4. Sometimes, the desired probabilities are obtained *analytically* from other related and more fundamental probabilities. The latter are called the *assumptions*, which in turn can be the equal probability assumptions, or be obtained empirically or subjectively. In fact, this is the major role of probability theory. It teaches us how to calculate the probability of having to toss a die five times before 6 is obtained. This probability is dependent on our assumption about the probability of obtaining 6 in one toss.

Of the possible paths to reach the desired probabilities as shown in Figure 1.4.1, the subjective approach is the least scientific and hence the least reliable. The equal probability assumption is applicable only in simple cases. The empirical method is valid but limited only to existing and repeated phenomena. The analytical approach is the most desirable.

FIGURE 1.4.1 How probabilities are obtained.

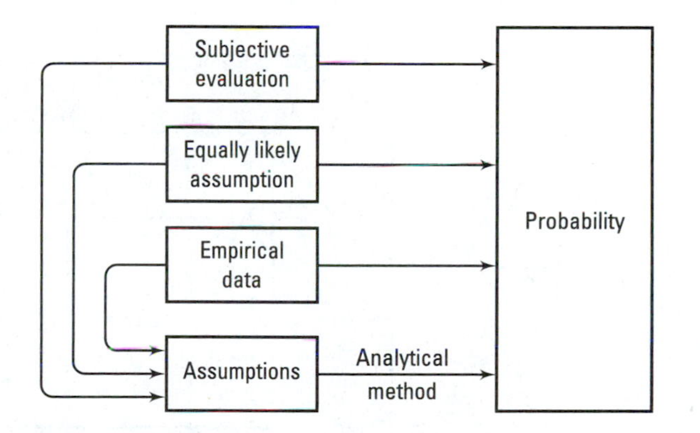

The analytical approach requires us first to formulate a *mathematical model*—that is, to make assumptions about the basic elements of the problem and mathematically describe the way they interact. The model idealizes a practical situation. It abstracts the essence of the problem to facilitate analysis. This means that, on the one hand, we cannot incorporate every aspect of the problem into our model, thus making it too complicated and hence intractable or unsolvable. On the other hand, we cannot make the model too simple, thus ignoring important parameters and arriving at results bearing no correlation to reality. Given our ability and resources, there is an optimal level of model complication, and we should strive for it.

1.5 PROBABILITIES OF COMBINED EVENTS

We first note that it is cumbersome, if not impossible, to assign probabilities to all events of a sample space. What we normally do is assign probabilities to a sufficient number of events only and then use the axioms to calculate the probabilities of the other events when needed. The following few results, which can be derived from the axioms but are stated here without proof, may be helpful when calculating the probabilities of the combined events:

1. $\Pr\{\emptyset\} = 0$
2. If $A_1, A_2, \ldots, A_N$ are mutually exclusive events, then

$$\Pr\{A_1 \cup A_2 \cup \cdots \cup A_N\} = \Pr\{A_1\} + \Pr\{A_2\} + \cdots + \Pr\{A_N\}$$

3. $\Pr\{A^c\} = 1 - \Pr\{A\}$
4. If $B_1, B_2, \ldots$ are mutually exclusive events such that $\bigcup_{n=1}^{\infty} B_n = S$, then, as shown in Figure 1.5.1,

$$\Pr\{A\} = \sum_{i=1}^{\infty} \Pr\{A \cap B_i\} \tag{1.5.1}$$

FIGURE 1.5.1 Equation (1.5.1).

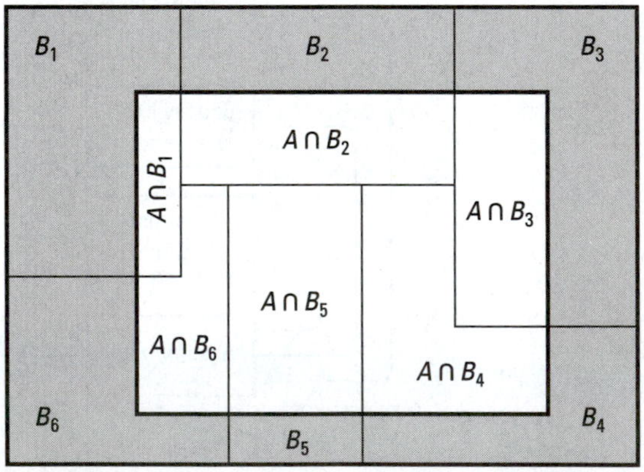

5. For any two events A and B,

$$\Pr\{A \cup B\} = \Pr\{A\} + \Pr\{B\} - \Pr\{A \cap B\} \tag{1.5.2}$$

If A and B are mutually exclusive, then

$$\Pr\{A \cup B\} = \Pr\{A\} + \Pr\{B\}$$

Equation (1.5.2) can be generalized into any number of events. For example, for three events A, B, and C,

$$\begin{aligned}\Pr\{A \cup B \cup C\} = \Pr\{A\} &+ \Pr\{B\} + \Pr\{C\} \\ &- \Pr\{A \cap B\} - \Pr\{B \cap C\} - \Pr\{C \cap A\} \\ &+ \Pr\{A \cap B \cap C\}\end{aligned}$$

1.6 CONDITIONAL PROBABILITIES

Axiom 2 underlines the important requirement that all probabilities need to be defined *with respect to a sample space*. If there is no possible confusion, then there is no need to spell out the sample space, and we can denote the probability of event A by $\Pr\{A\}$ as before. Otherwise, to make it clear, we write the probability as $\Pr\{A \mid S\}$ and read it as "the probability of A, *given* S."

A particularly important application of this requirement arises when *additional information* about the experiment becomes known, revealing with certainty that some outcomes must occur or cannot occur. This situation results in a *reduction of the original sample space*.

Let A and B be any two events in a sample space S, as in Figure 1.6.1. If it is now known that only outcomes in B can occur, the sample space S is reduced to B. The probability of A, defined with respect to this reduced sample space B, is called the *conditional probability* and can be denoted either by $\Pr\{A \mid B\}$ or by $\Pr\{A \mid \text{a description of } B\}$.

Consider a family having two children. Without any information, the sample space for the children's gender is $S \equiv \{bb, bg, gb, gg\}$, and the probability that both are girls is $\Pr\{gg \mid S\}$. However, if we know that at least one of them is a girl, then bb is no longer a possible outcome, and the sample space is therefore reduced to $B \equiv \{bg, gb, gg\}$.[2]

FIGURE 1.6.1 The reduced sample space *B*.

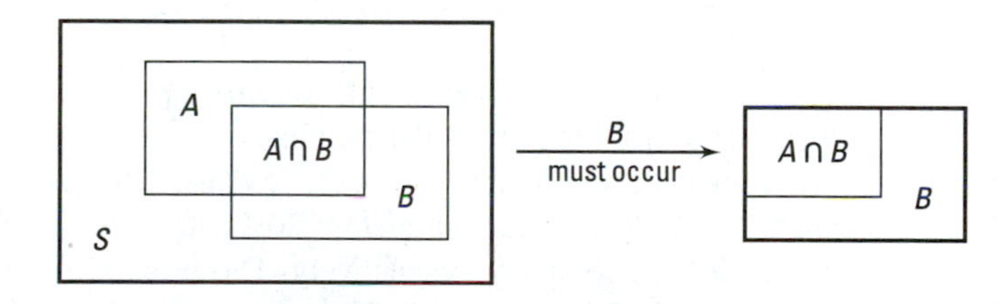

[2]The *equivalent relation* "≡" means that this is a definition.

The probability that both are girls now becomes the conditional probability $\Pr\{gg \mid B\}$, or $\Pr\{gg \mid \text{one of the children is a girl}\}$.

The additional information that event B must occur affects $\Pr\{A\}$ in two ways. First, this means that not every outcome in A, but only those in $A \cap B$, can occur. Thus, $\Pr\{A \mid S\}$ is reduced to $\Pr\{A \cap B \mid B\}$. Second, Axiom 2 now requires that $\Pr\{B \mid B\} = 1$. Hence,

$$\Pr\{A \mid B\} = \Pr\{A \cap B \mid B\} = \frac{\Pr\{A \cap B \mid B\}}{\Pr\{B \mid B\}}$$

Since we want to keep the *relative value* of $\Pr\{A \cap B\}$ and $\Pr\{B\}$ the same, regardless of whether the sample space is B or S, we *define*

$$\Pr\{A \mid B\} \equiv \frac{\Pr\{A \cap B \mid S\}}{\Pr\{B \mid S\}} \tag{1.6.1}$$

For the gender example, assuming all possible outcomes occur with equal probability, we have $\Pr\{gg \mid S\} = 1/4$ and $\Pr\{B \mid S\} = 3/4$. Thus,

$$\Pr\{gg \mid B\} = \frac{\Pr\{gg \cap B \mid S\}}{\Pr\{B \mid S\}} = \frac{\Pr\{gg \mid S\}}{\Pr\{B \mid S\}} = \frac{1/4}{3/4} = \frac{1}{3}$$

which is consistent with the fact that $B \equiv \{bg, gb, gg\}$ and all its elements occur with equal probability.

Defining the conditional probabilities in this manner, the three axioms are satisfied with both the original and the reduced sample spaces.

1.7 CONDITIONAL ARGUMENTS

Return now to Equation (1.5.1) and Figure 1.5.1, where we divided the sample space S into many mutually exclusive events $B_1, B_2, \ldots$ such that $\bigcup_{n=1}^{\infty} B_n = S$. Replacing $\Pr\{A \cap B_i\}$ with $\Pr\{A \mid B_i\}\Pr\{B_i\}$ in Equation (1.5.1), we obtain the following *law of total probability*:

$$\Pr\{A\} = \sum_{i=1}^{\infty} \Pr\{A \mid B_i\} \Pr\{B_i\} \tag{1.7.1}$$

$\Pr\{A\}$ can be considered the *weighted average* of all conditional probabilities $\Pr\{A \mid B_i\}$s, with $\Pr\{B_i\}$s as the weights.

In many cases, B_is are the possible *causes* and A is the *effect*. For example, three machines A, B, and C manufacture 20%, 30%, and 50% of the total number of items produced by a company, respectively. The probability that machine A produces a defective item is .05, machine B .03, and machine C .04. Let D be the event that an item selected at random is defective. What is $\Pr\{D\}$?

Equation (1.7.1) gives

$$\begin{aligned}\Pr\{D\} &= \Pr\{D \mid A\}\Pr\{A\} + \Pr\{D \mid B\}\Pr\{B\} + \Pr\{D \mid C\}\Pr\{C\}\\ &= (.20)(.05) + (.30)(.03) + (.50)(.04)\\ &= .039\end{aligned}$$

This calculation recognizes that there are three causes for a defective item: machines A, B, and C. Given the cause is A, then the probability of having a defective item is .05. Since machine A produces 20% of all items, we give cause A a weight of 0.20.

Graphically, we represent cause B_i as a *branch* in a *tree diagram*. The probability $\Pr\{B_i\}$ is written under branch B_i. The conditional probability $\Pr\{A \mid B_i\}$ is written at the end of it. For this example, the tree diagram is shown in Figure 1.7.1.

Following is an example in which causes are artificially created: Suppose we need to solicit answers to a question that might be too personal, such as: Have you been unfaithful to your spouse?

Asking this question without modification, we might not even get an answer, let alone an honest one. To make it easier to swallow, we conditionally mix it with another inoffensive question as in the following manner: In the first stage, we ask the interviewee to toss a coin and observe its outcome without telling us. In the second stage,

1. if the coin showed a head, then he or she would answer yes or no to the above question.
2. if it showed a tail, then yes or no is the answer to: Were you born in January?

Not having to reveal which question he or she answered, the interviewee might be more willing to participate in the survey.

Now let

1. $\Pr\{H\}$ be the probability of a head, which is $1/2$
2. $\Pr\{T\}$ be the probability of a tail, which is $1/2$
3. $\Pr\{Y \mid H\}$ be the probability that a person has been unfaithful, which is what we are looking for
4. $\Pr\{Y \mid T\}$ be the probability that a person was born in January, which is $1/12$
5. $\Pr\{Y\}$ be the probability that we have a yes to either question, which is the result of our survey

FIGURE 1.7.1 Tree diagram for the probability of a defective item.

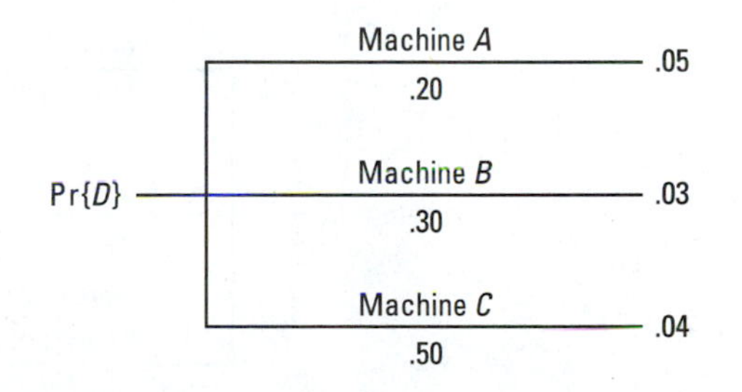

From the law of total probability (1.7.1)

$$\begin{aligned}\Pr\{Y\} &= \Pr\{Y \mid H\}\Pr\{H\} + \Pr\{Y \mid T\}\Pr\{T\} \\ &= \Pr\{Y \mid H\}(1/2) + (1/12)(1/2)\end{aligned}$$

Our desired probability can now be calculated as

$$\Pr\{Y \mid H\} = 2\Pr\{Y\} - (1/12)$$

In the example of the defective items, the desired probability of a defect is on the left-hand side of the law of total probability (1.7.1); in the survey example, the desired probability of unfaithfulness is on its right-hand side. In many cases, we see the desired probability on both sides of Law (1.7.1). We shall refer to the arguments leading to the last kind of relationship as the *conditional arguments*.

As an example, let us consider an experiment that has two mutually exclusive events A and B besides other events. We repeat this experiment until either A or B occurs. We are interested in the probability that A occurs before B.

In the first trial, either A occurs, B occurs, or neither.

1. The probability that A occurs in the first trial is $\Pr\{A\}$. If this happens, the probability that A occurs before B is 1.
2. The probability that B occurs in the first trial is $\Pr\{B\}$. If this happens, the probability that A occurs before B is 0.
3. The probability that neither A nor B occur in the first trial is $1 - \Pr\{A\} - \Pr\{B\}$. This does not cause any change in the probability that A occurs before B.

The tree diagram for this problem is in Figure 1.7.2.

From this

$$\begin{aligned}\Pr\{A \text{ before } B\} = {} & \Pr\{A\} \times (1) + \Pr\{B\} \times (0) \\ & + (1 - \Pr\{A\} - \Pr\{B\}) \times \Pr\{A \text{ before } B\}\end{aligned}$$

which has $\Pr\{A \text{ before } B\}$ on both sides. Hence,

$$\Pr\{A \text{ before } B\} = \frac{\Pr\{A\}}{\Pr\{A\} + \Pr\{B\}} \tag{1.7.2}$$

a very reasonable result, as it states that the higher the value of $\Pr\{A\}$, the higher the value of $\Pr\{A \text{ before } B\}$.

FIGURE 1.7.2 Tree diagram for Pr{*A* before *B*}.

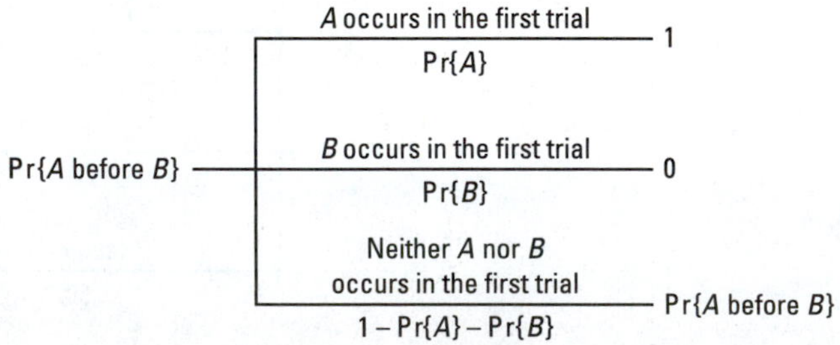

Pr{*A* before *B*}
A occurs in the first trial
Pr{*A*}
1
B occurs in the first trial
Pr{*B*}
0
Neither *A* nor *B* occurs in the first trial
1 – Pr{*A*} – Pr{*B*}
Pr{*A* before *B*}

1.8 INDEPENDENT EVENTS

Two events A and B are said to be *independent* if

$$\Pr\{A \cap B\} = \Pr\{A\} \Pr(B\}$$

Comparing this equation with Equation (1.6.1), we obtain

$$\Pr\{A \mid B\} = \Pr\{A\}$$

meaning that events A and B are independent if they have no cause-and-effect relationship: The probability of A is not influenced by the knowledge that B occurred, and vice versa.

The notion of independence should not be confused with that of exclusiveness. In fact, if A and B are mutually exclusive, $\Pr\{A \cap B\} = 0$; on the other hand, if A and B are independent, $\Pr\{A \cap B\} = \Pr\{A\} \Pr\{B\}$. Thus, the only way that A and B can be both independent and mutually exclusive is that one of them is impossible.

We note that, if events A and B are independent, then each of the pairs $\{A$ and $B^c\}$, $\{A^c$ and $B\}$, and $\{A^c$ and $B^c\}$ is also independent.

To generalize the concept of independence into more than two events, we say the n events $E_1, E_2, \ldots, E_n$ are *mutually independent* if the joint probability of any of their combinations is equal to the product of the individual probabilities. For example, three events E_1, E_2, and E_3 are mutually independent if all of the following relationships hold:

$$\begin{aligned}
\Pr\{E_1 \cap E_2 \cap E_3\} &= \Pr\{E_1\} \Pr\{E_2\} \Pr\{E_3\} \\
\Pr\{E_1 \cap E_2\} &= \Pr\{E_1\} \Pr\{E_2\} \\
\Pr\{E_2 \cap E_3\} &= \Pr\{E_2\} \Pr\{E_3\} \\
\Pr\{E_3 \cap E_1\} &= \Pr\{E_3\} \Pr\{E_1\}
\end{aligned}$$

As we shall see, a substantial part of this book is devoted to sequences of experiments that are dependent on each other. In preparation, we now introduce a sequence of identical experiments in which the outcomes are independent of each other. Here, we repeat the same experiment and observe the outcome Z_i of the ith experiment. We say they are mutually independent if the outcome of one experiment is not dependent on those of the previous ones. In other words, the information about the outcomes of the first, the second, $\ldots$, or the nth experiment is useless and does not help us to predict that of the $(n + 1)$th, or

$$\Pr\{Z_{n+1} = j \mid Z_1 = i_1, Z_2 = i_2, \ldots, Z_n = i_n\} = \Pr\{Z_{n+1} = j\}$$

1.9 RANDOM VARIABLES

Note that the outcome of an experiment can be anything, including a number $(1/2/3/\ldots)$, a color (*black/white/red/*...), a gender (*male/female*), or a logical *yes/no*. For the rest of this chapter, we shall restrict ourselves to the most important class of outcomes, which finds its way virtually into every practical application.

We refer to the outcomes that can be described as, or translated into, *numbers*. Any outcome can be translated into a number if there is a *function* assigning to it a numerical value. Such a function is called a *random variable*.

Distances, numbers of red balls drawn from an urn, durations waiting for a bus are all random variables; *head/tail*, genders, colors are not. However, we shall have a random variable if

1. *head* is considered equivalent to 1 and *tail* to 0.
2. a boy is considered equivalent to 0 and a girl to 1.
3. colors are studied by scientists and only color frequencies count.
4. the value of a painting is measured by the price the current owner paid for it.

The sample space of a random variable is a subset of the real line $(-\infty, \infty)$.

1.10 DISCRETE RANDOM VARIABLES AND THEIR EXPECTED VALUES

A random variable X is *discrete* if it only takes isolated values. Organized in increasing order, we can label its values sequentially as x_i, with $i = 0, 1, 2, 3, \ldots$. Its sample space therefore is *countable*. Note that the values of X do not have to be integers. As an illustration, the random variable D, which only takes value as .75, $\sqrt{3}$, or 3, is discrete.

The following definitions are applicable to a discrete random variable X:

1. The probability $p_i \equiv \Pr\{X = x_i\}$ is called the *probability mass function* of X. Since all axioms in §1.3 must be satisfied, especially $\sum_{k=0}^{\infty} p_k = 1$, not all p_is can be zero.

For the aforementioned random variable D, let us assume that

$$p_1 \equiv \Pr\{D = .75\} = .3 \qquad p_2 \equiv \Pr\{D = \sqrt{3}\} = .2 \qquad p_3 \equiv \Pr\{D = 3\} = .5$$

2. The *(cumulative)*[3] *distribution function* of X is defined as

$$F(x) \equiv \Pr\{X \leq x\} \quad \text{for all } -\infty < x < \infty \tag{1.10.1}$$

If we define

$$F_i \equiv \Pr\{x \leq x_i\} = \sum_{k=0}^{i} p_k \quad \text{for all } i = 0, 1, 2, \ldots \tag{1.10.2}$$

then $F(x)$ is a function that is *right-continuous* at all x_i, with $F(x) = F_i$ for all $x_i \leq x < x_{i+1}$.

For the random variable D, we now have

$$F_1 \equiv p_1 = .3 \qquad F_2 \equiv p_1 + p_2 = .5 \qquad F_3 \equiv p_1 + p_2 + p_3 = 1$$

Its distribution function $F(x)$ is shown in Figure 1.10.1a.

---

[3] Some definitions in this book have words enclosed within parentheses. These words may be dropped if there is no possible confusion.

FIGURE 1.10.1 The cumulative distribution function.

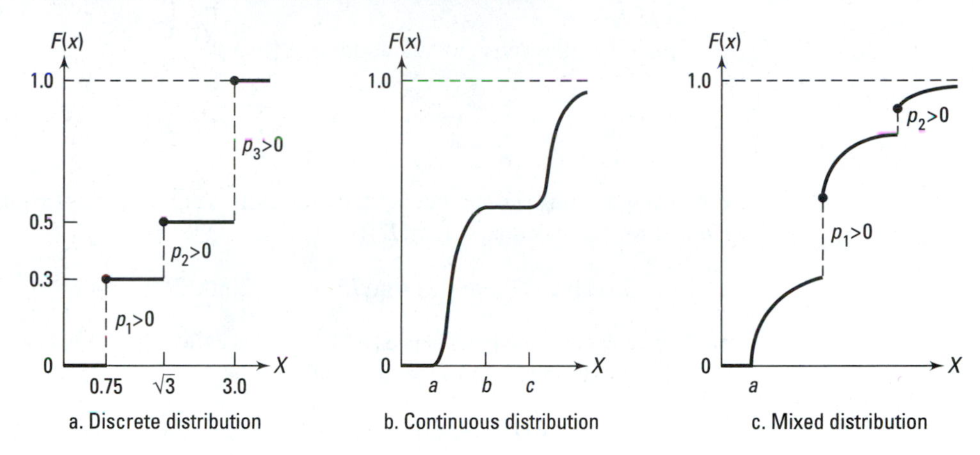

3. The *expected value*—or *mean* or *average* or *first moment (about the origin)*—of X is defined as

$$\mathcal{E}[X] \equiv \sum_{k=0}^{\infty} p_k x_k \tag{1.10.3}$$

This gives the expected value of the random variable D as

$$\mathcal{E}[D] \equiv (.3)(.75) + (.2)\sqrt{3} + (.5)(3) = 2.071$$

Equation (1.10.3) gives $\mathcal{E}[X]$ as a *weighted average* of all the possible values of X, in which the weight given to x_i is its probability p_i and the total weight is 1. The more probable a value, the more weight it has.

If we repeatedly observe the values of X indefinitely, we will see that the number of times X takes a particular value is proportional to its probability. Thus, $\mathcal{E}[X]$ can be considered the average of an infinite number of observations of X.

When we analyze a random variable X, we want to know its distribution function $F(x)$. This kind of result, however, may not be what the practitioners are looking for. In most instances, they do not want a function or a table, but only a single number. The expected value $\mathcal{E}[X]$ is one such important number. It measures the variable's *central tendency*; that is, it gives the general size of a random variable, or a value around which X tends to assume its values.

For any constant k and any two discrete random variables V and W, which can be dependent or independent, we can show that

$$\begin{aligned} \mathcal{E}[kV] &= k\mathcal{E}[V] \\ \mathcal{E}[V+k] &= \mathcal{E}[V] + k \\ \mathcal{E}[V+W] &= \mathcal{E}[V] + \mathcal{E}[W] \end{aligned} \tag{1.10.4}$$

4. Consider a discrete random variable X, taking values as x_i, with $i = 0, 1, 2, 3, \ldots$. As defined earlier, X is a function mapping an outcome with a real value. Consider now another real-value function $g(x)$. Because $g(X)$ is also a function mapping an outcome with a real value, it is also a discrete random variable, taking values as $g(x_i)$, with $i = 0, 1, 2, 3, \ldots$.

Thus, it makes sense to talk about of $\mathcal{E}[g(X)]$:

$$\mathcal{E}[g(X)] \equiv \sum_{k=0}^{\infty} p_k g(x_k) \tag{1.10.5}$$

For example, $\sin(D)$ is a random variable, taking values as $\sin(.75)$, $\sin(\sqrt{3})$, and $\sin(3)$. Hence, we can calculate $\mathcal{E}[\sin(D)]$ as

$$\mathcal{E}[\sin(D)] \equiv (.3)\sin(.75) + (.2)\sin(\sqrt{3}) + (.5)\sin(3) = .472$$

5. A special case of $\mathcal{E}[g(X)]$ is $\mathcal{E}[X^k]$, which is called the kth *moment (about the origin)* of X.[4] For example, the second moment of D is

$$\mathcal{E}[D^2] \equiv (.3)(.75)^2 + (.2)(\sqrt{3})^2 + (.5)(3)^2 = 5.268$$

6. If X takes a particular value x_i, then its *deviation* $x_i - \mathcal{E}[X]$ measures the distance from x_i to the expected value of X. Letting $k = \mathcal{E}[X]$ in the second equation of (1.10.4) yields the expected value of the deviations as $\mathcal{E}[X - \mathcal{E}[X]] = \mathcal{E}[X] - \mathcal{E}[X] = 0$.

7. The *variance* of X is the expected value of the squares of the deviations; that is,

$$\text{Var}[X] \equiv \sigma_X^2 \equiv \mathcal{E}[(X - \mathcal{E}[X])^2] \equiv \sum_{k=0}^{\infty} (x_k - \mathcal{E}[X])^2 \, p_k$$

The squares of the deviations can only be positive. As their expected value, $\text{Var}[X]$ measures the *dispersion* of the variable. The higher its value, the more scattered are the values of X away from its expected value.

A shortcut for the calculation of $\text{Var}[X]$ is

$$\text{Var}[X] = \mathcal{E}[X^2] - \mathcal{E}^2[X] \tag{1.10.6}$$

For random variable D

$$\text{Var}[D] = \mathcal{E}[D^2] - \mathcal{E}^2[D] = 5.26 - (2.07)^2 = .978$$

For any constant k and any two independent random variables V and W, we can show that

$$\begin{aligned} \text{Var}[kV] &= k^2\text{Var}[V] \\ \text{Var}[V + k] &= \text{Var}[V] \\ \text{Var}[V + W] &= \text{Var}[V] + \text{Var}[W] \end{aligned} \tag{1.10.7}$$

The expected value $\mathcal{E}[X]$ and the variance $\text{Var}[X]$ are the two most important single-value descriptions of a random variable X.

8. The positive square root of $\text{Var}[X]$ is called the *standard deviation* of X.

---

[4]Note the difference between this notation $\mathcal{E}[X^k]$ and $\mathcal{E}^k[X] \equiv (\mathcal{E}[X])^k$.

1.11 BERNOULLI RANDOM VARIABLES

A *Bernoulli trial* is an experiment, the sample space of which has only two outcomes, namely, *success* and *failure*. We normally write $\Pr\{Success\} = p$ and $\Pr\{Failure\} = 1 - p \equiv q$.

If a success occurs in a trial, we count the number of success in this trial as $I = 1$; otherwise, $I = 0$. The random variable I is called a *Bernoulli random variable*, or an *indicator function*.

Clearly, $\Pr\{I = 1\} = p$ and $\Pr\{I = 0\} = q$. Also,

$$\mathcal{E}[I^k] = (0)^k q + (1)^k p = p \quad \text{for all } k = 0, 1, 2, \ldots \tag{1.11.1}$$

and

$$\text{Var}[I] = \mathcal{E}[I^2] - \mathcal{E}^2[I] = p - p^2 = p(1-p) = pq \tag{1.11.2}$$

Simple as it is, the Bernoulli trials can be considered the building blocks for the construction of many important discrete distributions as we shall see in the following sections.

1.12 GEOMETRIC DISTRIBUTION

Consider now a *sequence* of independent Bernoulli trials having p as the probability of success in each trial. Let T be the *number of trials up to and including the first success*. Its sample space is $\{1, 2, 3, \ldots\}$. Since the first success occurs at the ith trial if and only if all first $i - 1$ trials result in failures and the ith trial in a success, we have

$$\Pr\{T = i\} = \begin{cases} p(1-p)^{i-1} & \text{for all } i = 1, 2, 3, \ldots \\ 0 & \text{otherwise} \end{cases} \tag{1.12.1}$$

We say that the random variable T has the *geometric distribution* with parameter p.

As defined in (1.10.2), with the help of Appendix (A.7),[5] we have

$$F_i \equiv \Pr\{T \le i\} = \sum_{k=1}^{i} pq^{k-1} = p\frac{1-q^i}{1-q} = 1 - q^i \quad \text{for all } i = 1, 2, 3, \ldots \tag{1.12.2}$$

Also, using Appendices (A.10) and (A.11), we now obtain

$$\mathcal{E}[T] = \sum_{i=1}^{\infty} pq^{i-1} i = \frac{p}{q}\frac{q}{(1-q)^2} = \frac{1}{p} \tag{1.12.3}$$

and

$$\mathcal{E}[T^2] = \sum_{i=1}^{\infty} pq^{i-1} i^2 = \frac{p}{q}\frac{q(1+q)}{(1-q)^3} = \frac{2-p}{p^2} \tag{1.12.4}$$

Hence,

$$\text{Var}[T] = \frac{2-p}{p^2} - \frac{1}{p^2} = \frac{1-p}{p^2} = \frac{q}{p^2} \tag{1.12.5}$$

---

[5] Appendix (A.7) is shorthand for Equation (7) in Appendix A in the back of this book.

Suppose a person buys a lottery ticket every week, and the probability of winning the weekly jackpot is always $p = 1/50{,}000{,}000$. According to Equation (1.12.3), on average, this person has to buy 50 million tickets before winning the jackpot for the first time.

Because the trials are mutually independent, one major property of a sequence of Bernoulli trials is that its development into the future is always independent of its development in the past. For this reason, we say that the sequence is *memoryless*.

Starting from the nth trial, let T_n be the number of trials until the next success. Because the probability of a success p remains the same and the sequence is memoryless, T_n is the same as the number of trials until the first success, starting from the first trial; that is, $T_n = T_1 = T$. It does not matter how many lottery tickets the buyer has bought in the past or how many times she has won so far, the expected number of extra tickets she still has to buy before winning the next jackpot is always 50 million.

Let S be the number of trials in between two consecutive successes. A corollary of the memoryless property is that the random variable S also has the same distribution as T. (Why?)

We should note that the term *geometric distribution* is also used for the distribution of the *number of failures* G before the first success, which is $T - 1$. Its sample space is $\{0, 1, 2, \ldots\}$ and its probability mass function is

$$\Pr\{G = i\} = p(1-p)^i \quad \text{for all } i = 0, 1, 2, \ldots \tag{1.12.6}$$

Since $G = T - 1$,

$$\mathcal{E}[G] = \mathcal{E}[T] - 1 = q/p \qquad \text{Var}[G] = \text{Var}[T] = q/p^2 \tag{1.12.7}$$

1.13 BINOMIAL DISTRIBUTION

Consider now n independent and identical Bernoulli trials, where n is a fixed integer. Let B be the number of successes in these n trials. Its sample space is $\{0, 1, 2, \ldots, n\}$.

There are many ways we can have i successes in n trials. One such way is to have all i successes in the first i trials and $n - i$ failures in the last $n - i$ trials. The other way is to have all $n - i$ failures in the first $n - i$ trials and i successes in the last i trials. It can be shown that the number of different ways we can have i successes in n trials is

$$\binom{n}{i} \equiv \frac{n!}{i!(n-i)!}$$

Thus, with p as the probability of success in each trial, we have

$$\Pr\{B = i\} = \begin{cases} \binom{n}{i} q^{n-i} p^i & \text{for all } i = 0, 1, 2, \ldots, n \\ 0 & \text{otherwise} \end{cases} \tag{1.13.1}$$

The random variable B is said to have a *binomial distribution* with parameters (n, p).

If I_i is an indicator function representing the number of success in the ith trial, then $B = I_1 + \cdots + I_n$. From Equations (1.10.4), (1.10.7), (1.11.1), and (1.11.2),

$$\begin{aligned} \mathcal{E}[B] &= \mathcal{E}[I_1] + \cdots + \mathcal{E}[I_n] = np \\ \mathrm{Var}[B] &= \mathrm{Var}[I_1] + \cdots + \mathrm{Var}[I_n] = npq \end{aligned} \tag{1.13.2}$$

1.14 NEGATIVE BINOMIAL DISTRIBUTION

In a sequence of independent and identical Bernoulli trials, let N_n be the *number of trials up to and including the nth success*, where n is a fixed number. The sample space of N_n is $\{n, n+1, n+2, \ldots\}$.

The aforementioned geometric random variable T is a special case of this random variable, with $T \equiv N_1$.

With p as the probability of success in each trial, we have shown that the probability of having exactly n successes in i independent Bernoulli trials is $\binom{i}{n} p^n q^{i-n}$. This can occur in two ways:

1. The nth success occurs at the ith trial. This happens with probability $\Pr\{N_n = i\}$.
2. The nth success occurs before the ith trial. In this case, there must be exactly n successes in the first $i-1$ trials, and the ith trial results in a failure. The probability of the former is $\binom{i-1}{n} p^n q^{i-1-n}$; that of the latter is $1-p$.

Thus,

$$\binom{i}{n} p^n q^{i-n} = \Pr\{N_n = i\} + (1-p)\binom{i-1}{n} p^n q^{i-1-n}$$

Hence,

$$\begin{aligned} \Pr\{N_n = i\} &= \frac{i!}{n!(i-n)!} p^n q^{i-n} - (1-p)\frac{(i-1)!}{n!(i-1-n)!} p^n q^{i-1-n} \\ &= \left[\frac{i}{n} - \frac{(i-p)}{q}\frac{i-n}{n}\right]\frac{(i-1)!}{(n-1)!(i-n)!} p^n q^{i-n} \\ &= \binom{i-1}{i-n} p^n (1-p)^{i-n} \qquad \text{for all } i = n, n+1, n+2, \ldots \end{aligned}$$

We say N_n has a *negative binomial distribution* with parameters (n, p).

Now let T_i be the number of trials between the $(i-1)$th and the ith successes, excluding the former but including the latter. All T_is are independent and have a geometric distribution with parameter p. Since $N_n = T_1 + \cdots + T_n$, from Equations (1.12.3) and (1.12.5) we have

$$\begin{aligned} \mathcal{E}[N_n] &= \mathcal{E}[T_1] + \cdots + \mathcal{E}[T_n] = n/p \\ \mathrm{Var}[N_n] &= \mathrm{Var}[T_1] + \cdots + \mathrm{Var}[T_n] = (nq)/p^2 \end{aligned} \tag{1.14.1}$$

The lottery ticket buyer has to buy an average of $(2)(50{,}000{,}000) = 100{,}000{,}000$ tickets to win the jackpot twice.

Let $M_n \equiv N_n - n$ be the number of *failures* up to and including the nth success. Then

$$\Pr\{M_n = i\} = \Pr\{N_n = i + n\} = \binom{n+i-1}{i} p^n (1-p)^i \quad \text{for all } i = 0, 1, 2, \ldots \tag{1.14.2}$$

M_n is also said to have the negative binomial distribution.

1.15 CONTINUOUS AND MIXED RANDOM VARIABLES

We said a random variable X is discrete if it only takes value from a set of countable $\{x_i\}$. Its distribution function $F(x) \equiv \Pr\{X \le x\}$ has a jump at x_i, the size of which is $p_i \equiv \Pr\{X = x_i\}$. In between the jumps, $F(x)$ is constant.

Suppose we allow the set of all x_i that have $p_i > 0$ to increase in number. The distribution function $F(x)$ now has more and more jumps. As this happens, the sizes of some jumps (that is, the values of some p_is) must decrease because we must always have $F(x) \le 1$. This should result in a smoother graph of $F(x)$.

Let the set of all distinct x_i that have $p_i > 0$ continue to increase in number until it becomes so "everywhere dense" within some interval (a, b) that its members become *uncountable* in that interval. In this case, except for a possible finite number of jumps, all jump sizes p_i should degenerate to zero, and $F(x)$ should increase continuously within (a, b).

1. If *all* jump sizes p_is degenerate to zero, the distribution function $F(x)$ becomes continuous at all points in $(-\infty, \infty)$, as shown in Figure 1.10.1b. The random variable X is no longer discrete but is said to be *continuous*.
2. If a *finite number* of jump sizes p_is retains a positive value, then, as shown in Figure 1.10.1c, the distribution function $F(x)$ has a finite number of jumps in $(-\infty, \infty)$ but otherwise increases continuously within (a, b). X is termed a *mixed* random variable.

1.16 CONTINUOUS RANDOM VARIABLES AND THEIR EXPECTED VALUES

Consider now a continuous random variable Y. Although its distribution function $F(x)$ is continuous, examples have been constructed to show that its derivative might not exist. In most cases, however, there exists a nonnegative function $f(x)$, which is defined for all $x \in (-\infty, \infty)$ and continuous at all but a finite number of points, such that

$$F(x) = \int_{-\infty}^{x} f(t)\, dt \quad \text{for all } -\infty < x < \infty \tag{1.16.1}$$

Assuming from now that this is the case, then $f(x)$ is called the *probability density function* of Y, and the distribution function $F(x)$ is said to be *absolutely continuous*.

Note that:

1. $F(y)$ is monotonically nondecreasing; that is, $\lim_{y\to-\infty} F(y) = 0$, $F(a) \le F(b)$ if $a < b$ and $\lim_{y\to\infty} F(y) = 1$ (Figure 1.10.1b).
2. For every pair of real numbers a and b ($a \le b$), it is not the value of $f(x)$ but the *area* under the curve of $f(x)$ between a and b that gives the probability $\Pr\{a \le Y \le b\}$; that is

$$\Pr\{a \le Y \le b\} = F(b) - F(a) = \int_a^b f(t)\,dt$$

3. The probability that a continuous random variable takes exactly the value a is

$$\Pr\{Y = a\} = F(a) - F(a) = \int_a^a f(t)\,dt = 0$$

Strange as it seems, this concept is closely related to the concept of a point having no width. By itself, it has no width; however, combined with an infinite number of other points, it may form a line segment having a positive width. Similarly, while $\Pr\{Y = a\} = 0$, we still can have $\Pr\{a \le Y \le b\} > 0$ for some interval (a, b).

For a continuous random variable Y, as a limiting form of Definition (1.10.3), we define its *expected value*—or *mean* or *average* or *first moment (about the origin)*—as

$$\mathcal{E}[Y] \equiv \int_{-\infty}^{\infty} t f(t)\,dt$$

For any real-value function $g(Y)$ of a continuous random variable Y, equivalent to Equation (1.10.5), we can also define

$$\mathcal{E}[g(Y)] \equiv \int_{-\infty}^{\infty} g(t) f(t)\,dt \qquad (1.16.2)$$

Similar to the discrete case, the kth *moment* of Y is $\mathcal{E}[Y^k]$ and its *variance* is $\mathrm{Var}[Y] \equiv \mathcal{E}[(Y - \mathcal{E}[Y])^2] = \mathcal{E}[Y^2] - \mathcal{E}^2[Y]$.

It can be shown that Equations (1.10.4) and (1.10.7) are also applicable when the random variables U and V are continuous.

We only present two continuous random variables in the next sections. More are presented later.

1.17 UNIFORM DISTRIBUTION

A continuous random variable U is *uniformly distributed* in the interval (a, b) if its probability density function $f(x)$ is a constant in that interval; that is

$$f(x) = \begin{cases} 1/(b-a) & \text{for all } a \le x \le b \\ 0 & \text{otherwise} \end{cases}$$

The probability of U taking a value in the interval $(c, d) \subset (a, b)$ is the area under the curve $f(x)$ from c to d, which is proportional to $d - c$.

The distribution function of U is

$$F(x) = \begin{cases} 0 & \text{for all } x < a \\ (x-a)/(b-a) & \text{for all } a \le x < b \\ 1 & \text{for all } x \ge b \end{cases}$$

Its expected value is the midpoint between a and b; that is

$$\mathcal{E}[U] = \int_a^b t \frac{1}{b-a}\, dt = \frac{1}{b-a} \frac{b^2 - a^2}{2} = \frac{a+b}{2}$$

Also

$$\mathcal{E}[U^2] = \int_a^b t^2 \frac{1}{b-a}\, dt = \frac{1}{b-a} \frac{b^3 - a^3}{3} = \frac{1}{3}\left(a^2 + ab + b^2\right)$$

Hence

$$\mathrm{Var}[U] = \frac{1}{3}\left(a^2 + ab + b^2\right) - \left(\frac{a+b}{2}\right)^2 = \frac{1}{12}(a-b)^2$$

1.18 NORMAL DISTRIBUTION

The probability density function $f(x)$ of a random variable Z having a *normal distribution* with mean μ and variance σ^2 is

$$f(x) = \frac{1}{\sigma\sqrt{2\pi}} \exp\left\{-\frac{1}{2}\left(\frac{x-\mu}{\sigma}\right)^2\right\} \quad \text{for all } -\infty < x < \infty$$

If $\mu = 0$ and $\sigma^2 = 1$, this probability density function becomes

$$f(x) = \frac{1}{\sqrt{2\pi}} \exp\left\{-\frac{x^2}{2}\right\} \quad \text{for all } -\infty < x < \infty \tag{1.18.1}$$

and the variable is said to have the *standard normal distribution*.

1.19 CONDITIONAL ARGUMENTS FOR A RANDOM VARIABLE

In §1.7, we presented the conditional arguments as a very useful tool to calculate some probabilities. In this section, we return to the law of total probability (1.7.1) with further examples to demonstrate its usefulness in obtaining the distribution of some random variables.

Consider the following problem: A bowl has three fish. For each fish, if a person randomly passes a net through the bowl, the probability of its being caught is p. We want to find the probability that exactly two fish are caught in the second pass.

To start, the probability of catching i fish in any pass is dependent on how many fish are in the bowl. If there are n fish, the number of fish caught is a random variable having a binomial distribution; that is

$$\Pr\{i \text{ out of } n \text{ fish caught}\} = \binom{n}{i} p^i (1-p)^{n-i}$$

The probability of catching two fish in the second pass is therefore dependent on how many were caught in the first pass, as depicted in the tree diagram in Figure 1.19.1.

From this

$$\begin{aligned}\Pr\{2 \text{ fish in the second pass}\} &= (1-p)^3\{3p^2(1-p)\} + 3p(1-p)^2\{p^2\} \\ &= 3p^2(1-p)^4 + 3p^3(1-p)^2\end{aligned}$$

1.20 CONDITIONAL EXPECTATIONS

Consider now a discrete random variable X. Let us assume that its distribution function is dependent on the occurrence of a certain event B: If B occurs, then $\Pr\{X = x_i\}$ will become $\Pr\{X = x_i \mid B\}$. In this case, the expected value $\mathcal{E}[X]$ as defined in Equation (1.10.3) will change into the *conditional* expected value

$$\mathcal{E}[X \mid B] \equiv \sum_{k=0}^{\infty} x_k \Pr\{X = x_k \mid B\}$$

Furthermore, the expected value $\mathcal{E}[g(X)]$ as defined in Equation (1.10.5) will change into

$$\mathcal{E}[g(X) \mid B] \equiv \sum_{k=0}^{\infty} g(x_k) \Pr\{X = x_k \mid B\}$$

Especially, we have

$$\mathrm{Var}[X \mid B] \equiv \mathcal{E}[(X - \mathcal{E}[X \mid B])^2 \mid B] = \mathcal{E}[X^2 \mid B] - \mathcal{E}^2[X \mid B]$$

FIGURE 1.19.1 Tree diagram for the probability of two fish caught in the second pass.

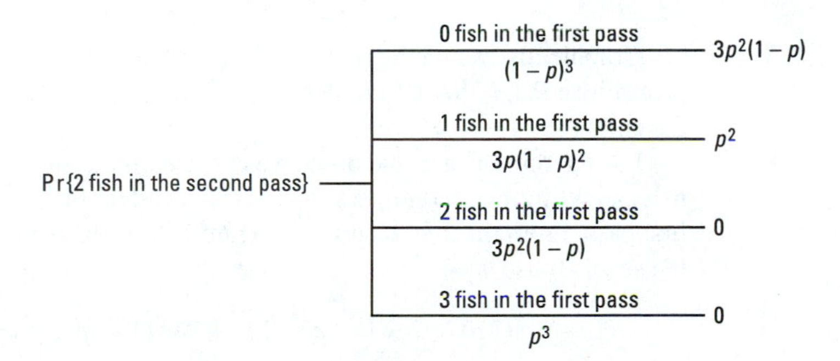

Let us now divide the sample space S into many mutually exclusive events $B_1, B_2, \ldots$ such that $\bigcup_{n=1}^{\infty} B_n = S$. From Definition (1.10.5) and Equation (1.7.1),

$$\begin{aligned}
\mathcal{E}[g(X)] &= \sum_{k=0}^{\infty} g(x_k) \Pr\{X = k\} \\
&= \sum_{k=0}^{\infty} g(x_k) \left(\sum_{i=1}^{\infty} \Pr\{X = k \mid B_i\} \Pr\{B_i\} \right) \\
&= \sum_{k=0}^{\infty} \sum_{i=1}^{\infty} g(x_k) \Pr\{X = k \mid B_i\} \Pr\{B_i\} \qquad (1.20.1) \\
&= \sum_{i=1}^{\infty} \left(\sum_{k=0}^{\infty} g(x_k) \Pr\{X = k \mid B_i\} \right) \Pr\{B_i\} \\
&= \sum_{i=1}^{\infty} \mathcal{E}\,[g(X) \mid B_i] \Pr\{B_i\}
\end{aligned}$$

Similarly, If the distribution function of a continuous random variable Y is conditioned on the occurrence of B, then its probability density function $f(y)$ will be written as $f(y \mid B)$, and the expected value $\mathcal{E}[g(Y)]$ as defined in Equation (1.16.2) will change into the *conditional* expected value

$$\mathcal{E}[g(Y) \mid B] \equiv \int_{-\infty}^{\infty} g(t) f(t \mid B)\, dt$$

Using the kind of manipulations found in the derivation of Equation (1.20.1), we should obtain a similar result for the continuous random variable Y. Thus, for a general random variable Z, discrete or continuous, we now have the following result, which is equivalent to Equation (1.7.1) and may be called the *law of total probability for expectations*:

$$\mathcal{E}[g(Z)] = \sum_{i=1}^{\infty} \mathcal{E}[g(Z) \mid B_i] \Pr\{B_i\} \qquad (1.20.2)$$

Graphically, we can represent this equation as a tree diagram. Each branch has the probability $\Pr\{B_i\}$ written under it and the conditional expected value $\mathcal{E}\{g(X) \mid B_i\}$ at the end of it.

For the three fish in the bowl, we can use this result to find the expected number of fish caught in the second pass, which is dependent on the number of fish caught in the first pass. From the tree diagram in Figure 1.20.1, the expected number of fish caught in the second pass is

$$(1-p)^3(3p) + 3p(1-p)^2(2p) + 3p^2(1-p)(p) = 3p(1-p)(1+4p^2)$$

FIGURE 1.20.1 **Tree diagram for the expected number of fish caught in the second pass.**

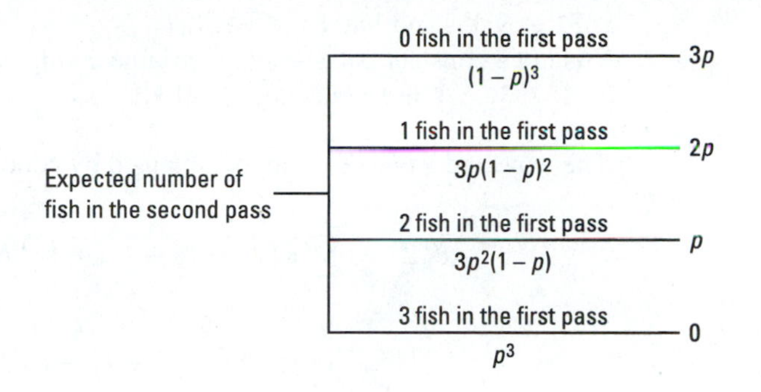

As an example of the *conditional expectation arguments* that result in the desired expected value being on both sides of Law (1.20.2), let us consider a man who is trapped in a mine with three doors. If he takes the first door, then he will wander around and return to the same position after 3 hours. If he takes the second door, he will be free after 4 hours. If he takes the third door, he will return to the same position after 5 hours. He always chooses the doors with equal probability each time. Let T be the duration before he can reach freedom. What is $\mathcal{E}[T]$?

Conditioning on the door taken, the tree diagram is presented in Figure 1.20.2. From this

$$\mathcal{E}[T] = (1/3)\{\mathcal{E}[T] + 3\} + (1/3)(4) + (1/3)\{\mathcal{E}[T] + 5\}$$

Hence, $\mathcal{E}[T] = 12$.

1.21 RANDOM SUM

Following is a simple yet important application of Law (1.20.2): We call Z a *random sum* if

$$Z \equiv Y_1 + Y_2 + \cdots + Y_N$$

where

FIGURE 1.20.2 **Tree diagram for the expected duration to freedom.**

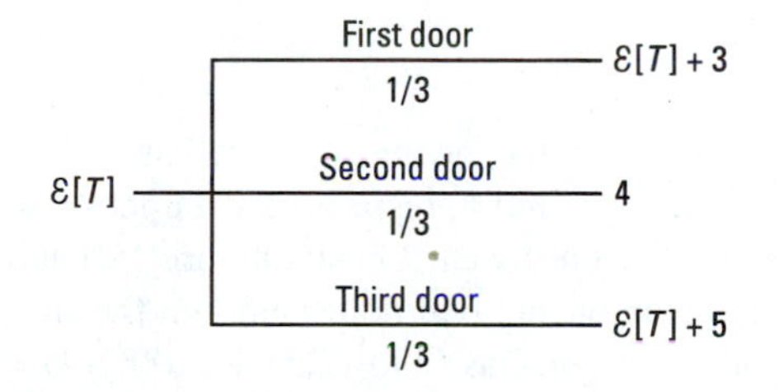

1. $Y_1, Y_2, Y_3, \ldots, Y_N$ are independent and identically distributed random variables, with $\mathcal{E}[Y] \equiv \mathcal{E}[Y_1]$ and $\mathrm{Var}[Y] \equiv \mathrm{Var}[Y_1]$
2. N is not a constant but a random variable taking value on the set of discrete integers $\{0, 1, 2, 3, \ldots\}$, independently of all Y_is

The expected value of Z can be obtained by conditioning on the value of N as

$$\begin{aligned}\mathcal{E}[Z] &= \sum_{i=0}^{\infty} \mathcal{E}[Y_1 + Y_2 + \cdots + Y_i \mid N = i] \Pr\{N = i\} \\ &= \sum_{i=0}^{\infty} i\mathcal{E}[Y] \Pr\{N = i\} = \mathcal{E}[Y] \sum_{i=0}^{\infty} i \Pr\{N = i\} \\ &= \mathcal{E}[Y]\mathcal{E}[N]\end{aligned} \tag{1.21.1}$$

Note that Equation (1.21.1) is also applicable even when Y_is are dependent with each other because we derive it by using Equation (1.10.4), which does not require that all Y_is are mutually independent.

Also by conditioning on the value of N, $\mathcal{E}[Z^2]$ can be obtained as

$$\begin{aligned}\mathcal{E}[Z^2] &= \sum_{i=0}^{\infty} \mathcal{E}[Z^2 \mid N = i] \Pr\{N = i\} \\ &= \sum_{i=0}^{\infty} \left(\mathrm{Var}\{Z \mid N = i\} + \mathcal{E}^2[Z \mid N = i]\right) \Pr\{N = i\} \\ &= \sum_{i=0}^{\infty} \left(i\mathrm{Var}[Y] + i^2\mathcal{E}^2[Y]\right) \Pr\{N = i\} \\ &= \mathrm{Var}[Y] \sum_{i=0}^{\infty} i \Pr\{N = i\} + \mathcal{E}^2[Y] \sum_{i=0}^{\infty} i^2 \Pr\{N = i\} \\ &= \mathrm{Var}[Y]\mathcal{E}[N] + \mathcal{E}^2[Y]\mathcal{E}[N^2]\end{aligned}$$

Thus,

$$\begin{aligned}\mathrm{Var}[Z] &= \mathcal{E}[Z^2] - \mathcal{E}^2[Z] \\ &= \mathrm{Var}[Y]\mathcal{E}[N] + \mathcal{E}^2[Y]\mathcal{E}[N^2] - (\mathcal{E}[Y]\mathcal{E}[N])^2 \\ &= \mathrm{Var}[Y]\mathcal{E}[N] + \mathcal{E}^2[Y]\left(\mathcal{E}[N^2] - \mathcal{E}^2[N]\right) \\ &= \mathrm{Var}[Y]\mathcal{E}[N] + \mathcal{E}^2[Y]\mathrm{Var}[N]\end{aligned} \tag{1.21.2}$$

Suppose that the mean and variance of the number of customers entering a store each hour are 2.5 and 4, respectively. Suppose also that the mean and variance of the dollar amount spent by each customer are 100 and 250, respectively. Equation (1.21.1) gives the mean of the total dollar amount the store takes in each hour as $(100)(2.5) = \$250$ and its variance as $(250)(2.5) + (100^2)(4) = 40{,}625$.

1.22 SUMMARY

This chapter reviews only the materials necessary for the later part of this book. The reader must have a good understanding of the following topics before advancing further:

1. the conditional probabilities and the conditional arguments, especially the law of total probability (1.7.1)
2. the random variables and their distributions, especially the indicator function, the geometric, binomial, negative binomial, and uniform random variables
3. the expected values of a random variable and the conditional expectation arguments, especially the law of total probability for expectations (1.20.2)

PROBLEMS

1.1 Suppose that each animal is classified according to the pair of genes it carries in one position in the chromosomes and that the gene is either a or b. Thus, each animal must have one of the following pairs of genes, which are called *genotypes*: aa, ab, or bb. (Note that ab is the same as ba.) If two animals are mated, there are six possible combinations of genotypes: (aa, aa), (bb, bb), (aa, ab), (ab, ab), (ab, bb), (aa, bb). Suppose the genotype of the offspring is a combination of one gene randomly selected from the father and one from the mother. For each of the combinations, find the probability that the genotype of the offspring is aa, ab, or bb.

1.2 Consider a game of tennis between two players A and B. Assume that A can win at any given point with probability p and B with probability $q \equiv 1 - p$. As shown in Figure 1.23.1, the game can be divided into two stages: the *preliminary stage* and the *final stage*. The final stage includes the five scores: (B) game B, (AB) advantage B (or 30–40), (D) deuce, (AA) advantage A (or 40–30), and (A) game A. The game is

FIGURE 1.23.1 The tennis games.

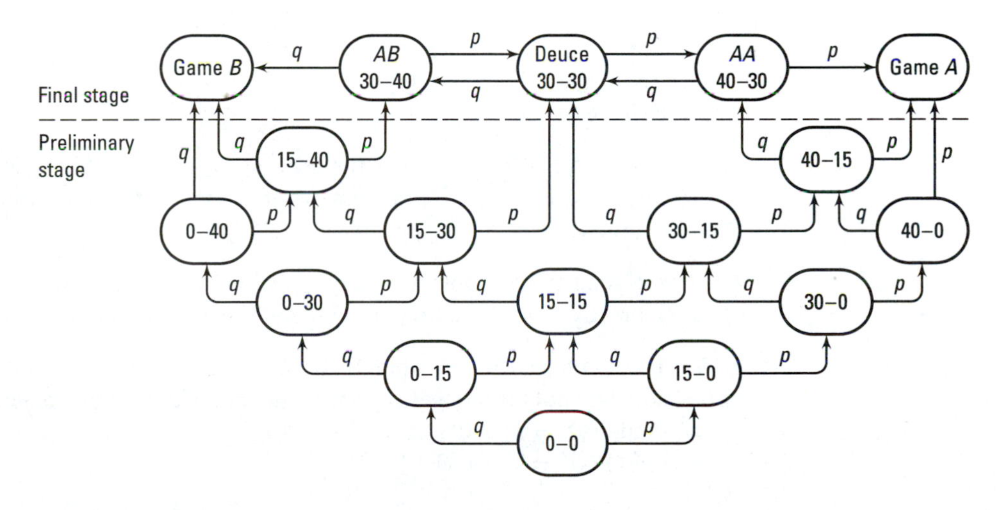

in the preliminary stage if it has not reached one of the foregoing scores. Show that immediately after the preliminary stage,

a. $\Pr\{B\} = (1+4p)q^4$
b. $\Pr\{AB\} = 4p^2q^3$
c. $\Pr\{D\} = 6p^2q^2$
d. $\Pr\{AA\} = 4p^3q^2$
e. $\Pr\{A\} = p^4(1+4q)$

1.3 Refer to the man trapped in a mine in §1.20. In the first attempt, he picks one of the three doors at random. However, if he takes a door that returns him to the same location, he will mark that door and not take it again.

a. What is the expected duration before he reaches freedom?
b. What is the expected number of doors he tries before he reaches freedom?

1.4 In a war, there are three surviving tanks A, B, and C. Tank A can hit its target with probability 1/3; tank B with probability 1/4; and tank C with probability 1/5. In a fight, they shoot simultaneously and randomly pick one of the other surviving tanks as a target. Show that, after the first round,

a. $\Pr\{\text{no tank survives}\} = 1/240$
b. $\Pr\{\text{all tanks survive}\} = 24/60$
c. $\Pr\{\text{only } A \text{ survives}\} = \Pr\{\text{only } B \text{ survives}\} = \Pr\{\text{only } C \text{ survives}\} = 1/24$
d. $\Pr\{A \text{ and } B \text{ survive}\} = 11/60$
e. $\Pr\{B \text{ and } C \text{ survive}\} = 1/8$
f. $\Pr\{A \text{ and } C \text{ survive}\} = 39/240$

1.5 A group of children has four boys under 10, six boys over 10, six girls under 10, and x girls over 10. What is the value of x if age and gender are independent?

1.6 An integer X between 1 and N is randomly selected. Let A be the event "X is prime" and B be the event "$X \geq 11$." Show that A and B are independent if $N = 20$ and dependent if $N = 21$.

1.7 Two dice are tossed. Let A be the event that the first die gives a 4, B be the event that the sum is 6, and C be the event that the sum is 7.

a. Are A and B independent?
b. Are A and C independent?
c. Explain the difference.

1.8 Let $p_i = (1-\rho)^2(i+1)\rho^i$ for all $i = 0, 1, 2, \ldots$. Show that p_i can be the probability mass function of a random variable. Hence, find the expected value of this random variable. *Hint:* Use Appendices (A.10) and (A.12).

1.9 For a geometric random variable G defined in §1.12, find $\mathcal{E}[G]$ and $\mathrm{Var}[G]$ in (1.12.7) directly from Definition (1.10.3) and Equation (1.10.6).

1.10 An urn contains $N-1$ white balls and one red ball. Let X and Y be the number of balls drawn until the red ball is drawn with and without replacement, respectively.

a. Find $\Pr\{X = k\}$ and $\mathcal{E}[X]$.
b. Find $\Pr\{Y = k\}$ and $\mathcal{E}[Y]$.

1.11 Dave repeatedly tosses three dice together. He stops only if the same number appears on exactly two dice.

a. What is the probability that he stops after the fifth toss?

b. What is the expected number of tosses he will make?

1.12 We perform the same experiment having two mutually exclusive events A and B again and again. Let N be the number of experiments performed until either A or B occurs. Find in terms of $\Pr\{A\}$ and $\Pr\{B\}$ the distribution of N and $\mathcal{E}[N]$.

1.13 A and B play the following dice game. A tosses two dice. If the sum is 7 or 11, A wins. If the sum is 2, 3, or 12, A loses. Otherwise, A continues to toss the two dice until she either obtains a 7 and loses or obtains the same number as in the first toss and wins.

a. What is the probability that A wins? *Hint:* Use Equation (1.7.2).

b. What is the average number of tosses in each game? *Hint:* Use Problem 1.12.

1.14 An urn contains N balls, either black or white. A ball is drawn at random. If it is black, there is a probability $1 - p$ that it will be returned to the urn; there is a probability p that a white ball will be returned to the urn instead. If it is white, it will be returned to the urn. This procedure is repeated indefinitely.

a. Given i white balls in the urn, let m_i be the expected number of further black balls drawn before the urn contains all white balls. Explain why

$$m_i = (1/p) + m_{i+1}$$

b. Hence, find m_0.

c. Given i white balls in the urn, let n_i be the expected number of further balls, either black or white, drawn before the urn contains all white balls. Explain why

$$n_i = \frac{N}{(N-i)p} + n_{i+1}$$

d. Hence, find n_0.

1.15 A zoo wishes to catch some animals of a certain species. For each animal caught, the probability that it is a male is p and a female is $q \equiv 1 - p$. Each animal roams around the search area independently.

a. Find the probability that the first male/female pair is caught exactly after the zoo captures the fifth animal. *Hint:* Condition on the gender of the first animal caught.

b. How many animals are expected to be caught before the first male/female pair is found? What is the value of p such that this expected value is a minimum? Can you guess this number?

c. How many animals are expected to be caught before the second male/female pair is found? What is this value when $p = q = .5$? *Hint:* Condition on the genders of the first two animals caught.

1.16 Consider a random variable Y such that $\Pr\{Y > n\} = q^n$ for all $n = 1, 2, 3, \ldots$. Show that Y has a geometric distribution.

1.17 Let X and Y be two independent geometric random variables having the same parameter ρ. Show that $\min(X, Y)$ is also geometric. *Hint:* Use Problem 1.16.

1.18 If a random variable T has a geometric distribution, show that

$$\Pr\{T > n + m \mid T > n\} = \Pr\{T > m\}$$

Show that the converse is also true: If a discrete random variable satisfies this equation for all n and m, then it has a geometric distribution.

1.19 Consider an infinite sequence of independent and identically distributed Bernoulli trials. Assume that we have to pay \$$c$ whenever a failure occurs. Also assume a discount rate of α $(0 \le \alpha \le 1)$. This means that a cost of \$1 incurred at the nth trial is equivalent to a present value of \$$\alpha^n$. Find the expected present value of the total cost.

1.20 For a binomial random variable B, find $\mathcal{E}[B]$ and Var$[B]$ in Equation (1.13.2) directly from Definition (1.10.3) and Equation (1.10.6).

1.21 When in flight, one engine of an airplane may fail with probability p, independent of the other engines. Suppose that an airplane will continue to fly if more than 50% of its engines operate. For what value of p is a four-engine plane safer than a two-engine plane?

1.22 In a sequence of independent and identically distributed Bernoulli trials with probability of success p, let $P_{n,i}$ be the probability of having i successes in n trials.

a. Explain why

$$P_{n,i} = pP_{n-1,i-1} + (1-p)\,P_{n-1,i} \quad \text{for all } n = 1, 2, 3, \ldots \text{ and } i = 1, 2, \ldots$$

b. Hence, use induction[6] to prove Equation (1.13.2).

1.23 Let X_i $(i = 1, 2, 3, \ldots)$ be a sequence of independent and identically distributed random variables with $\Pr\{X_1 = 1\} \equiv p$ and $\Pr\{X_1 = -1\} = 1 - p$. Let Z be a sequence of random variables such that $Z_n \equiv \sum_{i=1}^{n} X_i$. Find $\mathcal{E}[Z_n]$ and Var$[Z_n]$. *Hint:* Define $Y_i \equiv (X_i + 1)/2$.

1.24 A random variable X has a binomial distribution with parameters (N, p). Parameter N in turn is a random variable having a binomial distribution with parameters (M, r). Verify that X has a binomial distribution with parameters (M, pr).

1.25 For a negative binomial random variable N_n, find $\mathcal{E}[N_n]$ and Var$[N_n]$ in Equation (1.14.1) directly from Definition (1.10.3) and Equation (1.10.6).

1.26 In certain $n + m$ Bernoulli trials, it is known that there are k successes. Use the indicator function to show that the expected number of successes in the first n trials is $nk/(n + m)$.

1.27 N people write their names on N pieces of paper. These papers are then mixed together before each person randomly draws one (with replacement). Use the indicator function to show that the expected number of people picking their own names is always 1 regardless of the value of N.

---

[6]Mathematical induction can be used to prove that a certain mathematical proposition is true for all positive integers n greater than i. The proof requires two steps: (a) proving that the proposition is true for i; and (b) proving that the proposition is true for n, assuming that it is true for $n - 1$.

1.28 Assume that, if a monkey sits in front of a typewriter having 26 different letter keys, it will touch any key with equal probability.

a. Compute the expected number of different letters if the monkey touches ten keys.

b. On the average, how many keys must the monkey touch to get five different letters?

1.29 Let $Y_1, Y_2, \ldots$ be a sequence of independent and identically distributed continuous random variables. We say that a *record* occurs at time n if $Y_n = \max(Y_1, Y_2, \ldots, Y_n)$.

a. Show that $\Pr\{\text{A record occurs at time } n\} = 1/n$.

b. Find the expected number of records obtained by time 5.

c. Find the variance of the number of records obtained by time 5.

1.30 Let X and Y be two independent random variables with means μ_X and μ_Y and variances σ_X^2 and σ_Y^2, respectively. Show that

$$\text{Var}[XY] = \sigma_X^2\sigma_Y^2 + (\mu_Y)^2\sigma_X^2 + (\mu_X)^2\sigma_Y^2$$

1.31 Two players A and B take turns to play on a gaming machine. All plays are independent. A plays first. On his turn, A can score a success against the machine with probability a; B with probability b. The first to score a success wins.

a. Find the probability that A wins the game

i. by obtaining the probability that A wins at his nth turn

ii. by using conditional arguments

b. Find the expected number of games A plays

i. by showing that the number of games A plays has a geometric distribution

ii. by using conditional arguments

c. Find the expected number of games B plays

i. by showing that the number of games B plays has a geometric distribution

ii. by using conditional arguments

1.32 We perform the same experiment having two mutually exclusive events A and B again and again. Find in terms of $\Pr\{A\}$ and $\Pr\{B\}$ the probability that A occurs at least once before B occurs twice.

1.33 A biased coin, having $\Pr\{H\} = p$ and $\Pr\{T\} = 1 - p$, is tossed repeatedly. Let α_n be the probability that an even number of Hs was obtained up to and including the nth toss. (Zero is considered an even number.)

a. Show that $\alpha_0 = 1$ and

$$\alpha_n = p + (1 - 2p)\,\alpha_{n-1} \quad \text{for all } n > 0$$

b. Hence, prove by induction that

$$\alpha_n = \frac{1 + (1 - 2p)^n}{2} \quad \text{for all } n > 0$$

1.34 Find the expected value and the variance of a geometric random variable with parameter p by conditioning on the outcome of the first trial.

1.35 A *multinomial* trial is an experiment having $N + 1$ outcomes, conveniently labeled $0, 1, 2, \ldots, N$. We denote the probability that outcome i occurs in each trial by p_i.

($\sum_{j=0}^{N} p_i = 1$.) In a sequence of independent and identically distributed multinomial trials, let v_i be the number of times outcome i occurs before outcome 0. Show that $\mathcal{E}[v_i] = p_i/p_0$ for all $i \neq 0$.

1.36 Refer to the man trapped in a mine in §1.20.

a. What is the variance of the duration before he reaches freedom?

b. What is the expected number of doors he tries before he reaches freedom?

1.37 A woman is trapped in a mine having two doors. If she takes the left door, then she will wander around and return to the same position after 3 hours. If she takes the right door, the probability that she will be free after 4 hours is 1/3; the probability that she will return to the same position after 2 hours is 2/3. In the first attempt, she picks a door at random. However, if she takes one door and returns, she will try the other door the next time.

a. What is the expected duration before she reaches freedom? *Hint:* Find two equations for $\mathcal{E}[T \mid R]$ and $\mathcal{E}[T \mid L]$.

b. What is the expected number of doors she tries before reaching freedom?

1.38 A biased coin, having $\Pr\{H\} = p$ and $\Pr\{T\} = 1 - p$, is tossed repeatedly. Let m be the number of tosses of this coin to get two consecutive Hs or two consecutive Ts. Use conditional arguments to show that

$$\mathcal{E}[m] = \frac{2 + pq}{1 - pq}$$

Hint: Find two equations for $\mathcal{E}[m \mid H]$ and $\mathcal{E}[m \mid T]$.

1.39 A biased coin, having $\Pr\{H\} = p$ and $\Pr\{T\} = 1 - p$, is tossed repeatedly. Let m be the number of tosses of this coin to get two consecutive Hs. Find $\mathcal{E}[m]$ and $\text{Var}[m]$ by conditioning on whether the first T was at the first, second, or third toss.

1.40 A biased coin, having $\Pr\{H\} = p$ and $\Pr\{T\} = 1 - p$, is tossed repeatedly. Let n_k be the expected number of tosses until k consecutive Hs are obtained.

a. By conditioning on the time of the first T, show that

$$n_k = (1 - p) \sum_{j=1}^{k} p^{j-1} (j + n_k) + kp^k$$

Hence, show that

$$n_k = \frac{1}{p} + \frac{1}{p^2} + \cdots + \frac{1}{p^k} = \frac{1 - p^k}{p^k (1 - p)}$$

Hint: Use Appendix (A.9).

b. By conditioning on the outcome of the toss immediately after $k - 1$ consecutive Hs have been obtained, show that

$$n_k = (n_{k-1} + 1) / p$$

Hence, prove the foregoing result for n_k by induction.

c. Let v_i be the number of further tosses needed to have k consecutive Hs given that i consecutive Hs have already been obtained. Use conditional arguments to show that

$$v_i = 1 + (1-p)v_0 + pv_{i+1} \quad \text{for all } 0 \le i < k$$

Starting from $v_k = 0$, find $v_0 = n_k$ recursively.

1.41 There are $N+1$ points; each is assigned a value. They are ordered according to their values, the point with the lowest value is labeled 0, and that with the highest value is labeled N. An algorithm starts with point i and moves to one of the lower points $0, 1, \ldots, i-1$ with equal probability. By conditioning on the first transition, prove that the expected number of transitions to move from point i to point 0 is $\sum_{j=1}^{i-1} 1/j$.

1.42 Three people A, B, and C are playing a series of games as follows. In the first turn, A and B play against each other; C is watching. Whoever loses in the nth turn will watch the other two play in the $(n+1)$th turn. Whoever wins in two successive turns wins the game. Let p_{ij} be the probability that player i beats player j $(i, j = A, B, C;$ $p_{ij} + p_{ji} = 1)$. Assume the games are fair; that is, $p_{ij} = .5$ for all $i, j = A, B, C$.

a. Find the probability that A, B, or C wins the game. Does it matter who plays in the first turn?

b. Find the expected number of turns in each game.

1.43 Let $x_1, x_2, x_3, \ldots$ be a sequence of independent and identically distributed bids on a certain asset that is offered for sale. Suppose the successful bid is the one that equals or exceeds a predetermined amount M. Let $f(x)$ be the probability density function of x_1 and $\alpha \equiv \Pr\{x_1 \ge M\}$. Let μ be the expected number of bids before the asset is sold. By conditioning on whether the first bid is a success or not, show that $\mu = 1/\alpha$.

1.44 Find the conditions for α, β, and γ so that the following function is a probability density function

$$g(x) = \begin{cases} e^{-\alpha x}(\beta + \gamma x) & \text{for all } x > 0 \\ 0 & \text{otherwise} \end{cases}$$

Hint: $\int_0^\infty e^{-\alpha x}(\beta + \gamma x)\,dx = \beta\alpha + \gamma/\alpha^2$.

1.45 A point A is chosen according to a uniform distribution on the interval [0, 1]. After its coordinate x has been observed, another point B is chosen according to a uniform distribution on the interval $[0, x]$. What is the expected coordinate of B?

1.46 Let $Z \equiv X^3$, where X is uniformly distributed over [0, 1]. Find $\Pr\{Z \le t\}$ and $\mathcal{E}[Z]$.

1.47 Let X and Y be two independent and uniformly distributed random variables on [0, 1]. Calculate $\Pr\{X < Y\}$. Can the result be generalized into a distribution other than the uniform distribution?

1.48 For a discrete random variable X taking values on the set of nonnegative integers,

a. Show that

$$\mathcal{E}[X] = \sum_{n=0}^{\infty} \Pr\{X > n\}$$

b. Hence, prove Equation (1.12.3) using Problem 1.16.

1.49 For a discrete random variable X taking values on the set of nonnegative integers,

a. Show that

$$\mathcal{E}[(X)(1 + X)] = 2\sum_{n=0}^{\infty} n \Pr\{X > n\}$$

b. Hence, prove Equation (1.12.4).

1.50 An urn contains N balls labeled $1, 2, \ldots, N$. Let X be the number of balls drawn with replacement until the same ball is drawn twice.

a. Explain why $X < N$.

b. Find $\Pr\{X > k\}$.

c. Hence, find $\Pr\{X = k\}$.

d. Hence, find $\mathcal{E}[X]$ by using Problem 1.48.

e. Apply the problem when $N = 5$.

1.51 Consider a nonnegative continuous random variable X having distribution function $F(x)$.

a. Use integration by parts to show that

$$\mathcal{E}[X] = \int_0^{\infty} \Pr\{X > t\}\, dt$$

b. Hence, show that

$$\mathcal{E}[X^r] = r\int_0^{\infty} t^{r-1} \Pr\{X > t\}\, dt$$

c. Let T be a constant. Show that

$$\mathcal{E}[\min(X, T)] = T - \int_0^{T} F(t)\, dt$$

d. Show that

$$\mathcal{E}[X|X < T] = T - \frac{1}{F(T)}\int_0^{T} F(t)\, dt$$

1.52 Let X and Y be two independent random variables having distribution functions $F_X(x)$ and $F_Y(x)$, respectively. Let $U \equiv \max(X, Y)$ and $V \equiv \min(X, Y)$.

a. Express the distribution functions $F_U(x)$ and $F_V(x)$ of U and V, respectively, in terms of $F_X(x)$ and $F_Y(x)$.

b. Hence, show that

$$\mathcal{E}[\min\{X, Y\}] = \int_0^{\infty} [1 - F_X(t)][1 - F_Y(t)]\, dt$$

and

$$\mathcal{E}[\max(X, Y)] = \int_0^\infty [1 - F(t)G(t)]\, dt$$

1.53 Let $Y_1, Y_2, \ldots, Y_n$ be n independent and uniformly distributed random variables on [0,1]. Let $V \equiv \min(Y_1, Y_2, \ldots, Y_n)$.

a. Show that $\Pr\{V \le t\} = 1 - (1 - t)^n$ for all $0 \le t \le 1$.

b. Hence, show that $\mathcal{E}[U] = 1/(n + 1)$.

1.54 Let $Y_1, Y_2, \ldots, Y_n$ be n independent and uniformly distributed random variables on [0,1]. Let $U \equiv \max(Y_1, Y_2, \ldots, Y_n)$.

a. Show that $\Pr\{U \le t\} = t^n$ for all $0 \le t \le 1$.

b. Hence, show that $\mathcal{E}[U] = n/(n + 1)$.

1.55 Suppose an instrument is functional at time 0. Its *reliability* $R(t)$ at time t is the probability that it survives up to time t or that it functions *continually* in the interval $(0, t]$. Its mean-time-to-system-failure *MTSF* is the expected duration from time $t = 0$ to the time the system fails. Show that

$$MTSF = \int_0^\infty R(u)\, du$$

1.56 Let $G(x)$ and $H(x)$ be any two functions of x having derivatives $g(x)$ and $h(x)$, respectively. A *convolution* of $G(x)$ and $H(x)$ is defined as:

$$G(x) * H(x) \equiv \int_0^x G(x - s)h(s)\, ds = \int_0^x H(x - s)g(s)\, ds$$

Let X and Y be two independent, nonnegative random variables having distribution functions $F_X(t)$ and $F_Y(t)$, respectively. Explain why the distribution function of $Z \equiv X + Y$ is $F_Z(t) = F_X(t) * F_Y(t)$.

CHAPTER 2

Discrete-Time Markov Chains

2.1 STOCHASTIC PROCESSES

We discussed sequences of independent outcomes of identical experiments in §1.8. A more general concept is that of a *stochastic process*, which is a sequence of outcomes X_t, where t runs over some *parameter space* T. These outcomes may be dependent on earlier ones in the sequence.

If the parameter space T is *countable*, the sequence is called a *discrete-time* process and is denoted by $\{X_n\}_{n=0,1,2,\dots}$. The index n identifies the *steps* of the process.

If the parameter space T is not countable, the sequence is called a *continuous-time* process and is denoted by $\{X(t)\}_{t\in[0,\infty)}$. The parameter t is called *time*, although it may be any continuous quantity such as distance. If t is true time, then it is referred to as "real time." To illustrate, the sequence of temperatures at some fixed location at time t, $\{T(t)\}_{t\in[0,\infty)}$, is a continuous-time process in real time; the sequence of temperatures ℓ feet from the end of a heat conduit at a fixed point in time, $\{T(\ell)\}_{\ell\in[0,\infty)}$, is also a continuous-time process, though not in real time.

Note that:

1. An outcome of a stochastic process is a sequence of outcomes obtained from a sequence of experiments. For example, let X_n be the number obtained in the nth toss of a die. Then $\{X_n\}_{n=0,1,2,\dots}$ is a stochastic process, and one of its possible outcomes could be $1 - 6 - 3 - 2 - 6 - \cdots$. To distinguish between the outcomes of each experiment and the outcomes of the process, we refer to the latter as the *(sample) paths*.
2. The sample space of a stochastic process is a collection of all possible paths.
3. The set of all possible and distinct outcomes of all experiments in a stochastic process is called its *state space* and normally is denoted by S. Its elements are called the *states*. The state space of a sequence of tossing a die is $\{0, 1, 2, 3, 4, 5, 6\}$.

Table 2.1.1 Four Types of Stochastic Processes

	Discrete Parameter Space T (step)	Continuous Parameter Space T (time)
Discrete State Space S (chain)	discrete-time chain (finite/infinite) $\{X_n\}_{n=0,1,2,\ldots}$	continuous-time chain (finite/infinite) $\{N(t)\}_{t\in[0,\infty)}$
Continuous State Space S	discrete-time process $\{W_n\}_{n=0,1,2,\ldots}$	continuous-time process $\{L(t)\}_{t\in[0,\infty)}$

If the state space of a stochastic process is countable, then the process is called a *chain*. Furthermore, if a chain has a finite number of states, it is said to be *finite*; otherwise, it is *infinite*.

As given in Table 2.1.1, we thus have the following four types of stochastic processes:

1. **Discrete-time chain:** An example of a discrete-time chain is $\{X_n\}_{n=0,1,2,\ldots}$, where X_n is the number of matches in the nth box of matches produced by a manufacturing operation. If the maximum capacity of the boxes is 100 matches, then this chain is finite and its state space is $\{0, 1, 2, \ldots, 100\}$.
2. **Continuous-time chain:** Let $N(t)$ be the total number of customers in a restaurant at time t; then $\{N(t)\}_{t\in[0,\infty)}$ would be a continuous-time chain.
3. **Discrete-time process:** Suppose W_n is the length of time the nth program has to wait before being executed by a computer; then $\{W_n\}_{n=0,1,2,\ldots}$ would be a discrete-time process.
4. **Continuous-time process:** As an example of a continuous-time process, we could have $\{L(t)\}_{t\in[0,\infty)}$, where $L(t)$ is the level of water in a dam at time t. If the maximum dam level is 50 feet, then the state space for this process would be the interval $[0,\ 50]$.

2.2 INDEPENDENT AND IDENTICALLY DISTRIBUTED RANDOM VARIABLES

The simplest kind of discrete-time chain is a sequence of independent and identically distributed discrete random variables $\{X_n\}_{n=0,1,2,\ldots}$, as introduced in §1.8. If we think of $X_0, X_1, \ldots, X_{n-1}$ as belonging to the past, X_n to the present, and X_{n+1} to the future, then because successive states are independent, the values of the past and current states $X_1, X_2, \ldots, X_n$ do not have any bearing on the future state X_{n+1}. In other words, the values of the former need not be known for the prediction of the latter, or $\Pr\{X_{n+1} = j\}$ remains the same regardless of the values of $X_1, X_2, \ldots, X_n$. More formally,

$$\Pr\{X_{n+1} = j \mid X_0 = i_0, X_1 = i_1, \ldots, X_n = i\} = \Pr\{X_{n+1} = j\}$$

Suppose a die is tossed repeatedly and independently. Let us assume that the last 100 tosses have all yielded 4. This accumulated knowledge would be helpful in checking the assumption that the die is fair; but if we continue to believe that it is fair, the probability of another 4 on the next toss is still 1/6. In other words, the probability of a 4 does not decrease, as some might think, to compensate for earlier outcomes.

2.3 MARKOV CHAINS OF ORDER *i*

A general discrete-time chain would be a sequence of discrete random variables $\{X_n\}_{n=0,1,2,\ldots}$, in which X_{k+1} is dependent on all previous outcomes $X_0, X_1, \ldots, X_k$. Calculation of $\Pr\{X_{k+1} = j\}$ would require the complete information about the outcomes of the process from the initial to the current step.

Analysis of this type of chain can easily become unmanageable, especially for long-term forecasts. Fortunately, in many practical situations, the influence of the chain's earlier outcomes on its future one tends to diminish rapidly as time passes. Thus, for mathematical tractability, even when reality may argue moderately against it, we nearly always can assume that X_{n+1} *is dependent only on* i *previous outcomes*, where $i \geq 1$ is a fixed and finite number. In this case, obtaining $\Pr\{X_{n+1} = j\}$ requires only the information about the previous i outcomes—that is, those from step $n - i + 1$ to step n—all information about the process before that having been judged and assumed irrelevant:

$$\begin{aligned}\Pr\{X_{n+1} = j \mid X_0 = i_0, X_1 = i_1, \ldots, X_n = i\}\\ = \Pr\{X_{n+1} = j \mid X_{n-i+1} = i_{n-i+1}, X_{n-i+2} = i_{n-i+2}, \ldots, X_n = i\}\end{aligned}$$

We call this kind of chain an *ith-order Markov chain*, or a *Markov chain of order i*.

Note that when the next outcome depends on none of the earlier ones, the chain is a sequence of independent outcomes, and we may refer to it as the Markov chain of order 0.

As a simple example, suppose a person's emotional mood on a particular day is dependent only on his or her mood the previous 2 days:

1. If happy both yesterday and today, the person's mood is moderately likely to change for the worse, so there is a only a 40% chance of happiness for tomorrow.
2. If happy today but not yesterday, the happy mood is more likely to last, and there is a 70% chance the person will be happy tomorrow.
3. If happy yesterday but not today, the sad mood is likely to continue, and the person's probability of returning to the happy state is only 35%.
4. Finally, if sad both previous days, the sad mood is likely to subside, and happiness has a 80% chance for tomorrow.

The chain representing this person's mood, with state space $\{s \equiv \text{sad},\ h \equiv \text{happy}\}$, would be Markovian of order 2.

A Historical Note: A word or two about Markov seem appropriate before we go on. Andrei Andreyevich Markov (1856–1922) was a professor at Saint Petersburg University in Russia from 1886 and a full member of the Saint Petersburg Academy of Sciences from 1896. He began studying simple chains in 1906, subsequently published many articles on the subject, and later gained a reputation as a pioneer in the field. The expression *Markov chain* appeared no later than 1926 (Sheynin, 1988). ■

Markov Chains in Action: Which Order for Daily Precipitation? Rainfall patterns have been studied using Markov models. In a study of water resources, Chin (1977) investigated daily precipitation records from 1948 to 1973 at more than 100 locations in the United States. He found that

> the proper Markov order of the daily precipitation process depends primarily on the season of the year and to a lesser degree on the geographical location. For the winter months of January and February, second- or higher-order conditional dependence in the daily precipitation process is prevalent for the majority of stations .... On the other hand, for the summer months of July-August a majority of stations displayed first-order conditional dependence in the daily precipitation occurrence process. ■

Markov Chains in Action: Which Order for Group Dynamics? Another example of a Markov chain of order i is given by Parker (1988), who observed nine four-man groups engaged in discussion and noted the order in which the persons spoke. He selected the second-order as the best Markov model, arguing that who will speak next depends on who the previous two speakers were. For instance, "the most likely speaker after Andrew and then Bill would be Andrew again." ■

To predict the behavior of a Markov chain of order i at step $n + 1$, we need to remember the chain's behavior from step $n - i + 1$ to step n. We show later how this memory can all be stored in an equivalent Markov chain of order 1. So let us first concentrate on the first-order Markov chains.

2.4 FIRST-ORDER MARKOV CHAINS

We usually refer to the first-order Markov chains simply as *Markov chains*, without mentioning the order. For these chains, only their present (at time n), not their past (at times $0, 1, 2, \ldots, n-1$), has any influence on their future (at time $n+1$). In other words, for all $n > 0$,

$$\Pr\{X_{n+1} = j \mid X_0 = i_0, X_1 = i_1, \ldots, X_n = i\} = \Pr\{X_{n+1} = j \mid X_n = i\}$$

Such chains can be thought of as *memoryless*.

Since the behavior of a Markov chain at time $n + 1$ is dependent only on its behavior at time n, the chain can be described completely by its *(one-step) transition probability* at step n as

$$p_{ij}(n) \equiv \Pr\{X_{n+1} = j \mid X_n = i\} \quad \text{for all } i, j \in S, \text{ and } n \geq 0 \tag{2.4.1}$$

These probabilities can be grouped together into a *(one-step) transition matrix* as

$$\mathbf{P}(n) \equiv \left[p_{ij}(n)\right]_{i,j \in S} \tag{2.4.2}$$

Since the process must visit some state in the state space S at each step, the law of total probability requires

$$\sum_{j \in S} p_{ij}(n) = 1 \quad \text{for all } i \in S, \text{ and } n \geq 0$$

or *the sum along each row of a transition matrix must be 1.*[1]

Consider a drunkard who is in the middle of a bridge and is attempting to finish crossing it. However, the drunkard's current state of inebriation makes any attempt to step forward, toward the land on the other side, succeed only in a staggering sidestep 1 foot long to the left or to the right. Assume the bridge is 4 feet wide. Let X_n be the distance, measured in 1-foot segments, from the left edge of the bridge to the drunkard at step n. The state space of the chain $\{X_n\}_{n=0,1,2,\ldots}$ is thus $S = \{0, 1, 2, 3, 4\}$. At any time, if not at an edge of the bridge, the drunkard can sidestep 1 foot to the right with probability p or 1 foot to the left with probability $q \equiv 1 - p$. However, when an edge of the bridge is reached ($X_n = 0$ or $X_n = 4$), the drunkard immediately falls into the water and remains there.

We have, for example,

$$p_{23}(n) = \Pr\{X_{n+1} = 3 \mid X_n = 2\} = p \quad \text{(sidestep to the right from 2 to 3)}$$
$$p_{21}(n) = \Pr\{X_{n+1} = 1 \mid X_n = 2\} = q \quad \text{(sidestep to the left from 2 to 1)}$$

These probabilities describe a first-order Markov chain at time $n \geq 0$ with the following one-step transition matrix:

$$\mathbf{P}(n) \equiv \begin{array}{c} \\ 0 \\ 1 \\ 2 \\ 3 \\ 4 \end{array} \begin{array}{c} \begin{array}{ccccc} 0 & 1 & 2 & 3 & 4 \end{array} \\ \begin{pmatrix} 1 & 0 & 0 & 0 & 0 \\ q & 0 & p & 0 & 0 \\ 0 & q & 0 & p & 0 \\ 0 & 0 & q & 0 & p \\ 0 & 0 & 0 & 0 & 1 \end{pmatrix} \end{array} \tag{2.4.3}$$

Let us denote $\Pr\{X_n = i\}$ by $\pi_i^{(n)}$ and define the *state distribution* of X_n at step n as

$$\boldsymbol{\pi}^{(n)} \equiv [\Pr\{X_n = i\}]_{i \in S} \equiv \left[\pi_i^{(n)}\right]_{i \in S} \quad \text{for all } n \geq 0 \tag{2.4.4}$$

For example, if the state space is $\{1, 2, 3, \ldots\}$, this distribution is the vector

$$\boldsymbol{\pi}^{(n)} = (\pi_1^{(n)} \quad \pi_2^{(n)} \quad \pi_3^{(n)} \quad \ldots)$$

From the law of total probability (1.7.1), we now have

$$\pi_i^{(n)} = \sum_{j \in S} \Pr\{X_n = i | X_{n-1} = j\} \Pr\{X_{n-1} = j\} = \sum_{j \in S} \pi_j^{(n-1)} p_{ji}(n-1)$$

[1] Due to some round-off error, this sum might not be exactly 1 in some examples in this book.

In matrix form, this equation can be written as:

$$\pi^{(n)} = \pi^{(n-1)}\mathbf{P}(n-1) \tag{2.4.5}$$

2.5 MATRIX NOTATION FOR MARKOV CHAINS

In this book, bold capital letters denote a *matrix*. If we write $\mathbf{A} \equiv [a_{ij}]_{i,j\in S}$, as in Equation (2.4.2), we mean that its elements are scalars a_{ij} corresponding to *states* i and j of the sample space S in a Markov chain. When appropriate or necessary, the states are displayed along the rows and columns of this matrix, as in Equation (2.4.3).

Note that the states of a chain do not have to be numerical. They could be colors, brands, conditions, or other qualitative variables. Suppose now that the states are numbers. Then when we refer to the (i, j)-element of $\mathbf{A}$, or $a_{i,j}$, we do not mean the entry in the ith row from the top and jth column from the left of the matrix, but rather the entry in the row and column corresponding to states i and j, respectively. For example, the (4, 3)-element in Equation (2.4.3) is 0, while the element in the fourth row and the third column in that matrix is q. We would therefore say "the 4-row" and "the 3-column" rather than "the fourth row" and "the third column."

Likewise, we use a bold lowercase letter to denote a *vector*. The equivalent relation $\mathbf{a} \equiv [a_i]_{i\in S}$, as in Equation (2.4.4), means the i-element of $\mathbf{a}$ is scalars a_i corresponding to state i of a sample space S.

2.6 TRANSITION DIAGRAMS

The drunkard's walk across the bridge that we have been discussing can be represented by a *transition diagram*, as shown in Figure 2.6.1. Note that each *node* in this diagram represents a state in the state space. If the chain can go from state i to state j in one step, then an *arrow* is drawn from node i to node j with the corresponding probability $p_{ij}(n)$ written on it.

For small problems, this type of diagram represents the process more clearly than does a transition matrix.

2.7 THE IMPORTANCE OF FIRST-ORDER MARKOV CHAINS

First-order Markov chains are important because they seem to strike a middle ground in the art of model building.

FIGURE 2.6.1 **Transition diagram for the drunkard's walk example.**

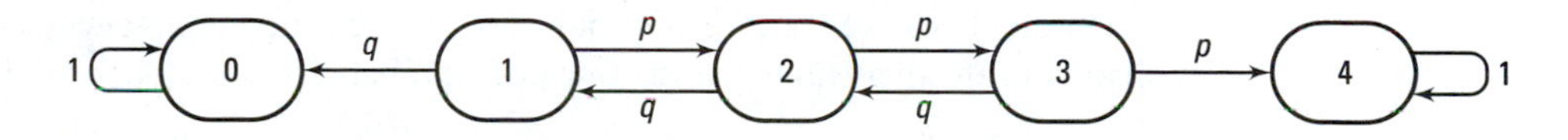

For example, consider the net worth of a person at age n ($n = 0, 1, 2, \ldots$) as a discrete-time chain. It is too simplistic to say that this chain is just a sequence of independent outcomes. In other words, the order of the chain must be at least one, or we need to know at least how much wealth a person has this year before we can determine the probability that he or she will have a given amount of wealth next year.

Is the first-order assumption sufficient then? Probably not. Although two people may have precisely the same wealth this year, their future accumulations almost certainly are influenced by the general wiseness (or foolishness) of their decision-making strategies, as reflected by their wealths in the past.

A higher-order dependence is probably more appropriate, and the results obtained would probably be more realistic. The difficulty is that, the higher the order of dependence, the more complicated the mathematics. Things are actually worse than they seem in this regard because, as we shall see later, the dimensionality of the problem increases much faster than the dependence order.

We are faced here with a trade-off problem: Should we settle for a first-order assumption, which is solvable but less realistic, or a higher-order assumption, which is mathematically more complicated although more defensible? As a general rule, the first-order dependence is a good first choice and often gives sufficiently satisfactory solutions. It is not as simplistic as the zero-order assumption, yet is simple enough so that many elegant mathematical tools are available for analysis.

It is also good to remember that even the best models are only idealizations of, and thus can never completely faithfully represent, real-world situations that are subject to many influencing factors. Even with a higher-order assumption, we will never achieve accurate results. So why spend that much more effort just to be a little closer to what basically is only an estimation?

Adding to the usefulness of first-order Markov chains is the fact that, as we will later see, higher-order Markov chains can be studied as equivalent, though more statewise complex, first-order chains.

2.8 TIME-HOMOGENEITY

To illustrate time-homogeneity, suppose the probability that the CEO of a large corporation will be a billionaire next year depends not only on a current net worth of $900 million, but also on the fact that the economy is unusually strong this year. The Markovian assumption still holds here because only current-year information (current wealth and current economic health) is needed to predict the future. Because the state of the economy fluctuates from year to year, the foregoing probability would also vary from year to year.

Generally, the Markovian assumption allows the transition probability at step n, $p_{ij}(n)$, to be a function of n. If $p_{ij}(n)$ does not depend on n, the chain is said to be *time-homogeneous*; otherwise, it is *time-inhomogeneous*.

In this book we shall restrict ourselves to the time-homogeneous cases; that is, we shall henceforth assume that the transition probabilities do not depend on time, or

$$p_{ij}(n) = p_{ij} \quad \text{for all } i, j \in S, \text{ and } n \geq 0$$

or

$$\mathbf{P}(n) = \mathbf{P} \quad \text{for all } n \geq 0$$

Recalling the drunkard on the bridge, is the first-order assumption realistic? For drunkards who are memoryless (get it?), it might well be. Is time-homogeneity a realistic assumption? Probably not. Matrix (2.4.3) might hold true for some initial values of n, but after a few steps (say, ten) the drunkard could cease to function entirely and become completely immobile. In that case, the transition matrix would change at step 11 so that

$$\mathbf{P}(n) = \begin{array}{c} \\ 0 \\ 1 \\ 2 \\ 3 \\ 4 \end{array} \begin{array}{c} \begin{array}{ccccc} 0 & 1 & 2 & 3 & 4 \end{array} \\ \begin{pmatrix} 1 & 0 & 0 & 0 & 0 \\ 0 & 1 & 0 & 0 & 0 \\ 0 & 0 & 1 & 0 & 0 \\ 0 & 0 & 0 & 1 & 0 \\ 0 & 0 & 0 & 0 & 1 \end{pmatrix} \end{array} \quad \text{for all } n \geq 11$$

There are many applications in which transition matrices do change with time, and hence, the time-homogeneity assumption would prove inappropriate.

Markov Chains in Action: Changes of Town Sizes. Analyzing the sizes of small towns in the state of Wisconsin, as reported by the *U.S. Census of Population* from 1880 to 1960 inclusive, Fuguitt (1965) concluded that "the simple Markov chain did not actually describe the data . . . since transition probability matrices were quite different for different decades." ■

If the transition matrix of a time-inhomogeneous chain changes in an arbitrary manner, there is little we can do to predict its future development. However, prediction may still be possible if its transition matrix is changing in a way that is describable as a function of the step n.

For example, consider the following *Pólya urn model* for describing the spread of an epidemic: An urn contains R red and G green balls at step 0. At each step, we draw a ball at random from the urn and then return it with k additional balls of the same color. The total number of balls in the urn after the nth step is thus $R + G + nk$. If we let X_n be the number of red balls after the nth step, it can be shown that $\{X_n\}_{n=0,1,2,\ldots}$ is an infinite Markov chain having state space $\{0, 1, 2, 3, \ldots\}$ and transition probabilities

$$p_{ij}(n) = \begin{cases} \dfrac{i}{R+G+nk} & \text{for } j = i + k \\ 1 - \dfrac{i}{R+G+nk} & \text{for } j = i \\ 0 & \text{otherwise} \end{cases} \tag{2.8.1}$$

This chain is time-inhomogeneous, but theoretical prediction is possible because its transition probabilities are a function of n.

Although much theoretical work has been done on the subject, to date we are not aware of any practical applications with empirical data reported in the literature.

2.9 SEASONAL CHAINS

Other than models having transition probabilities describable as a function of n, as with the previous urn model, prediction is also possible for time-inhomogeneous chains having *seasonal variation*. Here, the transition matrix $\mathbf{P}(n)$ repeats itself after a fixed number of steps d, or

$$\mathbf{P}(n+d) = \mathbf{P}(n) \quad \text{for all } n \geq 0$$

Following is an example of a seasonal chain having $d = 2$ and state space $\{A, B, C\}$:

$$\mathbf{P}(2n) = \begin{array}{c} \\ A \\ B \\ C \end{array}\!\!\begin{array}{c} \begin{array}{ccc} A & B & C \end{array} \\ \begin{pmatrix} .3 & .5 & .2 \\ .1 & .4 & .5 \\ .2 & .1 & .7 \end{pmatrix} \end{array} \qquad \mathbf{P}(2n+1) = \begin{array}{c} \\ A \\ B \\ C \end{array}\!\!\begin{array}{c} \begin{array}{ccc} A & B & C \end{array} \\ \begin{pmatrix} .8 & .1 & .1 \\ .3 & .3 & .4 \\ .1 & .8 & .1 \end{pmatrix} \end{array} \qquad \text{for all } n \geq 0 \tag{2.9.1}$$

We return to these chains in §2.14.

Markov Chains in Action: Calendar Variability in Solar Radiation. As an example of chains with seasonal variation, consider the global solar radiation received on a particular horizontal section of the earth's surface on day n. Although this may be described as a time-homogeneous chain, Aguiar et al. (1988) argued that "the probability function for the summer period is much different from that for the winter time, and treating the whole year in this fashion would clearly be too simple minded."

Another approach might have been to have 365 different transition matrices, one for each day of the year. That was clearly too many, and such a great amount of detail was not really necessary. Aguiar et al. finally settled on a seasonal chain with 12 monthly transition matrices (that is, $d = 12$). ■

Markov Chains in Action: Is Asthmatic Change Seasonal? Jain (1986) classified chronic bronchial asthma patients into three states: self-care, intermediate care, and intensive care. The condition of asthmatic patients is normally assumed to be dependent on the daily airborne allergenic pollen count, which is highly seasonal. However, using the 1-year statistics of 47 patients, Jain could not reject the null hypothesis that the transition matrix is time-homogeneous. ■

2.10 THE RAT IN THE MAZE

Consider an experiment in which a rat is wandering inside a maze. The maze has six rooms labeled F, 2, 3, 4, 5, and S, as in Figure 2.10.1. If a room has k doors, the probability that the rat selects a particular door is $1/k$. However, if the rat reaches room F, which contains food, or room S, which gives it an electrical shock, then it is kept there, and the experiment stops.

FIGURE 2.10.1 The maze.

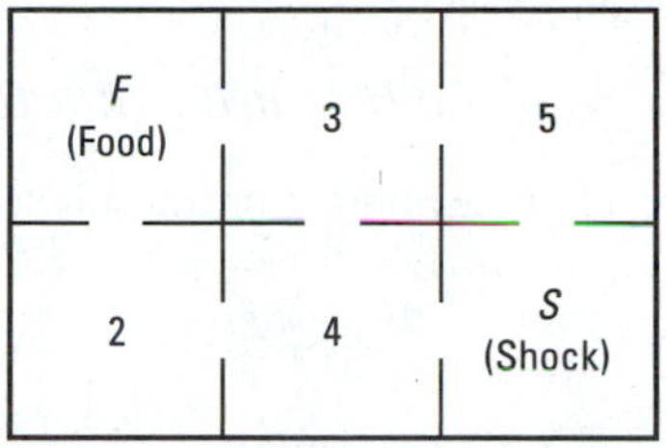

Let R_n be the room that the rat is in at step n. Chain $\{R_n\}_{n=0,1,2,\ldots}$, with state space $\{F, 2, 3, 4, 5, S\}$, is Markovian having transition matrix

$$
\mathbf{P} = \begin{array}{c} \\ F \\ 2 \\ 3 \\ 4 \\ 5 \\ S \end{array}
\begin{array}{c}
\begin{array}{cccccc} F & 2 & 3 & 4 & 5 & S \end{array} \\
\begin{pmatrix}
1 & 0 & 0 & 0 & 0 & 0 \\
1/2 & 0 & 0 & 1/2 & 0 & 0 \\
1/3 & 0 & 0 & 1/3 & 1/3 & 0 \\
0 & 1/3 & 1/3 & 0 & 0 & 1/3 \\
0 & 0 & 1/2 & 0 & 0 & 1/2 \\
0 & 0 & 0 & 0 & 0 & 1
\end{pmatrix}
\end{array}
$$

Its transition diagram is shown Figure 2.10.2.

2.11 EMPIRICAL TRANSITION MATRICES

In the rat-maze example, the transition matrix was obtained from an equal-probability assumption: If a room has k doors, the rat will have no preference for any particular door and thus will choose a door at random with probability $1/k$. In some applications, especially where humans are involved, this simple assumption may not be applicable. (Suppose you were in a similar maze. Would your choice be a random one?)

Lacking equal probabilities, the standard approach is to use statistical techniques to infer the values of the transition probabilities empirically from past data. To see how this might be accomplished, suppose the weather on a certain day is a Markov chain having

FIGURE 2.10.2 Transition diagram for the rat in the maze.

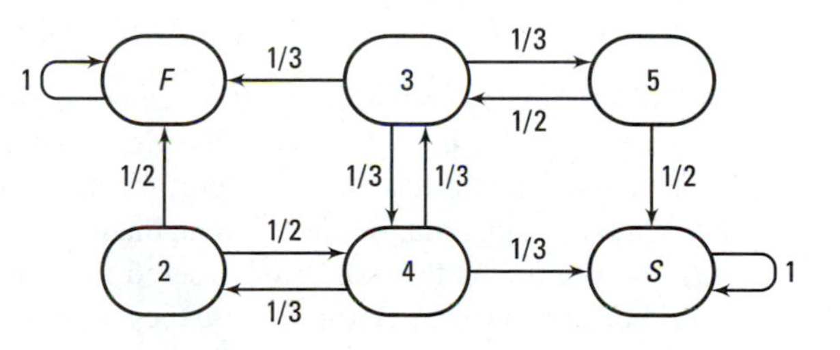

state space $\{R \equiv \text{rainy}, U \equiv \text{sunny}\}$. Furthermore, assume that the weather for the past 21 days has been

$$R\ U\ U\ R\ R\ U\ R\ R\ U\ U\ U\ R\ R\ U\ R\ R\ R\ U\ U\ U$$

Counting the number of transitions N_{ij} from state i to state j gives

$$[N_{ij}]_{i,j\in S} = \begin{matrix} & R & U \\ R & 6 & 5 \\ U & 4 & 5 \end{matrix}$$

The transition probability p_{ij} can now be estimated as

$$\hat{p}_{ij} = \frac{N_{ij}}{\sum_{k\in S} N_{ik}}$$

giving the estimation of **P** as

$$\widehat{\mathbf{P}} = \begin{matrix} & R & U \\ R & 6/(5+6) & 5/(5+6) \\ U & 4/(4+5) & 5/(4+5) \end{matrix} = \begin{matrix} & R & U \\ R & 6/11 & 5/11 \\ U & 4/9 & 5/9 \end{matrix}$$

Markov Chains in Action: Interpreting Data on Juvenile Crimes. The following application demonstrates that a critical *visual inspection* of empirically determined transition matrices may yield some very interesting insight into the process: There are three different conceptualizations of the lifetime careers of delinquent juveniles in the literature. Some studies stress the *situational* elements leading to the diversity of crimes. Some stress the *specializing* nature of crimes. Others suggest a *development* process in which offense type changes generally from less to more serious crimes over the course of a career. Smith and Smith (1984) divided the offenses into five states, or categories: nonindex (N), property (P), damage (D), robbery (R), and injury (I). Adding the state incarceration (IC), they obtained the following transition matrix based on a sample of all incarcerated juveniles (ages 13 to 18) who were institutionalized between October 1977 and December 1978 in New Jersey's correctional facilities:

$$\widehat{\mathbf{P}} = \begin{matrix} & IC & N & P & D & R & I \\ IC & 1 & 0 & 0 & 0 & 0 & 0 \\ N & .1242 & .4169 & .2999 & .0250 & .0371 & .0969 \\ P & .0615 & .2003 & .6026 & .0399 & .0278 & .0679 \\ D & .1048 & .2849 & .3602 & .1263 & .0242 & .0995 \\ R & .0996 & .1126 & .2944 & .0130 & .1840 & .2965 \\ I & .0678 & .2227 & .2984 & .0335 & .0898 & .2879 \end{matrix} \tag{2.11.1}$$

This estimated transition matrix seems to support the diversification perspective because the probability of a crime changing from one type to another is not insignificant in all cases. There is also some support for the specialization theory, as we can observe the relatively large magnitude in some diagonal values (for example, $p_{NN} = .4169$ or $p_{PP} = .6026$). As the states are ordered in an increasing degree of seriousness, this matrix offers no evidence for the tendency to move from a less serious to a more serious crime, since its values under the main diagonal are generally not smaller than those above it (Smith and Smith, 1984). ■

2.12 POPULATION-HOMOGENEITY

In obtaining transition matrices empirically, we have to keep in mind that they could be a function of many factors. Care should be taken to ensure that the population is *homogeneous*; that is, it does not comprise different *subpopulations*, and the transition probabilities associated with one subpopulation are sufficiently different from the other.

It is important that we neither over- nor underspecify the model. Overspecifying the model means that we divide our population too thinly into many subpopulations. This will lead to unnecessary computations and may cause statistical problems associated with small sample sizes from each subpopulation. On the other hand, underspecifying the model, or aggregating dissimilar subpopulations, can lead to misleading results.

Markov Chains in Action: Headaches in Male/Female Subpopulations. Headaches can be classified into four states: none, slight, moderate, and severe. Leviton et al. (1980) analyzed diaries of 177 female and 57 male patients for 28 days. "By observation of the stationary transition probabilities, it would appear that these probabilities were similar for males and females." ■

2.13 ENLARGING THE STATE SPACE

We now show that a Markov chain of any order can be studied as a Markov chain of order 1 if its state space is suitably *enlarged*. For illustration, consider the mood example introduced in §2.3. With the state space $\{s \equiv \text{sad},\ h \equiv \text{happy}\}$, this is a second-order Markov chain. However, if we expand the state space into $\{(s, s) \equiv (\text{sad, sad}), (h, s) \equiv (\text{happy, sad}), (s, h) \equiv (\text{sad, happy}), (h, h) \equiv (\text{happy, happy})\}$, we can represent the chain as a first-order one with the following transition matrix:

$$\mathbf{P} = \begin{array}{c} \\ (s,s) \\ (h,s) \\ (s,h) \\ (h,h) \end{array} \begin{array}{c} \begin{array}{cccc} (s,s) & (h,s) & (s,h) & (h,h) \end{array} \\ \begin{pmatrix} .20 & 0 & .80 & 0 \\ .65 & 0 & .35 & 0 \\ 0 & .30 & 0 & .70 \\ 0 & .60 & 0 & .40 \end{pmatrix} \end{array} \tag{2.13.1}$$

For a chain of order i, at the cost of having a larger transition matrix, the definition of the state space can be manipulated so as to include some memory of what happened within $i - 1$ steps immediately before the current step.

The difficulty that arises, however, is that the dimension of the transition matrix for the enlarged state space increases much more rapidly than the dependence order, especially when the number of states is large. For model builders, it is a challenge to find an optimal order in a given problem, striking a balance between the size of the transition matrix and the realness of the model. In most cases, low-order chains are preferred, thus keeping to a minimum the number of parameters to be estimated.

Markov Chains in Action: Old Faithful. Azzalini and Bowman (1990) suggested a second-order Markov chain to describe 299 eruptions of the Old Faithful geyser in Yellowstone National Park, Wyoming, from April through August 1985. Each eruption was classified as low (l) or high (h) depending on whether it was shorter or longer than 3 minutes. A second-order model was proposed, and then, with an enlarged state space, the following transition matrix was obtained:

$$\mathbf{P} = \begin{array}{c} \\ (l,h) \\ (h,l) \\ (h,h) \end{array} \begin{array}{c} \begin{array}{ccc} (l,h) & (h,l) & (h,h) \end{array} \\ \left(\begin{array}{ccc} 0 & .689 & .311 \\ 1 & 0 & 0 \\ 0 & .388 & .612 \end{array} \right) \end{array} \quad \blacksquare \tag{2.13.2}$$

2.14 USING SUPPLEMENTARY VARIABLES

Recall the rat-maze example and suppose now that the rat gets a shock only when it visits room S for the second time, not the first. When visiting this room for the first time, the rat is allowed to move to room 4 or 5, each with probability 1/2.

In this new incarnation, the chain $\{X_n\}_{n=0,1,2,\ldots}$, having state space $\{F, 2, 3, 4, 5, S\}$ as defined in §2.10 is no longer Markovian because knowing only that the rat is in room 5 is not sufficient to determine the probability that it receives a shock in the next step.

However, if we consider instead the *vector chain* $\{(X_n, Y_n)\}_{n=0,1,2}$, where Y_n is the number of times the rat has already visited room S, up to and including step n, then the redefined chain becomes Markovian again. The states of the new chain are $A \equiv (F, 0)$, $B \equiv (2, 0)$, $C \equiv (3, 0)$, $D \equiv (4, 0)$, $E \equiv (5, 0)$, $G \equiv (S, 1)$, $H \equiv (F, 1)$, $I \equiv (2, 1)$, $J \equiv (3, 1)$, $K \equiv (4, 1)$, $L \equiv (5, 1)$, $M \equiv (S, 2)$. Its transition matrix is:

	A	B	C	D	E	G	H	I	J	K	L	M
A	1	0	0	0	0	0	0	0	0	0	0	0
B	1/2	0	0	1/2	0	0	0	0	0	0	0	0
C	1/3	0	0	1/3	1/3	0	0	0	0	0	0	0
D	0	1/3	1/3	0	0	1/3	0	0	0	0	0	0
E	0	0	1/2	0	0	1/2	0	0	0	0	0	0
G	0	0	0	0	0	0	0	0	0	1/2	1/2	0
H	0	0	0	0	0	0	1	0	0	0	0	0
I	0	0	0	0	0	0	1/2	0	0	1/2	0	0
J	0	0	0	0	0	0	1/3	0	0	1/3	1/3	0
K	0	0	0	0	0	0	0	1/3	1/3	0	0	1/3
L	0	0	0	0	0	0	0	0	1/2	0	0	1/2
M	0	0	0	0	0	0	0	0	0	0	0	1

The transition diagram of this new chain is shown in Figure 2.14.1.

This example illustrates another method by which the memoryless restriction may be overcome. Here, the outcome X_n, which has no memory, is coupled with a *supplementary variable* Y_n, which records those events whose past occurrence is necessary to predict the future.

FIGURE 2.14.1 Transition diagram for the modified rat-maze example.

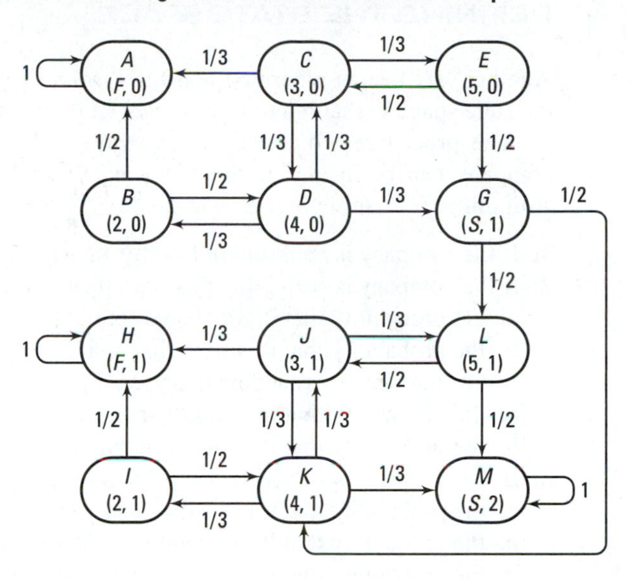

Markov Chains in Action: Wind Speed and Wind Direction. From the data obtained in their study of the wind speed variation during the course of a day, Roldan-Cañas et al. (1982) concluded that a knowledge of today's wind speed alone was insufficient for the prediction of tomorrow's wind speed. However, when today's wind speed was coupled with wind direction (that is, wind direction was used as a supplementary variable), the model became tenable.

> Although the dependence between wind speed and wind direction is not a general assumption, it seems a reasonable one for areas close to mountainous systems. In fact, it is widely supported by our observed data. ■

Supplementary variables would also help us to study Markov chains having seasonal transitions, as introduced in §2.9. For example, consider the chain with state space $\{A, B, C\}$ and two seasons in Equation (2.9.1). Coupled with a supplementary variable indicating whether the chain is in season 1 or 2, the state space now becomes $\{(A, 1), (B, 1), (C, 1), (A, 2), (B, 2), (C, 2)\}$, and we can study it as a time-homogeneous chain having transition matrix

$$\mathbf{P} = \begin{array}{c} (A,1) \\ (B,1) \\ (C,1) \\ (A,2) \\ (B,2) \\ (C,2) \end{array} \overset{\begin{array}{cccccc} (A,1) & (B,1) & (C,1) & (A,2) & (B,2) & (C,2) \end{array}}{\begin{pmatrix} 0 & 0 & 0 & .3 & .5 & .2 \\ 0 & 0 & 0 & .1 & .4 & .5 \\ 0 & 0 & 0 & .2 & .1 & .7 \\ .8 & .1 & .1 & 0 & 0 & 0 \\ .3 & .3 & .4 & 0 & 0 & 0 \\ .1 & .8 & .1 & 0 & 0 & 0 \end{pmatrix}} \qquad (2.14.1)$$

2.15 REFINING THE STATE SPACE

Another way that an otherwise non-Markovian chain can be *Markovized* is by refining the state space so that it has a sort of vestigial memory built into it.

The procedure is best explained with an example. Suppose each year a certain company can be in one of the following five financial states: $B \equiv$ bankrupt, $S \equiv$ struggling, $U \equiv$ surviving, $E \equiv$ expanding, and $H \equiv$ healthy. Suppose further that:

1. If the company is bankrupt or healthy, then it remains that way next year.
2. If the company is struggling this year, then
   a. the probability that it goes bankrupt next year is 1/2
   b. the probability that it is surviving and having a good year next year is 1/2
3. If the company is expanding this year, then
   a. the probability that it is healthy next year is 1/2
   b. the probability that it is surviving but having a bad year next year is 1/2
4. If the company is surviving and is having a good year, then
   a. the probability that it is healthy next year is 1/3
   b. the probability that it is struggling next year is 1/3
   c. the probability that it is surviving but having a bad year next year is 1/3
5. If the company is surviving but having a bad year, then
   a. the probability that it is expanding next year is 1/3
   b. the probability that it is bankrupt next year is 1/3
   c. the probability that it is surviving and having a good year next year is 1/3

With state space $\{B, S, U, E, H\}$, this chain is non-Markovian because, when the company is surviving, information about whether it is having a good or bad year is not included. However, if we simply *split* state U into two states $U_G \equiv$ surviving and having a good year and $U_B \equiv$ surviving and having a bad year, then we do have a Markov chain. The transition matrix for this chain would be

$$\mathbf{P} = \begin{array}{c} \\ B \\ S \\ U_G \\ U_B \\ E \\ H \end{array} \begin{array}{c} \begin{array}{cccccc} B & S & U_G & U_B & E & H \end{array} \\ \begin{pmatrix} 1 & 0 & 0 & 0 & 0 & 0 \\ 1/2 & 0 & 1/2 & 0 & 0 & 0 \\ 0 & 1/3 & 0 & 1/3 & 0 & 1/3 \\ 1/3 & 0 & 1/3 & 0 & 1/3 & 0 \\ 0 & 0 & 0 & 1/2 & 0 & 1/2 \\ 0 & 0 & 0 & 0 & 0 & 1 \end{pmatrix} \end{array} \qquad (2.15.1)$$

Its transition diagram is shown in Figure 2.15.1.

2.16 WHAT DO A LITTLE RAT AND A BIG COMPANY HAVE IN COMMON?

The rat-maze example introduced in §2.10 and the company just described arise from two different fields of application of Markov chains. Their transition matrices and transition diagrams seem to be different. However, if we exchange the positions of states U_B and

FIGURE 2.15.1 Transition diagram for the changing company.

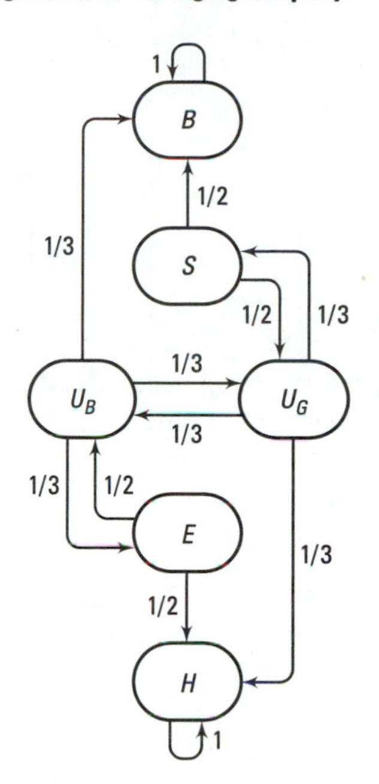

U_G in Matrix (2.15.1), then the two examples have identical transition matrices. Since the names of the states and the orders in which they are listed have no significance, the two chains are said to be *stochastically identical*—the behavior of the little rat and that of the big company in a competitive environment can be represented by the same model. The results obtained for one can be adapted to the other.

As you travel through this book, you will see that Markov chains have applications in many different fields, from biological science to business and from computer science to engineering. While these applications are different, many of them share essentially the same probability structure. We hope that, after seeing the similarity underlying these applications, you are encouraged to apply the techniques to your own field of interest.

2.17 LUMPABLE STATES

For the company studied in §2.15, we had to split state U into two states U_G and U_B to make its chain Markovian. If we now *lump* (*merge* or *combine*) two states U_G and U_B together into state U (that is, we reverse the splitting process), we again lose the Markov property of the chain.

Taking another example, consider a Markov chain having the following transition matrix

$$\mathbf{P} = \begin{array}{c} \\ 1 \\ 2 \\ 3 \\ 4 \\ 5 \end{array} \begin{array}{c} \begin{array}{ccccc} 1 & 2 & 3 & 4 & 5 \end{array} \\ \begin{pmatrix} .1 & .1 & .2 & .2 & .4 \\ .1 & .2 & .3 & .2 & .2 \\ .2 & .2 & .2 & .1 & .3 \\ .2 & .2 & .1 & .2 & .3 \\ .1 & .4 & .2 & .3 & 0 \end{pmatrix} \end{array} \tag{2.17.1}$$

Suppose we lump states 2 and 3 together. They then become *indistinguishable* in the sense that we can only tell if the chain is visiting state 2 or 3, but not which. Since the probability of moving from state i to the lumped state (2, 3) is $p_{i2} + p_{i3}$, we add the 2-column and the 3-column of matrix **P** together to get the (2, 3)-column of the lumped states:

$$\begin{array}{c} \\ 1 \\ 2 \\ 3 \\ 4 \\ 5 \end{array} \begin{array}{c} \begin{array}{cccc} 1 & (2,3) & 4 & 5 \end{array} \\ \begin{pmatrix} .1 & .3 & .2 & .4 \\ \hline .1 & .5 & .2 & .2 \\ .2 & .4 & .1 & .3 \\ \hline .2 & .3 & .2 & .3 \\ .1 & .6 & .3 & 0 \end{pmatrix} \end{array}$$

Now notice that the 2-row and the 3-row of the new matrix are different. For example, the transition probability $p_{2,1} = .1$ is different from the transition probability $p_{3,1} = .2$. Thus, the transition probability from the lumped state (2, 3) to state 1 cannot be determined because it requires the precise knowledge of which of the two states 3 or 4 the chain starts from. Hence, the chain is no longer Markovian if we lump states 2 and 3.

On the other hand, suppose we lump states 3 and 4 together. Adding the 3-column and the 4-column together yields:

$$\begin{array}{c} \\ 1 \\ 2 \\ 3 \\ 4 \\ 5 \end{array} \begin{array}{c} \begin{array}{cccc} 1 & 2 & (3,4) & 5 \end{array} \\ \begin{pmatrix} .1 & .1 & .4 & .4 \\ .1 & .2 & .5 & .2 \\ \hline .2 & .2 & .3 & .3 \\ .2 & .2 & .3 & .3 \\ \hline .1 & .4 & .5 & 0 \end{pmatrix} \end{array}$$

In this new matrix, the 3-row and the 4-row are identical. Now, if we want to determine the transition probability from the lumped state (3, 4) to state 1, there is no need to know whether the chain starts from state 3 or state 4 because this probability equals .2 in both cases. The transition probability from the lumped state (3, 4) to states 2 and 5 can be obtained in a like manner, and the Markov property is preserved.

States 3 and 4 are said to be *lumpable*. The resulting chain has transition matrix

$$\tilde{\mathbf{P}} = \begin{array}{c} \\ 1 \\ 2 \\ (3,4) \\ 5 \end{array} \begin{array}{c} \begin{array}{cccc} 1 & 2 & (3,4) & 5 \end{array} \\ \begin{pmatrix} .1 & .1 & .4 & .4 \\ .1 & .2 & .5 & .2 \\ .2 & .2 & .3 & .3 \\ .1 & .4 & .5 & 0 \end{pmatrix} \end{array}$$

Generally, to see whether the states in a subset of the state space are lumpable, we first add all the columns of the transition matrix corresponding to those states, thus forming a new column for the subset. The states are lumpable if, after this summation, *the resultant rows corresponding to all states in the subset are identical.*

A *partition* $\tilde{S}$ of the state space S of a Markov chain is defined as an exhaustive collection of disjoint subsets of S. For example, $\{(a, d, f), c, (b, e)\}$ is a partition of $S \equiv \{a, b, c, d, e, f\}$, whereas $\{(a, d, f), (b, e)\}$ is not. We say a Markov chain having state space S is *lumpable with respect to its partition* $\tilde{S}$ if the states within each subset are lumpable, when the lumpings are done simultaneously for all subsets. We have shown that Chain (2.17.1) is lumpable with respect to the partition $\{1, 2, (3, 4), 5\}$.

The reader can verify that, after lumping together states 3 and 4 of Chain (2.17.1), states 2 and 5 are also lumpable. Hence, this chain is also lumpable with respect to the partition $\{1, (2, 5), (3, 4)\}$. The resulting chain has transition matrix

$$\tilde{\mathbf{P}} = \begin{array}{c} \\ 1 \\ (2,5) \\ (3,4) \end{array} \begin{array}{c} \begin{array}{ccc} 1 & (2,5) & (3,4) \end{array} \\ \begin{pmatrix} .1 & .5 & .4 \\ .1 & .4 & .5 \\ .2 & .5 & .3 \end{pmatrix} \end{array}$$

Note that this chain is not lumpable with respect to the partition $\{1, (2, 5), 3, 4\}$ because states 2 and 5 are not lumpable unless states 3 and 4 are also lumped.

As you can see, lumpability in a Markov chain is rather restrictive. Furthermore, some information is lost when we lump states. However, when lumping is possible, it may have some analytical benefits because of the relative simplicity of the reduced model. Lumping also enables us to study the *macroscopic* behavior of the chain, where the transitions within a certain subset of the state space are not of interest.

2.18 DISCRETE-TIME CHAINS IN A CONTINUOUS-TIME CHAIN

There are at least three ways a discrete-time chain can be obtained from a continuous-time chain.

1. Embedded chains: Consider a chain developing along a continuous parameter space. Because its state space is discrete, there are *epochs* (points in time) at which the process changes values. Between these epochs, the chain remains in the same state. Figure 2.18.1 shows an example of a possible development of this type of process.

For the rat-maze example, there are epochs along the continuous-time axis when the rat moves from one room to another; in between, it stays in the same room.

FIGURE 2.18.1 A realization of an embedded chain.

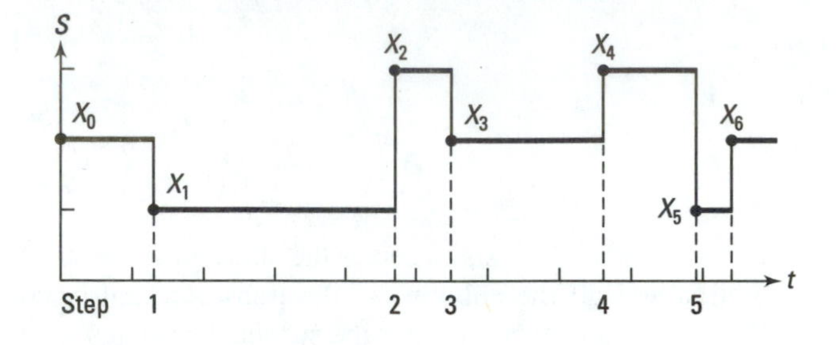

If we now observe this continuous-time process only when it changes values, we will have a discrete-time chain which is *embedded* in the continuous-time process. For the rat, if we are interested only in the sequence of rooms it visits, and thus only record its location immediately after it changes rooms, then we will have a discrete-time embedded chain, as in §2.10.

Note that, for this kind of chain, because no step occurs until it changes states, p_{ii} is zero for all states $i \in S$.

2. Equally spaced chains: The mood of a person as it develops in time should be a stochastic process having a continuous parameter space and a continuous state space. In the example in §2.3, however, we simplified this problem in two ways. First, we "discretized" the mood, classifying it only as sad or happy. Second, we observed the process only every 24 hours. We have here a chain *equally spaced* along the continuous-time axis.

Generally, an equally spaced chain is obtained from a continuous-time chain if the parameter space is divided into equal segments such as minutes, hours, or days. The process is then observed only at the beginning of these time segments (Figure 2.18.2).

3. Randomly spaced chains: If the chain is observed regularly, but not necessarily when it changes values or at equally spaced epochs, then it is a *randomly spaced* chain. An example of this kind of chain is $\{V_n\}_{n=0,1,2,\dots}$, where V_n is the number of code violations cited by an inspector who inspects a restaurant at random.

FIGURE 2.18.2 A realization of an equally spaced chain.

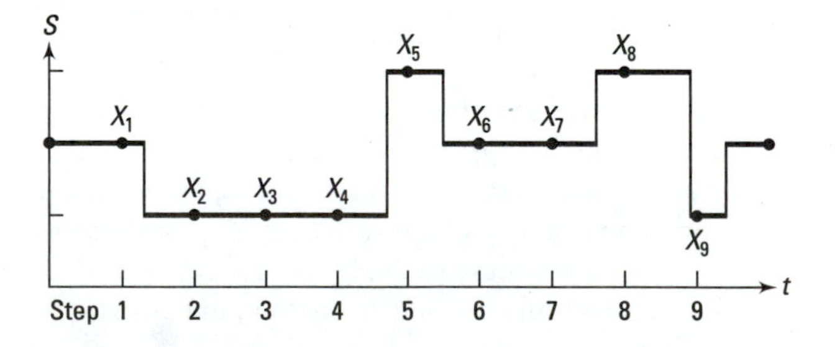

Markov Chains in Action: Discrete-Time Chains in Geology. We must emphasize again that the parameter space does not have to be real time. In geology, Markov chains have been used to study subsurface stratification, identifying different lithological types such as sandstone, shale, and siltstone. The parameter space is the vertical distance down from a certain height. Both the embedded and equally spaced discrete-time chains have been used as models. We use the embedded chain when the types are observed only at their transitions, disregarding the stratum's thickness in between. The equally spaced chain is appropriate when the types are observed at equally vertically spaced points (Kulatilake, 1987). ■

2.19 THE SOJOURN TIMES

We define the *sojourn time* of a chain at state i as the total number of steps it stays there at each visit. For the embedded chains discussed previously, sojourn times are always one. For the equally spaced chain, sojourn times have the *geometric distribution*; that is, the probability that the chain stays at state i for exactly n steps is $(1 - p_{ii})p_{ii}^{n-1}$, as in Equation (1.12.1).

It is important to note here that it is necessary that the sojourn times of an equally spaced chain be geometrically distributed; otherwise, the memoryless assumption is not valid.

Markov Chains in Action: Are Sleep Patterns Memoryless? A person's sleep pattern can be classified into five states, or categories, varying from awake to light sleep to deep sleep. Yang and Hursch (1973) obtained the distributions of the sojourn times in minutes from a sample of 34 nights of sleep for 20- to 39-year-old normal females. They found a significant difference between the sample distribution and the geometric distribution and concluded that the Markov chain model is inadequate. Similar conclusions were also obtained for the data of normal males (ages 20–39, 40–59, 60–69), pregnant women, and insomnia patients. ■

2.20 SUMMARY

1. A stochastic process can be classified according to whether its parameter space and its state space are discrete or continuous. It is called a chain if its state space is discrete.
2. A Markov chain of order i requires the information about the previous i outcomes (from step $n - i + 1$ to step n) for the prediction of the $n + 1$ outcome.
3. A first-order Markov chain can be fully described by either
   a. all its one-step transition probabilities $p_{ij}(n)$s
   b. its one-step transition matrix $\mathbf{P}(n)$
   c. its transition diagram
4. In this book, all Markov chains are assumed to be time-homogeneous (that is, the transition probabilities at step n do not depend on n) and population-homogeneous (that is, there are no subpopulations with differing transition matrices).

5. Sometimes a suitably defined state space can be used to convert a chain that lacks the Markov property to an equivalent one that possesses the Markov property. There are at least three different methods of conversion:
   a. enlarging the state space
   b. using the supplementary variables
   c. refining the state space
6. Normally, we cannot arbitrarily lump the states of a Markov chain and still preserve its Markovian property.
7. A discrete-time chain can be obtained from a continuous-time chain as:
   a. an embedded chain
   b. an equally spaced chain
   c. a randomly spaced chain

PROBLEMS

2.1 We perform a sequence of independent Bernoulli trials. Each has probability p of success (S) and $1 - p$ of failure (F). For each of the following chains, determine whether it is Markovian. If so, find its transition matrix; if not, modify the state space to make it Markovian.

a. $\{X_n\}_{n=0,1,2,\ldots}$, where X_n is the total number of Ss in the first n trials.

b. $\{Y_n\}_{n=0,1,2,\ldots}$, where Y_n is equal to 0, $1a$, $1b$, or 2 depending on whether the $(n-1)$st and the nth trials are SS, SF, FS, or FF, respectively.

c. $\{Z_n\}_{n=0,1,2,\ldots}$, where Z_n is the total number of Fs in the $(n-1)$st and the nth trials.

d. $\{T_n\}_{n=0,1,2,\ldots}$, where T_n is equal to 0 if the $(n-1)$st and the nth trials are SS and 1 otherwise.

e. $\{U_n\}_{n=0,1,2,\ldots}$, where $U_n = 0$ if the nth trial is S or $U_n = k$ if the nth trial is F and the last S was obtained at the $(n-k)$th trial. In other words, U_n is the accumulated number of Fs since the last S.

f. $\{V_n\}_{n=0,1,2,\ldots}$, where $V_n = i$ if the nth success is at the ith trial.

2.2 The following experiments are performed on two urns A and B containing a total of N balls. Let X_n be the number of balls in urn A immediately after the nth step. Show that $\{X_n\}_{n=0,1,2,\ldots}$ is Markovian and find its transition matrix.

a. At every step, one ball is equally likely to be drawn from among all N balls. It is then put into urn A with probability p or urn B with probability $1 - p$.

b. At every step, an urn is first selected with probability proportional to its contents; that is, if $X_n = i$, then A is selected with probability i/N. One ball is then drawn from urn A with probability p or from urn B with probability $1 - p$. The drawn ball is then put into the selected urn.

c. At every step, an urn is selected with probability proportional to its contents. One ball is equally likely to be drawn from among all N balls. The drawn ball is put into the selected urn.

2.3 The *Ehrenfest* model for heat exchange can be described as follows: Two urns, A and B, contain a total of $2N$ balls. At every step, one ball is equally likely to be drawn from among all $2N$ balls. It is then put into the other urn. Let X_n be the number of balls in urn A immediately after the nth step. Goldenberg (1988) uses this chain to represent futures and security prices, which tend to return to an equilibrium $N/2$. The greater the distance between the current price and $N/2$, the greater the probability that it moves toward $N/2$ rather than away from it. Show that $\{X_n\}_{n=0,1,2,...}$ is Markovian and find its transition matrix.

2.4 The *Bernoulli-Laplace* model to describe the flow of two incompressible liquids between two containers can be described as follows: N white and N black balls are distributed in two urns A and B, each containing N balls. At every step, one ball is drawn at random from each urn. The ball drawn from urn A is then put into urn B, and vice versa. Show that the number of white balls in urn A is a Markov chain and find its transition matrix.

2.5 An urn contains four balls, either black or white. A ball is drawn at random. If it is black, there is a probability $1-p$ that it will be returned to the urn; there is a probability p that a white ball will be returned to the urn instead. If it is white, it will be returned to the urn. This procedure is repeated indefinitely.

a. Let X_n be the number of white balls in the urn after the nth draw of a black ball. Find the transition matrix for the chain $\{X_n\}_{n=0,1,2,...}$.

b. Let Y_n be the number of white balls in the urn after the nth draw of a ball, either black or white. Find the transition matrix for the chain $\{Y_n\}_{n=0,1,2,...}$.

2.6 An urn contains three red and three white balls. A pair of balls is repeatedly drawn. If this pair has one red and one white ball, then two blue balls are put back into the urn instead; otherwise, they are returned into the urn. (This is a crude model of a chemical reaction in which a red and a white atom combine to form two blue atoms.) Let R_n be the number of red balls (and also the number of white balls) in the urn. Find the transition matrix for the chain $\{R_n\}_{n=0,1,2,...}$.

2.7 Balls are thrown one at a time and must land in one of the k cells with equal probability. Let X_n be the number of occupied cells after the nth throw. Find the transition matrix for the chain $\{X_n\}_{n=0,1,2,...}$.

2.8 Refer to Problem 1.1. Consider now the case of maximum *consanguinity*; that is, two animals are mated, and *from their offspring* two individuals of opposite sex are randomly chosen. These two are mated, and the procedure continues indefinitely. Find the transition matrix for the pairs of genotypes from one generation to the other.

2.9 Let $\{X_n\}_{n=0,1,2,...}$ be a sequence of independent and identically distributed discrete random variables with $\Pr\{X_1 = j\} \equiv p_j$ for all $j = 0, 1, 2, 3, \ldots$. Determine whether each of the following chains is Markovian or not. If so, find its transition probabilities.

a. $\{S_n\}_{n=0,1,2,...}$, where $S_n = \sum_{i=1}^{n} X_i$.

b. $\{M_n\}_{n=0,1,2,...}$, where $M_n = \max\{X_1, X_2, \ldots, X_n\}$.

c. $\{m_n\}_{n=0,1,2,...}$, where $m_n = \min\{X_1, X_2, \ldots, X_n\}$.

2.10 Let $\{X_n\}_{n=0,1,2,...}$ be a sequence of independent and identically distributed random variables taking values in $\{0, 1, 2, 3, 4\}$, with $\Pr\{X_1 = i\} \equiv p_i$ for all $0 \le i \le 4$. Let $Y_0 = 0$ and $Y_{n+1} = (Y_n + X_n) \bmod (5)$. Find the transition matrix for $\{Y_n\}_{n=0,1,2,...}$.[2]

2.11 Let Y_n be the sum of n independent rolls of a fair die. Find the transition matrix for $\{X_n\}_{n=0,1,2,...}$, where $X_n = Y_n \bmod (8)$.

2.12 Items are produced by a machine one at a time. Let the probability that an item is defective be p. Consider the following sampling plan. In stage A, each item is inspected. If s consecutive nondefective items are found, the plan changes to stage B, in which each item is inpected with probability α. As soon as a defective item is found, the sampling plan reverts back to stage A. This pattern repeats indefinitely.

If the nth item is found defective, let $X_n = 0$; otherwise, let X_n be the number of previously inspected items found to be nondefective. If $X_n > s$, let $X_n = s$. Explain why the transition matrix of $\{X_n\}_{n=0,1,2,...}$ is

$$\mathbf{P} = \begin{array}{c} \\ 0 \\ 1 \\ 2 \\ \cdots \\ s \end{array} \begin{array}{c} \begin{array}{cccccc} 0 & 1 & 2 & \cdots & s \end{array} \\ \begin{pmatrix} p & q & & & \\ p & & q & & \\ p & & & \cdots & \\ \cdots & \cdots & \cdots & \cdots & \cdots \\ \alpha p & & & \cdots & 1-\alpha p \end{pmatrix} \end{array}$$

2.13 John is married to Mary. Let H_n be the number of happy persons in the couple John and Mary during day n ($H_n = 0$, 1, or 2). Assume the probability that John is happy during day n is $(1 + H_{n-1})/4$. The probability that Mary is happy during day n is also assumed to be $(1+H_{n-1})/4$. What is the transtion matrix for the the chain $\{H_n\}_{n=0,1,2,...}$?

2.14 (Daykin et al., 1967) The game of *Snakes and Ladders* consists of a board having 100 squares labeled $1, 2, \ldots, 100$. On the board, there are ten snakes: $(27, 10)$, $(55, 16)$, $(61, 14)$, $(69, 50)$, $(79, 5)$, $(81, 44)$, $(87, 31)$, $(91, 25)$, $(95, 49)$, and $(97, 59)$. There are also ten ladders: $(6, 23)$, $(8, 30)$, $(13, 47)$, $(20, 39)$, $(33, 70)$, $(37, 75)$, $(41, 62)$, $(57, 83)$, $(66, 89)$, and $(77, 96)$. The player starts from square 1. In his turn, he rolls a die. If he is on square i and the number turned up is d, he moves to square $i + d$. However, if square $i + d = u$ is the start of a snake or a ladder (u, v), he has to move immediately to square v. Formulate the game as a Markov chain.

2.15 A number X_0 is first chosen at random from the integers 1, 2, 3, 4, 5. Let X_n be a number chosen at random from the integers $1, 2, \ldots, X_{n-1}$ for all $n = 1, 2, 3, \ldots$. Find the transition matrix for $\{X_n\}_{n=0,1,2,...}$.

2.16 A company has three machines. Each day, independent of each other, a machine breaks down with probability p. Each night, there is one repairperson who can repair at most one machine. Let X_n be the number of machines available at the beginning of the nth day. Find the transition matrix for the chain $\{X_n\}_{n=0,1,2,...}$.

2.17 Three items A, B, and C are stored separately in three locations 1, 2, and 3. When a request is made for an item, a *sequential search method* is used, which looks for the item

[2] The modulus operation $x \bmod (c)$ gives the remainder when dividing x by c. For example, $27 \bmod (5) = 2$.

at location 1 first, at location 2 next (if the item is not found at 1), and then at location 3 (if the item is not found at 2). Let $X_n \equiv (ijk)$, where i, j, and k are the names of the items in locations 1, 2, and 3, respectively, after the nth search. To minimize the number of searches, it is suggested that the requested item be moved to location 1. For example, if $X_{n-1} = (CBA)$ and item B is requested at time n, then $X_n = (BCA)$. Let a, b, and c be the probability that the request is for A, B, and C, respectively ($a+b+c=1$). Find the transition matrix for the chain $\{X_n\}_{n=0,1,2,\ldots}$.

2.18 The *Greenwood* model for the spreading of contagious disease describes a population of size N comprising two kinds of individuals: the infected and the susceptible (those not yet infected). Let p be the probability that a susceptible is infected in one day, independently of the others. Find the transition matrix for $\{X_n\}_{n=0,1,2,\ldots}$, where X_n is the number of susceptibles at day n (see also Minh, 1977).

2.19 Suppose that there are N fish in a bowl. For each fish, let p be the probability that it will be caught if a person passes a net through the bowl. Let X_n be the number of fish left in the bowl after the nth pass. Find the transition matrix for $\{X_n\}_{n=0,1,2,\ldots}$. How does this model compare with the Greenwood model in Problem 2.18?

2.20 Consider a sequence of genes. Each is made up of N subgenes, of which i are normal and $N-i$ abnormal. This is how the nth gene replicates. First, it doubles itself into $2N$ subgenes, with $2i$ normal and $2(N-i)$ abnormal. N subgenes among these $2N$ are then randomly selected; they form the $(n+1)$th gene. Let X_n be the number of normal subgenes in the nth gene. Find the transition probabilities of $\{X_n\}_{n=0,1,2,\ldots}$

a. in terms of N

b. when $N=3$

2.21 In each game, a gambler wins the dollars he bets with probability p and loses with probability $1-p$. If he has less than \$3, he will bet all he has. Otherwise, since his goal is to have \$5, he will only bet the difference between \$5 and what he has. He continues to bet until he either has \$0 or \$5. Let X_n be the amount he has immediately after the nth bet. Draw the transition diagram for this chain.

2.22 In a war, there are three surviving tanks, A, B, and C. Tank A can hit its target with probability 1/3, tank B with probability 1/4, and tank C with probability 1/5. In a fight, they repeatedly fire simultaneously until either only one tank survives or none. Assume a tank randomly picks one of the other surviving tanks as a target. Represent this problem as a Markov chain. *Hint:* Use Problem 1.4.

2.23 Do Problem 2.22 but assume now that a surviving tank picks its strongest opponent as target.

2.24 Let $\mathbf{P}_1$ and $\mathbf{P}_2$ be two transition matrices. A chain is constructed as follows. Before each transition, a Bernoulli trial having p as the probability of success is performed. If success, then $p_{ij} = [\mathbf{P}_1]_{ij}$; else, $p_{ij} = [\mathbf{P}_2]_{ij}$. Find the transition matrix of this chain.

2.25 Suppose that whether a person is happy or sad today depends on his or her mood in the previous three days. Show how this mood can be analyzed as a first-order Markov chain. How many states are needed?

2.26 Four boys (Tim, Sam, Bob, and Jim) play catch.

1. If Tim has the ball, he is equally likely to throw it to Sam or Jim.
2. If Sam has the ball, he will throw it to Tim with probability 1/3, and to Jim with probability 2/3.
3. If Bob has the ball, he will throw it to Sam.

Obtain the transition matrix for the following two cases:

a. If Jim has the ball, he will throw it to Tim with probability 2/3, or run away with the ball with probability 1/3.
b. If Jim has the ball for the first time, he will throw it to Tim; if he has the ball for the second time, he will run away with it.

2.27 Weather is observed at an observation point three times daily, 8 hours apart: 8:00 A.M., 4:00 P.M., and 12:00 A.M. The observer records whether it is clear (E), cloudy (C), or raining (R). Let $\mathbf{P}_A$ be the 8-hour transition matrix from 12:00 A.M. to 8:00 A.M., $\mathbf{P}_B$ from 8:00 A.M. to 4:00 P.M., and $\mathbf{P}_C$ from 4:00 P.M. to 12:00 A.M. Assume

$$\mathbf{P}_A = \begin{array}{c} \\ E \\ C \\ R \end{array} \begin{array}{c} \begin{array}{ccc} E & C & R \end{array} \\ \begin{pmatrix} .3 & .4 & .3 \\ .2 & .5 & .3 \\ .1 & .5 & .4 \end{pmatrix} \end{array} \qquad \mathbf{P}_B = \begin{array}{c} \\ E \\ C \\ R \end{array} \begin{array}{c} \begin{array}{ccc} E & C & R \end{array} \\ \begin{pmatrix} .2 & .6 & .2 \\ .1 & .3 & .6 \\ .3 & .3 & .4 \end{pmatrix} \end{array}$$

$$\mathbf{P}_C = \begin{array}{c} \\ E \\ C \\ R \end{array} \begin{array}{c} \begin{array}{ccc} E & C & R \end{array} \\ \begin{pmatrix} .1 & .7 & .2 \\ .4 & .2 & .4 \\ .3 & .1 & .6 \end{pmatrix} \end{array}$$

Let Z_n be the kind of weather observed every 8 hours starting from 8:00 A.M. of day 0. Use a supplementary variable to make $\{Z_n\}_{n=0,1,2,\ldots}$ Markovian. Obtain its transition matrix $\mathbf{P}_Z$.

2.28 Let X_n be the number of bombers a squadron has at 8:00 A.M. on day n. If $X_n > 0$, they are sent on a mission; otherwise, no mission is flown. If sent out during the day, each bomber has a probability p_D of being shot down. Let Y_n be the number of bombers left after the mission at 8:00 P.M. on day n. They will fly missions the following day. If $Y_n < 3$, then $3 - Y_n$ replacement bombers will be flown to the squadron at night to bring it to its full strength of 3. Since replacement bombers must fly over enemy lines, there is a probability p_N that each will be shot down on the way during the night. Let Z_n be the number of bombers observed every 12 hours starting from 8:00 A.M. of day 0. Use a supplementary variable to make $\{Z_n\}_{n=0,1,2,\ldots}$ Markovian. Obtain its transition matrix $\mathbf{P}_Z$.

2.29 The life L of an instrument is a discrete random variable with $\Pr\{L = i\} = \alpha_i$ for $i = 1, 2, \ldots$. When it fails, it will be replaced with an identical component. The *planned replacement policy* calls for replacing the component when it fails or when it reaches age $N + 1$, whichever comes first. We study the chain $\{a_n\}_{n=0,1,2,\ldots}$ having state space $\{0, 1, 2, \ldots, N - 1, N\}$, where $a_n = i$ is the age of the equipment at time n

$(0 \le i \le N)$. (In this way, we can distinguish between replacement-by-failure at 0 and planned replacement at N). Explain why

$$\mathbf{P} = \begin{array}{c} 0 \\ 1 \\ 2 \\ \cdots \\ N-1 \\ N \end{array} \overset{\begin{array}{cccccc} 0 & 1 & 2 & \cdots & N-1 & N \end{array}}{\begin{pmatrix} \frac{\alpha_1}{\Lambda_1} & \frac{\Lambda_2}{\Lambda_1} & & \cdots & & \\ \frac{\alpha_2}{\Lambda_2} & & \frac{\Lambda_3}{\Lambda_2} & \cdots & & \\ \frac{\alpha_3}{\Lambda_3} & & & \cdots & & \\ \cdots & \cdots & \cdots & \cdots & \cdots & \cdots \\ \frac{\alpha_N}{\Lambda_N} & & & \cdots & & \frac{\Lambda_{N+1}}{\Lambda_N} \\ \frac{\alpha_1}{\Lambda_1} & \frac{\Lambda_2}{\Lambda_1} & & \cdots & & \end{pmatrix}}$$

where $\Lambda_i \equiv \sum_{j=i}^{\infty} \alpha_j$.

2.30 Consider a time-inhomogeneous Markov chain $\{X_n\}_{n=0,1,2,\ldots}$ with state space $\{0, 1, 2, 3\}$. Suppose

$$\Pr\{X_{n+1} = j \mid X_n = i\} = \begin{cases} p_{ij} & \text{when } n \text{ is even} \\ r_{ij} & \text{when } n \text{ is odd} \end{cases}$$

Use a supplementary variable to make this a time-homogeneous chain.

2.31 A market survey of 1500 consumers having a choice of three brands A, B, and C shows the following results:

$$[N_{ij}]_{i,j\in S} = \begin{array}{c} A \\ B \\ C \end{array} \overset{\begin{array}{ccc} A & B & C \end{array}}{\begin{pmatrix} 250 & 175 & 75 \\ 100 & 450 & 250 \\ 50 & 100 & 50 \end{pmatrix}}$$

where N_{ij} is the number of consumers changing from brand i to brand j in 1 year. Estimate the transition matrix for this chain.

2.32 Consider a chain with transition matrix:

$$\mathbf{P} = \begin{array}{c} 1 \\ 2 \\ 3 \\ 4 \\ 5 \end{array} \overset{\begin{array}{ccccc} 1 & 2 & 3 & 4 & 5 \end{array}}{\begin{pmatrix} .1 & .1 & .3 & .3 & .2 \\ .2 & .2 & .3 & .3 & 0 \\ 0 & .1 & .4 & .3 & .2 \\ .1 & 0 & .1 & .4 & .4 \\ .2 & .3 & .1 & .3 & .1 \end{pmatrix}}$$

Is it lumpable with respect to the following partitions? If so, find the transition matrix of the lumped chain.

- **a.** $\{1, (2, 3), 4, 5\}$
- **b.** $\{(1, 5), 2, 3, 4\}$
- **c.** $\{(1, 5), (2, 3), 4\}$

2.33 Consider a chain with transition matrix:

$$\mathbf{P} = \begin{array}{c} \\ 1 \\ 2 \\ 3 \\ 4 \\ 5 \\ 6 \\ 7 \end{array} \begin{array}{c} \begin{array}{ccccccc} 1 & 2 & 3 & 4 & 5 & 6 & 7 \end{array} \\ \begin{pmatrix} .1 & .1 & .2 & .1 & .2 & .1 & .2 \\ .1 & .1 & .3 & 0 & 0 & .1 & .4 \\ .1 & .4 & .2 & .1 & 0 & .2 & 0 \\ .2 & .2 & .2 & 0 & .1 & 0 & .3 \\ 0 & .3 & .1 & .2 & .1 & .2 & .1 \\ .5 & .1 & .1 & .1 & .2 & 0 & 0 \\ .2 & 0 & .3 & .2 & .1 & .1 & .1 \end{pmatrix} \end{array}$$

Is it lumpable with respect to the following partitions? If so, find the transition matrix of the lumped chain.

a. $\{(1, 7), 2, 3, 4, 5, 6\}$
b. $\{(1, 2, 4), (3, 5), 6\}$
c. $\{(1, 7), (3, 5), (2, 4, 6)\}$

2.34 A chain has state space S (with N states) and transition matrix $\mathbf{P}$. Let

1. $\tilde{S} \equiv \{S_1, S_2, \ldots, S_M\}$ be a partition of S, where S_i has N_i states.
2. $\mathbf{W}$ be an $(M \times N)$ matrix, the i-row of which corresponds to $S_i \subset \tilde{S}$ and the j-column of which corresponds to state $j \in S$. Its (i, j)-element is equal to $1/N_i$ if $j \in S_i$ and 0 otherwise.
3. $\mathbf{V}$ be an $(N \times M)$ matrix, the i-row of which corresponds to state $i \in S$ and the j-column of which corresponds to $S_j \subset \tilde{S}$. Its (i, j)-element is equal to 1 if $i \in S_j$ and 0 otherwise.

For example, Chain (2.17.1) admitting partition $\{1, (2, 5), (3, 4)\}$ has $N = 5$, $M = 3$, $N_1 = 1$, $N_2 = 2$, $N_3 = 2$, and

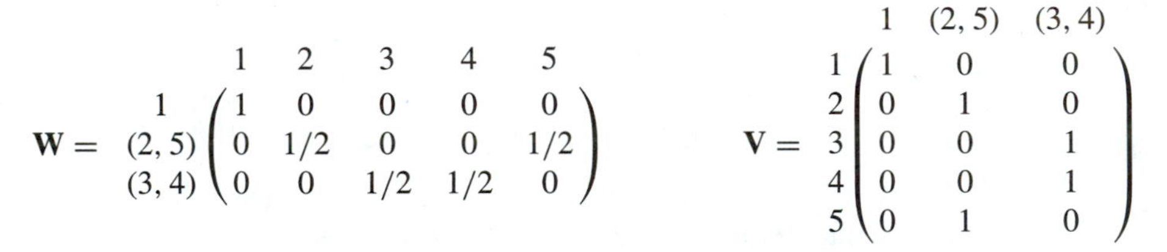

a. Explain why the condition of lumpability is equivalent to $\mathbf{VWPV} = \mathbf{PV}$.
b. Explain why the lumped matrix, if it exists, is $\tilde{\mathbf{P}} = \mathbf{WPV}$.
c. Show that $\mathbf{WV} = \mathbf{I}$. ($\mathbf{I}$ is an identity matrix having 1s along the main diagonal and 0s elsewhere.)
d. Hence, show that $\tilde{\mathbf{P}}^{(n)} = \mathbf{WP}^{(n)}\mathbf{V}$.
e. Apply the results to the chains in Problems 2.32 and 2.33.

CHAPTER 3

Transient and Limiting Results

3.1 PATHS

A *path of length k* in a Markov chain is a sequence of states it visits from step n to step $n+k$. A path both starting and ending with the same state i is also called a *loop* from state i. A loop from state i may contain another loop from state i of smaller length in it.

For the rat-maze example in §2.10, one path of length 5 could be

$$H \equiv 3 - 4 - 2 - 4 - 3 - 5$$

Path H includes a loop from state 3 of length 4 and a loop from state 4 of length 2.

Because a first-order Markov chain $\{X_n\}_{n=0,1,2,\ldots}$ is memoryless, we can write

$$\begin{aligned}
&\Pr\{X_{n+2} = i_{n+2}, X_{n+1} = i_{n+1}, X_n = i_n\} \\
&\quad = \Pr\{X_{n+2} = i_{n+2} \mid X_{n+1} = i_{n+1}, X_n = i_n\} \Pr\{X_{n+1} = i_{n+1}, X_n = i_n\} \\
&\quad = \Pr\{X_{n+2} = i_{n+2} \mid X_{n+1} = i_{n+1}\} \Pr\{X_{n+1} = i_{n+1} \mid X_n = i_n\} \Pr\{X_n = i_n\} \\
&\quad = p_{i_{n+1}, i_{n+2}} p_{i_n, i_{n+1}} \Pr\{X_n = i_n\}
\end{aligned}$$

In this manner, we can show that, starting from state i_n, the probability that the chain follows a particular path $i_n - i_{n+1} - i_{n+2} - \cdots - i_{n+k}$ is simply

$$\begin{aligned}
&\Pr\{X_{n+1} = i_{n+1}, X_{n+2} = i_{n+2}, \cdots, X_{n+k} = i_{n+k} \mid X_n = i_n\} \\
&\quad = p_{i_n, i_{n+1}} p_{i_{n+1}, i_{n+2}} \cdots p_{i_{n+k-1}, i_{n+k}}
\end{aligned}$$

Starting from state 3, the probability that the rat (or the chain) follows the foregoing path H is

$$p_{34} p_{42} p_{24} p_{43} p_{35} = (1/3)(1/3)(1/2)(1/3)(1/3)$$

3.2 k-STEP TRANSITION PROBABILITIES

Let us term

$$p_{ij}^{(k)} \equiv \Pr\{X_{n+k} = j \mid X_n = i\} \quad \text{for all } k \geq 1, n \geq 0, \text{ and } i, j \in S$$

the *k-step transition probabilities* of the chain. The matrix

$$\mathbf{P}^{(k)} \equiv \left[p_{ij}^{(k)}\right]_{i,j \in S} \qquad \text{for all } k \geq 1$$

is called its *k-step transition matrix*.

Note that, because we assume the chain is time-homogeneous, $p_{ij}^{(k)}$ is independent on the step n. Thus, it is the same as

$$p_{ij}^{(k)} \equiv \Pr\{X_k = j \mid X_0 = i\} \quad \text{for all } k \geq 1, \text{ and } i, j \in S \tag{3.2.1}$$

Furthermore,

$$p_{ij}^{(1)} = p_{ij}$$
$$\mathbf{P}^{(1)} \equiv \left[p_{ij}^{(1)}\right]_{i,j \in S} = \left[p_{ij}\right]_{i,j \in S} \equiv \mathbf{P}$$

For completeness, we define

$$p_{ij}^{(0)} \equiv \delta_{ij}$$
$$\mathbf{P}^{(0)} \equiv \left[p_{ij}^{(0)}\right]_{i,j \in S} = \left[\delta_{ij}\right]_{i,j \in S} = \mathbf{I}$$

where the *Kronecker delta* δ_{ij} is defined as

$$\delta_{ij} = \begin{cases} 1 & \text{for } i = j \\ 0 & \text{otherwise} \end{cases}$$

and the *identity matrix* **I** is a matrix having 1s along the main diagonal and 0s elsewhere.

In the mood example of §2.13, if a person is sad Monday, the probability that the person will be happy Tuesday is a one-step transition probability; the probability that the person will be sad the following Monday is a seven-step transition probability.

From the values of the one-step transition probabilities p_{ij}s, we shall now obtain the values of the k-step transition probabilities $p_{ij}^{(k)}$ for all $k > 1$ and $i, j \in S$.

3.3 PATH ANALYSIS

Note that $p_{ij}^{(k)}$ is the probability that the chain develops from state i at step n to state j at step $n+k$ *regardless of the path taken*. Therefore, one way to calculate that probability is to spell out all possible paths of length k from state i to state j and add their probabilities. This technique is called *path analysis*.

For example, for the rat in the maze to get from room 3 to room S in four steps, there are seven possible paths as shown in Table 3.3.1.

Hence,

$$p_{3,S}^{(4)} = \frac{1}{54} + \frac{1}{36} + \frac{1}{6} + \frac{1}{54} + \frac{1}{81} + \frac{1}{54} + \frac{1}{9} = \frac{121}{324}$$

3.4 MATRIX MULTIPLICATION METHOD

A drawback of path analysis is that it quickly becomes complicated as the length of the path or the number of states increases, making it easy to overlook paths. What we need is a way to determine the k-step transition probabilities without having to enumerate these paths. Matrix multiplication provides an answer, as it can be shown that

$$\mathbf{P}^{(k)} = \mathbf{P}^k \quad \text{for all } k \geq 0 \tag{3.4.1}$$

That is, *the k-step transition matrix is the one-step transition matrix raised to the power k*; or $p_{ij}^{(k)}$ is equal to the (i, j)-element of $\mathbf{P}^k$.

Table 3.3.1 Paths from room 3 to room 5

Path	Probability
3-5-3-4-S	(1/3)(1/2)(1/3)(1/3) = 1/54
3-5-3-5-S	(1/3)(1/2)(1/3)(1/2) = 1/36
3-5-S-S-S	(1/3)(1/2)(1)(1) = 1/6
3-4-2-4-S	(1/3)(1/3)(1/2)(1/3) = 1/54
3-4-3-4-S	(1/3)(1/3)(1/3)(1/3) = 1/81
3-4-3-5-S	(1/3)(1/3)(1/3)(1/2) = 1/54
3-4-S-S-S	(1/3)(1/3)(1)(1) = 1/9

We can prove this equation by induction.[1] First, we can see that it is true for $k = 0$ and $k = 1$. Second, let us now assume that it is true for $k - 1$; that is, $p_{i\ell}^{(k-1)} = [\mathbf{P}^{k-1}]_{i\ell}$. In this case, from the law of total probability (1.7.1),

$$\begin{aligned}\Pr\{X_k = j \mid X_0 = i\} &= \sum_{\ell \in S} \Pr\{X_k = j \mid X_{k-1} = \ell\} \Pr\{X_{k-1} = \ell \mid X_0 = i\} \\ &= \sum_{\ell \in S} p_{i\ell}^{(k-1)} p_{\ell j} = \sum_{\ell \in S} [\mathbf{P}^{k-1}]_{i\ell}\, p_{\ell j} = [\mathbf{P}^k]_{ij}\end{aligned}$$

Hence, this equation is also true for all $k > 1$.

As an illustration, consider the two-step transition probability $p_{32}^{(2)}$ in the rat-maze example. From path analysis, we see there is only one possible path, namely, $3 - 4 - 2$. Thus, $p_{32} = p_{34}p_{42} = (1/3)(1/3) = 1/9$. If we calculate the (3, 2)-element of $\mathbf{P}^2$, we obtain the same probability:

$$\begin{aligned}&p_{3F}p_{F2} + p_{32}p_{22} + p_{33}p_{32} + p_{34}p_{42} + p_{35}p_{52} + p_{3S}p_{S2} \\ &= \left(\frac{1}{3}\right) 0 + (0)(0) + (0)(0) + \left(\frac{1}{3}\right)\left(\frac{1}{3}\right) + \left(\frac{1}{3}\right)(0) + (0)(0) = \frac{1}{9}\end{aligned}$$

By multiplying matrices, we have to automatically take into consideration impossible paths such as $3 - S - 2$. However, the answer is still correct and the additional work is justified because the method produces a systematic way to avoid overlooking any paths.

In addition, by using matrices we can conveniently multiply higher powers of $\mathbf{P}$ together. To calculate $\mathbf{P}^{(10)}$, for example, we need not multiply $\mathbf{P}$ by itself ten times. Instead, we first find $\mathbf{P}^2$, then $\mathbf{P}^4 = \mathbf{P}^2\mathbf{P}^2$, next $\mathbf{P}^8 = \mathbf{P}^4\mathbf{P}^4$, and finally $\mathbf{P}^{10} = \mathbf{P}^8\mathbf{P}^2$, which involves only four multiplications.

It can be proved that all rows of $\mathbf{P}^n$ add to 1 for all $n \geq 1$.

Markov Chains in Action: Progress of Labor in Human Birthing. Nagamatsu et al. (1988) have applied Markov models in the study of the human birthing process. They divided human labor into eight stages depending on the size of the cervical dilatation: 4 cm, 5 cm, 6 cm, 7 cm, 8 cm, 9 cm, 10 cm, and delivery. From data on 625 primiparas (women who give birth for the first time), they obtained the following 30-minute transition matrix $\mathbf{P}$:[2]

	4 cm	5 cm	6 cm	7 cm	8 cm	9 cm	10 cm	Delivery
4 cm	.627	.218	.078	.027	.024	.014	.010	.002
5 cm		.405	.323	.112	.062	.053	.029	.016
6 cm			.306	.315	.171	.090	.088	.030
7 cm				.213	.317	.229	.187	.054
8 cm					.179	.336	.368	.117
9 cm						.192	.560	.248
10 cm							.520	.480
Delivery								1

[1] Mathematical induction can be used to prove that a certain mathematical proposition is true for all positive integers n greater than i. The proof requires two steps: (a) proving that the proposition is true for i and (b) proving that the proposition is true for n, assuming that it is true for $n - 1$.

[2] Sometimes we leave the (i, j)-element of a transition matrix blank if $p_{ij} = 0$.

This matrix could then be raised to the nth power to obtain the probabilities of transitions to the various states $(n/2)$ hours later. For example, the following 90-minute, or (3/2)-hour, transition matrix was obtained by calculating $\mathbf{P}^3$:

	4 cm	5 cm	6 cm	7 cm	8 cm	9 cm	10 cm	Delivery
4 cm	.247	.177	.147	.096	.081	.074	.106	.073
5 cm		.066	.123	.127	.127	.132	.231	.195
6 cm			.029	.064	.101	.141	.331	.335
7 cm				.010	.037	.090	.364	.499
8 cm					.006	.035	.313	.646
9 cm						.007	.228	.765
10 cm							.141	.859
Delivery								1

From the resulting matrix, the probability for a woman currently having a 4-cm cervical dilatation to deliver within 90 minutes is shown to be 7.3%, compared with 49.9% for a woman having a 7-cm cervical dilatation. ■

3.5 CHAPMAN-KOLMOGOROV EQUATION

From Equation (3.4.1), we can write

$$\mathbf{P}^{(m+n)} = \mathbf{P}^{(m)}\mathbf{P}^{(n)}$$

which is known as the *Chapman-Kolmogorov equation*. The following relationships represent special cases of this equation:

$$\begin{aligned} p_{ij}^{(n+1)} &= \left[\mathbf{P}\mathbf{P}^{(n)}\right]_{ij} = \sum_{k\in S} p_{ik}\, p_{kj}^{(n)} \\ p_{ij}^{(n+1)} &= \left[\mathbf{P}^{(n)}\mathbf{P}\right]_{ij} = \sum_{k\in S} p_{ik}^{(n)}\, p_{kj} \end{aligned} \tag{3.5.1}$$

3.6 INITIAL CONDITIONS

To understand the development of a chain fully, not only do we need to know how it develops from step n to step $n+1$. We also need to know the *initial conditions* of the chain—that is, the state it starts out in at time $t = 0$. The former is contained in the one-step transition matrix $\mathbf{P}$; the latter in the *initial distribution*

$$\boldsymbol{\pi}^{(0)} \equiv [\Pr\{X_0 = i\}]_{i\in S}$$

If the state space is $\{1, 2, 3, \ldots\}$, this initial condition is the vector

$$\boldsymbol{\pi}^{(0)} = (\,\Pr\{X_0 = 1\} \quad \Pr\{X_0 = 2\} \quad \Pr\{X_0 = 3\} \quad \ldots)$$

3.7 TRANSIENT RESULTS

As defined in Equation (3.2.1), the k-step transition probabilities $p_{ij}^{(k)}$s are conditioned on the initial conditions $\boldsymbol{\pi}^{(0)}$, assumed to be unknown. Thus, those $p_{ij}^{(k)}$s are rarely what practitioners seek. Of more practical interest is when the initial conditions $\boldsymbol{\pi}^{(0)}$ are known, and we want to find the unconditional state distribution of X_n, as defined in (2.4.4):

$$\boldsymbol{\pi}^{(n)} \equiv [\Pr\{X_n = i\}]_{i\in S} \equiv \left[\pi_i^{(n)}\right]_{i\in S} \quad \text{for all } n \geq 1$$

When n is finite, we call the values of $\Pr\{X_n = i\}$, or $\boldsymbol{\pi}^{(n)}$, the *transient results*.

In the mood example as modified in §2.13, one normally doesn't hear questions such as: "We do not know anything about our boss's feelings this Monday and Tuesday. What are the various probabilities that she will be happy Friday conditioned on how she feels on Monday and Tuesday?" A more likely question is: "Our boss was sad on Monday and happy Tuesday. What is the probability that she will still be happy on Friday?"

In answering this question, we start from Equation (2.4.5) that

$$\boldsymbol{\pi}^{(n)} = \boldsymbol{\pi}^{(n-1)}\mathbf{P} = \boldsymbol{\pi}^{(n-2)}\mathbf{PP} = \boldsymbol{\pi}^{(n-2)}\mathbf{P}^2 = \cdots \tag{3.7.1}$$

or

$$\boldsymbol{\pi}^{(n)} = \boldsymbol{\pi}^{(0)}\mathbf{P}^n \quad \text{for all } n \geq 0 \tag{3.7.2}$$

Note that the law of total probability (1.7.1) yields

$$\pi_j^{(n)} = \Pr\{X_n = j\} = \sum_{i\in S}\Pr\{X_n = j|X_0 = i\}\Pr\{X_0 = i\} = \sum_{i\in S}\pi_i^{(0)}p_{ij}^{(n)}$$

which can be written in matrix form as:

$$\boldsymbol{\pi}^{(n)} = \boldsymbol{\pi}^{(0)}\mathbf{P}^{(n)} \quad \text{for all } n \geq 0 \tag{3.7.3}$$

Comparing Equations (3.7.2) and (3.7.3), we now can prove the *matrix multiplication method* (3.4.1), which states that $\mathbf{P}^{(k)} = \mathbf{P}^k$.

To find the probability that the boss will be happy Friday, we first note that there are three steps from (Monday–Tuesday) to (Thursday–Friday): Step 1 from (Monday–Tuesday) to (Tuesday–Wednesday), step 2 from (Tuesday–Wednesday) to (Wednesday–Thursday), and step 3 from (Wednesday–Thursday) to (Thursday–Friday). We therefore calculate $\mathbf{P}^3$, where $\mathbf{P}$ is found in (2.13.1):

$$\mathbf{P}^3 = \begin{array}{c} \\ (s,s) \\ (h,s) \\ (s,h) \\ (h,h) \end{array} \begin{array}{c} \begin{array}{cccc} (s,s) & (h,s) & (s,h) & (h,h) \end{array} \\ \begin{pmatrix} .1640 & .3840 & .1160 & .3360 \\ .0943 & .3030 & .1408 & .4620 \\ .3120 & .1995 & .3030 & .1855 \\ .2340 & .1590 & .3960 & .2110 \end{pmatrix} \end{array}$$

From the information given in the question, we obtain the initial condition as

$$\pi^{(0)} = \begin{matrix} (s,s) & (h,s) & (s,h) & (h,h) \\ (\ 0 & 0 & 1 & 0\ ) \end{matrix}$$

Thus,

$$\pi^{(3)} = \pi^{(0)}\mathbf{P}^3 = \begin{matrix} (s,s) & (h,s) & (s,h) & (h,h) \\ (.3120 & .1995 & .3030 & .1855) \end{matrix}$$

giving the probability that the boss will be happy Friday as .4885, which is the sum of the probability that she is sad Thursday and happy Friday (.3030) and the probability that she is happy Thursday and happy Friday (.1855).

If the four-step transition matrix from (Monday–Tuesday) to (Friday–Saturday) is available instead; that is,

$$\mathbf{P}^4 = \begin{matrix} & (s,s) & (h,s) & (s,h) & (h,h) \\ (s,s) & .2824 & .2364 & .2656 & .2156 \\ (h,s) & .2158 & .3194 & .1815 & .2833 \\ (s,h) & .1921 & .2022 & .3194 & .2863 \\ (h,h) & .1502 & .2454 & .2429 & .3616 \end{matrix}$$

then the desired probability can be obtained as $.2022 + .2863 = .4885$, where .2022 and .2863 are the probabilities for transitions from (s, h) on (Monday–Tuesday) to (h, s) and (h, h) on (Friday–Saturday), respectively.

Markov Chains in Action: Planning for Dental Care. To evaluate dental care programs, Freed et al. (1979) analyzed records of 578 patients who entered the UCLA Venice Dental Clinic in Los Angeles between July 1, 1971, and June 30, 1972. The states were defined as: E (episodic care, a specific problem usually associated with pain and discomfort), I (initial care, a complete diagnostic evaluation and treatment plan), M (maintenance care, preventive and treatment service), and N (nonuse care or no visit). Following the patients for 3 years, they obtained the following 1-year transition matrix:

$$\mathbf{P} = \begin{matrix} & M & I & E & N \\ M & .346 & .014 & .003 & .637 \\ I & .372 & .110 & .010 & .508 \\ E & .190 & .152 & .095 & .562 \\ N & .003 & .053 & .012 & .933 \end{matrix}$$

Given that initially 14.5% of the patients are in state E and 85.5% are in state I, the reader may wish to verify the following table:

	Percentage Expected at End of Year				
	Year 1	Year 2	Year 3	Year 4	Year 5
Maintenance *M*	34.5	16.8	8.0	5.1	4.2
Initial *I*	11.6	4.8	5.0	5.4	5.5
Episodic *E*	2.3	1.1	1.1	1.2	1.2
Nonuse *N*	51.6	77.3	85.8	88.3	89.0

It can be seen that only 22.7% of the patients would be expected to be receiving treatment (in the M, I or E state) after 2 years. After 5 years, only 10.9% would be projected to be receiving some form of treatment. ■

Markov Chains in Action: Stages of Retinopathic Degeneration. Diabetic retinopathy (*DR*) is a common disease among people with insulin-dependent (type-I) diabetes. Patients with early *DR* tend to have progressive retinal changes, which may cause blindness. The illness can be classified into five stages depending on the progress of the retinopathic degeneration, ranging from best to worst: state 1 indicates no retinopathy, state 5–6 indicates preproliferative and proliferative retinopathy.

Garg et al. (1990) analyzed data obtained during 693 visits of 259 *DR* patients for about 2.4 years and obtained the following 1-year transition matrix:

$$\mathbf{P} = \begin{array}{c} \\ 1 \\ 2 \\ 3 \\ 4 \\ 5\text{–}6 \end{array} \begin{array}{c} \begin{array}{ccccc} 1 & 2 & 3 & 4 & 5\text{–}6 \end{array} \\ \begin{pmatrix} .65 & .27 & .06 & .03 & .00 \\ .16 & .59 & .20 & .04 & .01 \\ .04 & .40 & .44 & .12 & .01 \\ .00 & .13 & .38 & .38 & .13 \\ & & & & 1 \end{pmatrix} \end{array}$$

The 4-year transition matrix can be obtained as:

$$\mathbf{P}^4 = \begin{array}{c} \\ 1 \\ 2 \\ 3 \\ 4 \\ 5\text{–}6 \end{array} \begin{array}{c} \begin{array}{ccccc} 1 & 2 & 3 & 4 & 5\text{–}6 \end{array} \\ \begin{pmatrix} .31 & .41 & .20 & .07 & .03 \\ .22 & .42 & .23 & .08 & .06 \\ .18 & .42 & .26 & .09 & .08 \\ .12 & .35 & .24 & .10 & .24 \\ & & & & 1 \end{pmatrix} \end{array}$$

With the initial distribution of the patients as

$$\pi^{(0)} = \begin{array}{c} \begin{array}{ccccc} 1 & 2 & 3 & 4 & 5\text{–}6 \end{array} \\ \begin{pmatrix} .337 & .449 & .194 & .200 & 0 \end{pmatrix} \end{array}$$

the mean grade is 1.9. The product $\pi^{(0)}\mathbf{P}^4$

> . . . gives the distribution of grades that one would expect in this sample after 4 years of observation in the absence of treatment. This distribution is as follows: grade 1, 26.8%; grade 2, 42.9%; grade 3, 20.3%; grade 4, 6.4%; and grades 5 and 6, 3.6%. This distribution has a mean grade of 2.2, indicating about 0.3 grade units of increase during a 4-year period in the absence of treatment. ■

3.8 VALIDATING THE MODEL

Not only can Equation (3.7.2) be used for predicting future distributions, but it can also be used to *validate* the model, whereby the *predicted* state distribution obtained from Equation (3.7.2) can be compared with the *observed* one to make sure that the two are reasonably close.

Markov Chains in Action: A Model for Releasing of the Chronically Mentally Ill. Healthcare professionals usually avoid moving the chronically mentally ill directly from a hospital to an independent community setting. In Billings, Montana, for example, they are first placed in a 24-hour group home, which is staffed around the clock by mental health workers. If their progress in that facility is satisfactory, they are placed in a 40-hour group home, which is staffed for 8 hours each day, 5 days a week. If their progress continues, they are moved to a cooperative apartment, in which a degree of self-government can be exercised. Finally, when judged ready, they are allowed to move into the community.

Drachman (1981) analyzed data collected in Billings between January 1, 1978, and May 31, 1979, tracking patients' progress through the release program. Six states were used: H (hospital), I (inpatient unit), 24 (24-hour group home), 40 (40-hour group home), A (cooperative apartment), and C (community). Here is a 12-week (roughly 3-month) transition matrix:

$$\mathbf{P} = \begin{array}{c} \\ H \\ I \\ 24 \\ 40 \\ A \\ C \end{array} \begin{array}{c} \begin{array}{cccccc} H & I & 24 & 40 & A & C \end{array} \\ \begin{pmatrix} .7143 & .0714 & .0714 & .0000 & .0000 & .1429 \\ .1177 & .0588 & .2941 & .1177 & .0000 & .4118 \\ .0109 & .0109 & .7283 & .0652 & .0000 & .1848 \\ .0213 & .0213 & .0213 & .7660 & .0426 & .1277 \\ .0000 & .0172 & .0172 & .0172 & .7931 & .1552 \\ .0136 & .0442 & .0578 & .0034 & .0272 & .8537 \end{pmatrix} \end{array}$$

From this matrix we obtain

$$\mathbf{P}^6 = \begin{array}{c} \\ H \\ I \\ 24 \\ 40 \\ A \\ C \end{array} \begin{array}{c} \begin{array}{cccccc} H & I & 24 & 40 & A & C \end{array} \\ \begin{pmatrix} .1723 & .0424 & .2002 & .0585 & .0372 & .4894 \\ .0678 & .0359 & .2032 & .1010 & .0600 & .5323 \\ .0454 & .0323 & .2539 & .1167 & .0507 & .5010 \\ .0548 & .0330 & .1256 & .2373 & .1046 & .4447 \\ .0282 & .0313 & .1180 & .0592 & .2870 & .4762 \\ .0489 & .0374 & .1758 & .0548 & .0758 & .6073 \end{pmatrix} \end{array}$$

Let $\mathbf{o}_1$ be the vector of the observed number of clients in each of the six locations at the beginning of the studied period. From the data, $\mathbf{o}_1 = (1, 0, 15, 8, 10, 53)$.

Let $\mathbf{e}_7$ be the vectors of the estimated number of clients in each of the six locations at the end of the 17-month period, based on the transition matrix $\mathbf{P}$ (17 months are roughly six 12-week periods). From Equation (3.7.2), we have

$$\mathbf{e}_7 = \mathbf{o}_1\mathbf{P}^6 = (\,4.17 \quad 3.09 \quad 15.51 \quad 7.20 \quad 8.52 \quad 48.51\,)$$

> The actual end-of-period frequencies [$\mathbf{o}_7$] were found to be $(\,5 \quad 1 \quad 15 \quad 7 \quad 8 \quad 51\,)$. A chi-square test comparing $\mathbf{e}_7$ with $\mathbf{o}_7$ showed very close agreement between the expected and observed values. . . . Therefore, it appears that a first-order, stationary Markov model adequately fits the data. ■

3.9 LIMITING DISTRIBUTIONS

A taxicab driver, whom we call Tex, does business in three adjoining cities, A, B, and C. Let T_n be the city he is in at the beginning of the nth hour. Assume the initial condition

$$\pi^{(0)} = \begin{matrix} A & B & C \\ (.3 & .4 & .3) \end{matrix} \tag{3.9.1}$$

and the 1-hour transition matrix

$$\mathbf{P} = \begin{matrix} & \begin{matrix} A & B & C \end{matrix} \\ \begin{matrix} A \\ B \\ C \end{matrix} & \begin{pmatrix} .4 & .1 & .5 \\ .2 & .2 & .6 \\ .3 & .4 & .3 \end{pmatrix} \end{matrix} \tag{3.9.2}$$

In Table 3.9.1, we present some developments of $\mathbf{P}^{(n)}$ and $\pi^{(n)}$ of the chain $\{T_n\}_{n=0,1,2,\ldots}$ as the functions of n. Note that the sequence of $\Pr\{T_n = j\}$ seems to converge to a fixed value as $n \to \infty$ for all $j = A$, B, and C. In other words, the distributions $\{\pi^{(n)}\}_{n=0,1,2,\ldots}$ seem to approach a certain vector π as $n \to \infty$.

Generally, if the sequence of distributions $\{\pi_n\}_{n=0,1,2,\ldots}$ converges to a fixed vector π, we say that the chain $\{X_n\}_{n=0,1,2,\ldots}$ *converges in distribution* to X, where X is a random

Table 3.9.1 Development of $\mathbf{P}^{(n)}$ and $\pi^{(n)}$

n	$\mathbf{P}^{(n)}$	$\pi^{(n)}$		
		$\Pr\{T_n = A\}$	$\Pr\{T_n = B\}$	$\Pr\{T_n = C\}$
0		.3	.4	.3
1	$\begin{pmatrix} .4 & .1 & .5 \\ .2 & .2 & .6 \\ .3 & .4 & .3 \end{pmatrix}$	.29	.23	.48
2	$\begin{pmatrix} .33 & .26 & .41 \\ .30 & .30 & .40 \\ .29 & .23 & .48 \end{pmatrix}$	.306	.267	.427
4	$\begin{pmatrix} .3058 & .2581 & .4361 \\ .3050 & .2600 & .4350 \\ .3039 & .2548 & .4413 \end{pmatrix}$	.30491	.25787	.43722
6	$\begin{pmatrix} .3048 & .2572 & .4379 \\ .3048 & .2573 & .4378 \\ .3047 & .2569 & .4383 \end{pmatrix}$	.30477	.25720	.43802

variable having distribution $\boldsymbol{\pi}$. The vector $\boldsymbol{\pi}$ is called the *limiting distribution* of the chain.[3]

If, at any step n, the distribution of X_n is the same as $\boldsymbol{\pi}$, we say the chain has reached the *steady state*, or *equilibrium*.

Turning back to the rat-maze example, if the rat starts from room F, that is,

$$\boldsymbol{\pi}^{(0)} = \begin{matrix} F & 2 & 3 & 4 & 5 & S \\ (1 & 0 & 0 & 0 & 0 & 0) \end{matrix}$$

then it remains in room F forever, or

$$\lim_{n\to\infty} \boldsymbol{\pi}^{(n)} = \begin{matrix} F & 2 & 3 & 4 & 5 & S \\ (1 & 0 & 0 & 0 & 0 & 0) \end{matrix}$$

In this case, the rat has reached a steady state even at step 0.[4]

Additionally, if the rat starts from room S, or

$$\boldsymbol{\pi}^{(0)} = \begin{matrix} F & 2 & 3 & 4 & 5 & S \\ (0 & 0 & 0 & 0 & 0 & 1) \end{matrix}$$

then it remains in room S permanently and

$$\lim_{n\to\infty} \boldsymbol{\pi}^{(n)} = \begin{matrix} F & 2 & 3 & 4 & 5 & S \\ (0 & 0 & 0 & 0 & 0 & 1) \end{matrix}$$

Note that, in this example, although $\lim_{n\to\infty} \boldsymbol{\pi}^{(n)}$ exists, it is *dependent on the initial conditions*.

On the other hand, for Tex and his taxicab, if we now change the initial condition $\boldsymbol{\pi}^{(0)}$ in Equation (3.9.1) to another arbitrary distribution and carry out the calculations again, we will observe that the new sequence $\{\boldsymbol{\pi}_n\}_{n=0,1,2,...}$ seems to converge to the *same* limiting distribution as before. Here, the limiting distribution is independent of the initial conditions.

For general Markov chains, we want to find out the conditions that lead to the convergence of the sequence of distributions $\{\boldsymbol{\pi}_n\}_{n=0,1,2,...}$. If such a limiting distribution $\boldsymbol{\pi}$ exists, we also want to find out the conditions for its independence of the initial conditions $\boldsymbol{\pi}^{(0)}$. These issues are addressed in later chapters.

For now, we note that, if the limiting distribution exists and is independent of the initial conditions, the sequence of matrices $\{\mathbf{P}^{(n)}\}_{n=1,2,...}$ must converge to a matrix $\mathbf{P}^{(\infty)}$ in which *all rows are equal to* $\boldsymbol{\pi}$. (The reader may wish to determine why.)

Markov Chains in Action: Toward Educational Equality between Racial Groups. Based on the October 1960 *Current Population Survey* by the U.S. Bureau of the Census, Lieberson and Fuguitt (1967) examined the educational status of 20- to 24-year-old

[3] Generally, the adjective *limiting* means "as the time (or step) approaches ∞."

[4] We said in §2.10 that the experiment stops when $R_n = F$ or $R_n = S$. The discussion here assumes that the steps n can be arbitrarily defined after the rat reaches room F or S.

men, as compared to their father's status, and obtained the following intergenerational transition matrix:

$$\mathbf{P} = \begin{array}{c} \text{less than} \\ \text{high school} \\ \text{high school} \\ \text{more than} \\ \text{high school} \end{array} \begin{pmatrix} & \text{less than high school} & \text{high school} & \text{more than high school} \\ & .43 & .34 & .23 \\ & .10 & .36 & .54 \\ & .05 & .15 & .80 \end{pmatrix}$$

The authors argued that there are two kinds of disadvantages faced by members of racial and ethnic groups. One is based solely on their racial or ethnic origin, resulting in different intergenerational transition matrices for white and nonwhite groups. Even when this kind of discrimination is completely eradicated, there is still a second disadvantage that occurs because nonwhite groups currently occupy an inferior aggregate position. In other words, even if the intergenerational transition matrix **P** were the same for both groups, the initial conditions $\pi^{(0)}$s are different.

The reader may want to verify the results obtained in Table 3.9.2, assuming the matrix **P** is applicable for both white and nonwhite groups. The index of dissimilarity in the table is the percentage that one race would have to change educational categories if the two were to have identical distributions. For example, in 1960, $75 - 46 = 29\%$ of

Table 3.9.2 Educational Composition by Race

		Percentage Distribution			Index of Dissimilarity
		less than high school	high school	more than high school	
$\pi^{(1)}$ (1960)	white	46	31	23	29
	black	75	16	09	
$\pi^{(2)}$	white	24	30	46	13
	black	34	33	33	
$\pi^{(3)}$	white	16	26	58	6
	black	20	28	52	
$\pi^{(4)}$	white	12	23	64	3
	black	14	25	61	
$\pi^{(5)}$	white	11	22	67	1
	black	11	23	66	
$\pi^{(6)}$	white	10	22	68	1
	black	11	22	67	
$\pi^{(7)}$	white	10	21	69	1
	black	10	22	68	
$\pi^{(8)}$	white	10	21	69	0
	black	10	21	69	

nonwhites have to change away from less-than-high-school state for the two groups to have the same state distribution.

> We know that both . . . [nonwhite] and white occupational distributions will approach convergence at equilibrium, which is determined by the values of . . . [**P**] alone. What is of interest here is the changing pattern of . . . [nonwhite/white] differences over time until they become negligible, which depends both upon the . . . [**P**] matrix and the two initial vectors used.
>
> . . . In 1960, the index of dissimilarity in education between white and nonwhite employed men was 29. . . . This would drop to 13 in the second generation, 6 in the third generation, and would eventually reach 0 in the eighth generation.
>
> . . . These data do underscore the position that an end to discrimination will not result in occupational equality immediately. ■

3.10 EXPECTED NUMBERS OF VISITS

There is an important result that we would like to establish here before leaving this chapter. It is related to the total number of visits $v_{ij}^{(n)}$ the chain makes to state j in the first n steps, from step 0 to step $n-1$, starting from state i.

Given $X_0 = i$, we define the indicator function at step $k \geq 0$ as

$$I_{ij}^{(k)} = \begin{cases} 1 & \text{if } X_k = j \\ 0 & \text{otherwise} \end{cases}$$

Since the expected number of visits the chain makes to state j at step k is $\mathcal{E}[I_{ij}^{(k)}] = \Pr\{I_{ij}^{(k)} = 1\} = p_{ij}^{(k)}$ (Equation 1.11.1), the expected number of visits to state j from step 0 to step $n-1$ inclusive is therefore

$$\mathcal{E}\left[v_{ij}^{(n)}\right] = \sum_{k=0}^{n-1} p_{ij}^{(k)} \tag{3.10.1}$$

In matrix form,

$$\mathbf{V}^{(n)} \equiv \left[\mathcal{E}\left[v_{ij}^{(n)}\right]\right]_{i,j\in S} = \sum_{k=0}^{n-1} \mathbf{P}^{(k)} = \sum_{k=0}^{n-1} \mathbf{P}^{k} \tag{3.10.2}$$

For the taxicab example of §3.9,

$$\mathbf{V}^{(4)} = \mathbf{I} + \mathbf{P} + \mathbf{P}^2 + \mathbf{P}^3 = \begin{matrix} & A & B & C \\ A & 2.04 & .61 & 1.35 \\ B & .80 & 1.75 & 1.45 \\ C & .90 & .90 & 2.21 \end{matrix}$$

Suppose Tex is at city A at 8:00 A.M. Then the expected total number of visits he makes to cities A, B, and C from 8:00 A.M. to 11:00 A.M. are 2.04, .61, and 1.35, respectively.

Let $\mathcal{E}[v_{ij}] \equiv \lim_{n\to\infty} \mathcal{E}[v_{ij}^{(n)}]$ be the expected number of times the chain visits state j throughout its life, starting from state i. Then

$$\mathbf{V} \equiv \left[\mathcal{E}[v_{ij}]\right]_{i,j\in S} = \sum_{k=0}^{\infty} \mathbf{P}^k \tag{3.10.3}$$

3.11 SUMMARY

1. The k-step transition probability from state i to state j, $p_{ij}^{(k)}$, is the (i, j)-element of $\mathbf{P}^k$.
2. A first-order Markov chain is described completely by its transition matrix $\mathbf{P}$ and its initial condition $\boldsymbol{\pi}^{(0)}$.
3. Equation (3.7.2) gives the transient result $\boldsymbol{\pi}^{(n)}$ as $\boldsymbol{\pi}^{(0)}\mathbf{P}^n$.
4. Some chains have a limiting distribution $\boldsymbol{\pi}$; others do not. If this limiting distribution exists, it may be dependent on the initial condition.
5. If the chain's limiting distribution exists, $\mathbf{P}^{(n)}$ converges to a limit $\mathbf{P}^{(\infty)}$ as $n \to \infty$. If the limiting distribution $\boldsymbol{\pi}$ is independent on the initial condition, all rows of $\mathbf{P}^{(\infty)}$ are equal to $\boldsymbol{\pi}$.
6. Starting from state i, the expected total number of visits to state j in the first n steps is $\mathcal{E}[v_{ij}^{(n)}] = \sum_{k=0}^{n-1} p_{ij}^{(k)}$.

PROBLEMS

3.1 Consider a Markov chain with the following transition matrix:

$$\mathbf{P} = \begin{array}{c} \\ A \\ B \\ C \\ D \end{array} \begin{array}{c} \begin{array}{cccc} A & B & C & D \end{array} \\ \begin{pmatrix} .1 & .3 & .2 & .4 \\ .2 & .3 & .2 & .3 \\ .3 & .3 & .1 & .3 \\ .2 & .1 & .4 & .3 \end{pmatrix} \end{array}$$

a. Find the probability that the chain follows the path $B - A - B - C - B - A$.
b. Find $\Pr\{X_1 = A, X_3 = C, X_4 = B, X_5 = A \mid X_0 = B\}$.
c. Find $\mathbf{P}^5$.
d. Find $\Pr\{X_5 = A \mid X_0 = B\}$.
e. Find $\boldsymbol{\pi}^{(5)}$ given $\boldsymbol{\pi}^{(0)} = (.2 \quad .4 \quad .3 \quad .1)$.

3.2 Consider the chain studied in Problem 2.1b. Find $\mathbf{P}^n$ for all $n \geq 2$ and explain why they are independent of n and the initial condition.

3.3 (Jaggi and Lau, 1974) A firm has a three-rank hierarchy: Rank A is the lowest, rank B the middle, and rank C the highest. Initially, there are 100 employees at rank A, 80 at rank B, and 60 at rank C. Let the economic value of an employee at ranks A, B, and C be 1, 2, and 3, respectively. Assume that there is no demotion and each employee advances through the ranks according to the following 1-year transition matrix:

$$\mathbf{P} = \begin{array}{c} \\ A \\ B \\ C \\ X \end{array} \begin{array}{c} \begin{array}{cccc} A & B & C & X \end{array} \\ \begin{pmatrix} .4 & .4 & .1 & .1 \\ & .5 & .4 & .1 \\ & & .8 & .2 \\ & & & 1 \end{pmatrix} \end{array}$$

(State X represents the employee's exit from the firm.)

a. After 2 years, what is the expected economic value of the employees who were at rank A originally? At rank B originally? At rank C originally?

b. What is the expected total economic value of the firm after 2 years?

3.4 Verify the calculations by Nagamatsu et al. (1988) for human birthing in §3.4.

3.5 Verify the calculations by Freed et al. (1979) for dental care in §3.7.

3.6 Verify the calculations by Garg et al. (1990) for stages of retinopathic degeneration in §3.7.

3.7 Verify the calculations by Drachman (1981) for a release program of the chronically mental ill in §3.8.

3.8 Verify the calculations by Lieberson and Fuguitt (1967) for educational equality in §3.9.

3.9 Ticul, the largest community in western Yucatán Peninsula, Mexico, has two major ethnic groups: Mestizos and Catrines. Thompson (1970) found that, of 123 persons, 47 are poor Mestizos (state 1), 24 are ordinary Mestizos (state 2), 14 are "fine" Mestizos (state 3), 20 are poor Catrines (state 4), 12 are ordinary Catrines (state 5), and 6 are wealthy Catrines (state 6). The father-to-son transition matrix is:

$$\mathbf{P} = \begin{array}{c} \\ 1 \\ 2 \\ 3 \\ 4 \\ 5 \\ 6 \end{array} \begin{array}{c} \begin{array}{cccccc} 1 & 2 & 3 & 4 & 5 & 6 \end{array} \\ \begin{pmatrix} .62 & .11 & .05 & .15 & .05 & \\ & .47 & & .32 & .21 & \\ & .18 & .45 & .09 & .09 & .18 \\ & & & .50 & .50 & \\ & & & .33 & .67 & \\ & & & & & 1 \end{pmatrix} \end{array}$$

(Note the tendency of Mestizos to change to Catrines.) Find the distribution of the status after two generations.

3.10 Each year from 1961 to 1968, Anderson (1974) surveyed a total of 255 male undergraduate students at a private southern medical school. They were asked about their career preference in medicine, which can be in one of following six states: (1) general practice, (2) internal medicine, (3) general surgery, (4) academic medicine, (5) other specialty, and (6) undecided.

Of these 255 students, 118 (group 1) provided complete data over the six transitions, spanning from pretraining ($n = 0$) through internship ($n = 6$). In their first year, 11 students were in state 1, 8 in state 2, 17 in state 3, 13 in state 4, 30 in state 5, and 39 in state 6. The following transition matrix was obtained:

$$\mathbf{P} = \begin{array}{c} \\ 1 \\ 2 \\ 3 \\ 4 \\ 5 \\ 6 \end{array} \begin{array}{c} \begin{array}{cccccc} 1 & 2 & 3 & 4 & 5 & 6 \end{array} \\ \begin{pmatrix} .410 & .148 & .049 & .066 & .197 & .131 \\ .053 & .561 & .088 & .123 & .175 & \\ .045 & .091 & .545 & .159 & .159 & \\ .023 & .144 & .114 & .477 & .205 & .068 \\ .054 & .108 & .180 & .072 & .532 & .054 \\ .056 & .081 & .243 & .108 & .297 & .216 \end{pmatrix} \end{array}$$

From those who did not yield complete data, 80 students (group 2) provided data from $n = 0$ through $n = 4$. When $n = 4$, among these 80 students, 3 were in state 1, 14 were in state 2, 22 were in state 3, 18 were in state 4, 18 were in state 5, and 5 were in state 6. Assume the initial distribution and transition matrix for group 2 are the same as those for group 1. How do you compare the observed frequencies of group 2 with its expected frequencies when $n = 4$? Can you agree "that the model is a suitable descriptor of the specialty choice process and that it has predictive utility"?

3.11 Consider a Markov chain $\{X_n\}_{n=0,1,2,\ldots}$ in which X_n is the number of successes in n Bernoulli trials, and each trial has a probability of success p.

a. Find the transition matrix $\mathbf{P}$.

b. Obtain $\mathbf{P}^{(2)}$, $\mathbf{P}^{(3)}$. Guess the form of $\mathbf{P}^{(n)}$ for all $n > 1$ and then explain why.

3.12 Refer to Problem 2.18 in which X_n is the number susceptible to disease at day n. Let $\mu_{i,n} \equiv \mathcal{E}[X_n \mid X_0 = i]$.

a. Show that $\mu_{i,n} = (1 - p)\mu_{i,n-1}$ for all $n \geq 1$.

b. Hence, find $\mu_{i,n}$ and $\lim_{n\to\infty} \mu_{i,n}$.

3.13 Consider the Ehrenfest model in Problem 2.3 in which X_n is the number of balls in urn A immediately after the nth step. Let $\mu_{i,n} \equiv \mathcal{E}[X_n \mid X_0 = i]$.

a. By conditioning on the value of X_{n-1}, show that

$$\mu_{i,n} = 1 + (1 - 1/N)\,\mu_{i,n-1} \quad \text{for all } n > 0$$

b. Hence, show (by induction or some other method) that

$$\mu_{i,n} = N + (i - N)\,(1 - 1/N)^n \quad \text{for all } n > 0$$

c. Thus, find $\lim_{n\to\infty} \mathcal{E}[X_n]$. Can this be intuitive?

3.14 Consider the model studied in Problem 2.4 in which X_n is the number of white balls in urn A after the nth switch of balls. Define $\mu_{i,n} \equiv \mathcal{E}[X_n \mid X_0 = i]$.

a. Show that

$$\mu_{i,n} = 1 + (1 - 2/N)\,\mu_{i,n-1} \quad \text{for all } n > 0$$

b. Hence, find $\lim_{n\to\infty} \mathcal{E}[X_n]$. *Hint:* Use Problem 3.13.

3.15 Consider a Markov chain $\{X_n\}_{n=0,1,2,...}$ having state space $S \equiv \{0, 1, 2, \ldots, N\}$. At each transition, it always moves to a different state with equal probability. In other words, for all $0 \le i \le N$,

$$p_{ij} = \begin{cases} 0 & \text{for } j = i \\ 1/N & \text{for all } j \ne i \end{cases}$$

Let $\mu_{i,n} \equiv \mathcal{E}[X_n | X_0 = i]$.

a. By conditioning on the value of X_{n-1}, show that

$$\mu_{i,n} = (N+1)/2 - \mu_{i,n-1}/N \quad \text{for all } n > 0$$

b. Hence, show (by induction or some other method) that

$$\mu_{i,n} = N/2 + (i - N/2)(-1/N)^n \quad \text{for all } n > 0$$

c. Thus, find $\lim_{n\to\infty} \mathcal{E}[X_n]$.

3.16 Consider a Markov chain with state space $S \equiv \{0, 1, 2, \ldots, N\}$ that may remain at the same state at each transition or move to another state with equal probability p. In other words, for all $0 \le i \le N$,

$$p_{ij} = \begin{cases} 1 - Np & \text{for } j = i \\ p & \text{for all } j \ne i \end{cases}$$

where $0 \le Np \le 1$. Let $\mu_{i,n} \equiv \mathcal{E}[X_n | X_0 = i]$.

a. By conditioning on the value of X_{n-1}, show that

$$\mu_{i,n} = (N+1)Np/2 + [1 - (N+1)p]\,\mu_{i,n-1} \quad \text{for all } n > 0$$

b. Hence, show (by induction or some other method) that

$$\mu_{i,n} = N/2 + (i - N/2)[1 - (N+1)p]^n \quad \text{for all } n > 0$$

c. Let $N = 20$. Draw a graph of $\mu_{3,n}$ as a function of n when

i. $p = .015$

ii. $p = .05$ (which is Problem 3.15)

d. Thus, find $\lim_{n\to\infty} \mathcal{E}[X_n]$.

3.17 Consider a chain $\{X_n\}_{n=0,1,2,...}$ taking values from the set of integers. Suppose its transition probabilities p_{ij}s are such that

$$\sum_{j\in S} j p_{ij} = Ai + B$$

for all $i \in S$ and for some constants A and B. Let $\mu_{i,n} \equiv \mathcal{E}[X_n | X_0 = i]$.

a. By conditioning on the value of X_{n-1}, show that

$$\mu_{i,n} = A\mu_{i,n-1} + B \quad \text{for all } n > 0$$

b. Thus, if $A \ne 1$, show (by induction or some other method) that

$$\mu_{i,n} = A^n\left(i - \frac{B}{1-A}\right) + \frac{B}{1-A} \quad \text{for all } n > 0$$

c. Hence, if $A < 1$, find $\lim_{n\to\infty} \mathcal{E}[X_n]$.

3.18 Consider a population in which each individual produces i new offspring at the end of his lifetime with probability p_i, independent of the other individuals. Let μ and σ^2 be the expected value and the variance of the number of offspring, respectively, from each individual. Let X_n be the number of individuals at the nth generation. The Markov chain $\{X_n\}_{n=0,1,2,\ldots}$, taking values from the set of positive integers, is known as the *branching process*. Assume $x_0 = 1$.

a. Show that

$$\mathcal{E}[X_n] = \mu^n \quad \text{for all } n > 0$$

Hence, if $\mu < 1$, this population will be extinct with probability 1, or $\pi_0 \equiv \lim_{n\to\infty} \Pr\{X_n = 0\} = 1$.

b. Assume $\mu \geq 1$. Show that the probability of extinction π_0 is the smallest number that satisfies the equation

$$x = \sum_{j=0}^{\infty} x^j p_j$$

Hence, find π_0 when $p_0 = 1/5$, $p_1 = 1/5$, and $p_2 = 3/5$. *Hint:* Show by induction that, if x is a solution of the equation, then $x \geq \Pr\{X_n = 0\}$ for all $n > 1$.

c. Show that

$$\text{Var}[X_n] = \sigma^2\mu^{n-1} + \mu^2\text{Var}[X_{n-1}] \quad \text{for all } n > 0$$

Hence, show (by induction or some other method) that

$$\text{Var}[X_n] = \begin{cases} \sigma^2\mu^{n-1}\,(\mu^n - 1)\,/\,(\mu - 1) & \text{if } \mu \neq 1 \\ n\sigma^2 & \text{if } \mu = 1 \end{cases}$$

3.19 Refer to Problem 2.30. Show that the chain $\{X_{2n}\}_{n=0,1,2,\ldots}$ is time-homogeneous. Find its transition matrix.

3.20 Refer to the weather in Problem 2.27.

a. Let X_n be the kind of weather at 8:00 A.M. on day n. Find the transition matrix for $\{X_n\}_{n=0,1,2,\ldots}$.

b. Let Y_n be the kind of weather at 4:00 P.M. on day n. Find the transition matrix for $\{Y_n\}_{n=0,1,2,\ldots}$.

c. Let T_n be the kind of weather at 12:00 A.M. on day n. Find the transition matrix for $\{T_n\}_{n=0,1,2,\ldots}$.

3.21 Refer to the bombers in Problem 2.28. Let $\mu_{i,n} \equiv \mathcal{E}[X_n \mid X_0 = i]$, $\nu_{i,n} \equiv \mathcal{E}[Y_n \mid X_0 = i]$, and $r \equiv p_N(1 - p_D)$.

a. Find the transition matrix for the chains $\{X_n\}_{n=0,1,2,\ldots}$ and $\{Y_n\}_{n=0,1,2,\ldots}$.

b. Show that $\nu_{i,n} = (1 - p_D)\mu_{i,n}$ for all $n \geq 1$.

c. Hence, show that $\mu_{i,n+1} = r\mu_{i,n} + 3(1 - p_N)$ for all $n \geq 1$.

d. Hence, show (by induction or some other method) that

$$\mu_{i,n} = r^n i + 3q_N(1 - r^n)/(1 - r) \quad \text{for all } n \geq 0$$

e. From this, obtain $\lim_{n\to\infty} \mathcal{E}[X_n]$ and $\lim_{n\to\infty} \mathcal{E}[Y_n]$.
f. Does the chain $\{Z_n\}_{n=0,1,2,\ldots}$ converge in distribution?

3.22 Consider a chain $\{X_n\}_{n=0,1,2,\ldots}$ having transition matrix **P**.
a. Let ${}_\ell p_{ij}^{(k)}$ be the probability that it visits state j in step k but does not visit a certain state ℓ in between, given that it starts from state i ($\ell \neq i, j$); that is,

$$ {}_\ell p_{ij}^{(k)} \equiv \Pr\{X_1 \neq \ell, X_2 \neq \ell, \ldots, X_{k-1} \neq \ell, X_k = j \mid X_0 = i\} $$

We call ${}_\ell p_{ij}^{(k)}$ the k-step transition probability *under the taboo* ℓ. Let ${}_\ell\mathbf{P}$ be the matrix **P** without the ℓ-row and the ℓ-column. Explain why

$$ {}_\ell p_{ij}^{(k)} = [{}_\ell\mathbf{P}^k]_{ij} \quad \text{for all } \ell \neq i, j $$

b. In the mood example with **P** as in Equation (2.13.1), Alan observed that his boss was sad Monday but happy Tuesday. From §3.7, he knows that the probability that his boss will be happy this Friday is .4885. However, this will be a special Friday—he plans to ask for a raise. Not wanting to jeopardize his chance, Alan wants to know the probability that his boss not only will be happy this Friday, but also has had no two consecutive sad days in between. He will only hit his boss with the pay request under these conditions. Can you help by using taboo probabilities?

3.23 Suppose the transition matrix **P** can be *decomposed* into

$$ \mathbf{P} = \mathbf{H}\mathbf{J}\mathbf{H}^{-1} $$

where $\mathbf{H}^{-1}$ is the inverse of **H** and **J** is a diagonal matrix (that is, a matrix having zeros everywhere except along the diagonal).
a. Show that

$$ \mathbf{P}^k = \mathbf{H}\mathbf{J}^k\mathbf{H}^{-1} $$

where $\mathbf{J}^k$ is obtained simply by raising each of the diagonal elements of **J** to the power k.
b. For the taxicab transition matrix (3.9.2), verify that

$$ \mathbf{H} = \begin{pmatrix} 0.30476 & 0.63130 & 0.06394 \\ 0.30476 & -0.41424 & 0.10948 \\ 0.30476 & -0.19602 & -0.10874 \end{pmatrix} $$

$$ \mathbf{H}^{-1} = \begin{pmatrix} 1 & .84375 & 1.4375 \\ 1 & -.79129 & -.20871 \\ 1 & 3.7913 & -4.7913 \end{pmatrix} $$

and

$$ \mathbf{J} = \begin{pmatrix} 1 & 0 & 0 \\ 0 & 0.17913 & 0 \\ 0 & 0 & -0.27913 \end{pmatrix} $$

Hence, find

i. $\mathbf{P}^3$

ii. $\lim_{n\to\infty} \mathbf{P}^n$

iii. $\lim_{n\to\infty} \boldsymbol{\pi}^{(n)}$

3.24 Consider a two-state Markov chain having

$$\mathbf{P} = \begin{pmatrix} 1-\alpha & \alpha \\ \beta & 1-\beta \end{pmatrix}$$

with $0 < \alpha, \beta < 1$.

a. Show that

$$\mathbf{P}^n = \frac{1}{\alpha+\beta}\begin{pmatrix} \beta & \alpha \\ \beta & \alpha \end{pmatrix} + \frac{(1-\alpha-\beta)^n}{\alpha+\beta}\begin{pmatrix} \alpha & -\alpha \\ -\beta & \beta \end{pmatrix} \quad \text{for all } n \geq 0$$

i. by induction

ii. by using the result in Problem 3.23. *Hint:* Decompose $\mathbf{P}$ into

$$\mathbf{P} = \frac{1}{\alpha+\beta}\begin{pmatrix} \beta & \alpha \\ \beta & -\beta \end{pmatrix}\begin{pmatrix} 1 & 0 \\ 0 & 1-\beta-\alpha \end{pmatrix}\begin{pmatrix} 1 & \alpha/\beta \\ 1 & -1 \end{pmatrix}$$

b. Hence, find $\lim_{n\to\infty} \mathbf{P}^{(n)}$ and $\lim_{n\to\infty} \boldsymbol{\pi}^n$.

c. From Equation (3.10.2), find $\mathbf{V}^{(n)}$ for all $n \geq 0$. What is this value when $\alpha = \beta$?

3.25 Consider the following two-state *Markov-Bernoulli* chain $\{X_n\}_{n=0,1,2,\ldots}$ in which

$$\mathbf{P} = \begin{matrix} & \begin{matrix} 0 & & 1 \end{matrix} \\ \begin{matrix} 0 \\ 1 \end{matrix} & \begin{pmatrix} 1-(1-c)\alpha & (1-c)\alpha \\ (1-c)(1-\alpha) & c+(1-c)\alpha \end{pmatrix} \end{matrix}$$

with $0 < c < 1$. Let $p_n \equiv \Pr\{X_n = 1\}$ for all $n \geq 0$. Show that

$$p_n = \alpha + c^n(p_0 - \alpha) \quad \text{for all } n \geq 1$$

a. by using the result in Problem 3.24

b. by using the Chapman-Kolmogorov Equation (3.5.1) to show that

$$p_n = cp_{n-1} + (1-c)\,\alpha \quad \text{for all } n \geq 1$$

3.26 Consider a Markov chain with state space $S \equiv \{0, 1, 2, \ldots, N\}$. If it is at state i, it may remain at state i or move to another state with equal probability p_i; that is,

$$p_{ij} = \begin{cases} 1 - Np_i & \text{for } j = i \\ p_i & \text{for all } j \neq i \end{cases}$$

where $0 \leq Np_i \leq 1$ for all $0 \leq i \leq N$. Consider a particular state i.

a. Can you lump this chain with respect to the partition $\{i, \text{not-}i\}$, where not-i comprises all states in S except i?

b. If $p_i = p$ for all $0 \leq i \leq N$, can you lump this chain with respect to the partition $\{i, \text{not-}i\}$?

c. Thus, obtain the transient result $p_{ii}^{(n)}$ for all $n > 1$ when $p_i = p$ for all $0 \leq i \leq N$ by using Problem 3.24.

d. Hence, find $\lim_{n\to\infty} p_{ii}^{(n)}$.

3.27 Refer to Problem 1.1, where the genotype of the offspring of an animal in the $(n+1)$th generation has one gene randomly selected from its father and one from its mother. Assume now that the gene from its mother is no longer randomly selected but has the probability p of being a and $1-p$ of being b. Let X_n be the genotype of the father in the nth generation.

a. Show that the transition matrix for the chain $\{X_n\}_{n=0,1,2,\ldots}$ is

$$\mathbf{P} = \begin{array}{c} \\ aa \\ ab \\ bb \end{array} \begin{array}{c} \begin{array}{ccc} aa & ab & bb \end{array} \\ \begin{pmatrix} p & 1-p & 0 \\ p/2 & 1/2 & (1-p)/2 \\ 0 & p & 1-p \end{pmatrix} \end{array}$$

b. Use induction to show that

$$\mathbf{P}^n = \mathbf{Q} + \frac{1}{2^{n-1}}\mathbf{R}$$

where

$$\mathbf{Q} = \begin{pmatrix} p^2 & 2p(1-p) & (1-p)^2 \\ p^2 & 2p(1-p) & (1-p)^2 \\ p^2 & 2p(1-p) & (1-p)^2 \end{pmatrix}$$

$$\mathbf{R} = \begin{pmatrix} p(1-p) & (1-p)(1-2p) & -(1-p)^2 \\ p(1-2p)/2 & 1/2-2p(1-p) & -(1-p)(1-2p)/2 \\ -p^2 & p(2p-1) & p(1-p) \end{pmatrix}$$

c. Hence, find the limiting distribution of the genotypes.

3.28

a. Consider a Markov chain $\{X_n\}_{n=0,1,2,\ldots}$ having state space S and transition matrix $\mathbf{P}$. Suppose that there is a cost c_i incurred at step n whenever $X_n = i$ for all $i \in S$. Suppose also that costs are discounted at rate α $(0 \le \alpha \le 1)$. Thus, a cost of \$1 incurred at step n is equivalent to a cost of \$$\alpha^n$ at step 0. Given $X_0 = i$, let t_i be the expected present value of the total cost incurred over the entire life of the chain. Show that

$$\mathbf{t} \equiv [t_i]_{i\in S} = (\mathbf{I} - \alpha\mathbf{P})^{-1}\,\mathbf{c}$$

where $\mathbf{c} \equiv [c_i]_{i\in S}$ and $(\mathbf{I}-\alpha\mathbf{P})^{-1} \equiv \left(\mathbf{I} + \alpha\mathbf{P} + \alpha^2\mathbf{P}^2 + \cdots\right)$.

b. Each year, the weather can be either good (G), fair (F), or poor (P). The transition matrix is

$$\mathbf{P} = \begin{array}{c} \\ G \\ F \\ P \end{array} \begin{array}{c} \begin{array}{ccc} G & F & P \end{array} \\ \begin{pmatrix} .3 & .5 & .2 \\ .1 & .3 & .6 \\ .2 & .4 & .4 \end{pmatrix} \end{array}$$

If the weather is good, a company makes \$2 million; if fair, it makes \$1 million; if poor, it loses \$3 million. Let $\alpha = .8$. What is the expected present value of its total income if the company is having a good year now?

CHAPTER 4

Classification of Finite Chains

4.1 INTRODUCTION

At the end of the previous chapter, we noted that some Markov chains have a limiting distribution, and others do not. We also noted that the limiting distribution of a chain, if it exists, may be dependent on the initial conditions. Thus, to prepare for the analysis of the limiting behavior of Markov chains in future chapters, we use this chapter to explain how to classify them according to the manner in which they develop over time.

This chapter deals only with *finite* chains—that is, those having a finite number of states. We leave the classification of infinite chains until Chapter 7.

4.2 TRANSIENT AND RECURRENT STATES

First, let us look at each state individually. Starting from state i, let f_{ii} be the probability that the chain will eventually return to it. We call this the *return probability*.

If $f_{ii} = 1$, meaning that the chain will *surely* return to state i, state i is said to be *recurrent*. It is the nature of Markov chains that, once a chain returns to a recurrent state i, it will forget its past and thus will surely return to state i again. This means that state i is recurrent if and only if the expected number of visits to it throughout the chain's life is infinite. According to Equation (3.10.3), state i is therefore recurrent if and only if

$$\sum_{n=0}^{\infty} p_{ii}^{(n)} = \infty$$

For the taxicab example introduced in §3.9, we conjectured that all $p_{ii}^{(n)}$s converge to some positive numbers as $n \to \infty$. Thus, $\sum_{n=0}^{\infty} p_{ii}^{(n)} = \infty$ for all $i = A, B, C$, and all cities are therefore recurrent. Once Tex leaves city A, he will surely return to it.

On the other hand, if $f_{jj} < 1$, state j is said to be *transient*. Starting from a transient state j, there is a nonzero probability $1 - f_{jj}$ that the chain will never return to it.

Assume a Markov chain is visiting a transient state j in step 0. Let us say that we have a failure if the chain revisits state j at some later step (with probability $f_{jj} < 1$) and a success if the chain continues forever without returning to state j (with probability $1 - f_{jj}$). Because of the memoryless property, we now have a sequence of independent and identically distributed Bernoulli trials in which the number of visits v_{jj} to state j throughout the life of the chain is the number of failures before the first success. Thus, v_{jj} has a *geometric* distribution with finite mean $f_{jj}/(1 - f_{jj})$ (Equation 1.12.7).

In the rat-maze example, after leaving room 2, the rat might enter room F or S, causing the termination of the experiment and preventing a return to room 2. Thus, $f_{22} < 1$, and room 2 is a transient state.

Suppose now that there is a finite chain in which all states are transient. After some finite number of steps, one state becomes "unreturnable"; after another finite number of steps, another state becomes likewise; then another.... Eventually, the chain runs out of states and has nowhere to go. It follows that not all states in a finite chain can be transient. Every finite Markov chain must have at least one recurrent state.

4.3 REACHABILITY BETWEEN STATES

We now look at the relationships among the states. Consider three states i, k, and ℓ of a chain. Suppose the chain can go from state i to state k in one step but cannot go from state i to state ℓ in one step; that is, $p_{ik} > 0$ and $p_{i\ell} = 0$. Also suppose the chain can go from state k to state ℓ in one step; that is, $p_{k\ell} > 0$ (Figure 4.3.1). Then the chain can go from state i to state ℓ in two steps along the path $i - k - \ell$.

Regardless of the number of steps needed, as long as the chain can go from state i to state ℓ (that is, there is a positive probability that the chain goes from state i to state ℓ), we say state ℓ is *reachable from state* i and we denote this by $i \to \ell$. Otherwise, we write $i \not\to \ell$.

For the rat-maze example, although $p_{25} = 0$, $2 \to 5$ because the probability that the rat follows path $2 - 4 - 3 - 5$ is $(1/2)(1/3)(1/3) > 0$.

FIGURE 4.3.1 State ℓ is reachable from state i.

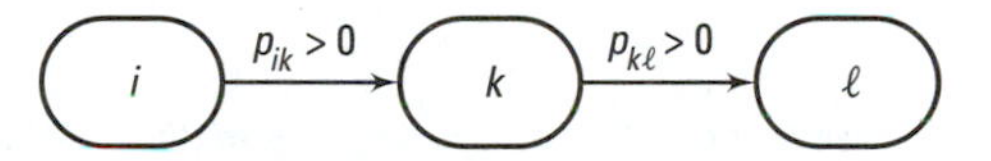

Note that $i \to \ell$ if and only if there is a path with positive probability of *any length* from state i to state ℓ. Thus, $i \to \ell$ if and only if

$$\sum_{n=0}^{\infty} p_{i\ell}^{(n)} > 0$$

From Equation (3.10.3), this also means that $i \to \ell$ if and only if, starting from state i, the expected number of times the chain visits state ℓ throughout its life is nonzero.

To be consistent with our definition that $p_{ii}^{(0)} = 1$ in §3.2, we say $i \to i$.

In a finite chain with N states, suppose there is a path from state i to state ℓ having length greater than N. Then there must be some loops (as defined in §3.1) in it. (Why?) If we eliminate all these loops, we obtain one reduced path from state i to state ℓ having length less than N. Now, if the original path has a nonzero probability, all one-step probabilities along it must also be nonzero because the probability of the original path is the product of these one-step probabilities. Hence, the reduced path also has a nonzero probability. Therefore, if ℓ is reachable from i, there must be a number $n < N$ such that $p_{i\ell}^{(n)} > 0$. Or $i \to \ell$ if and only if

$$\sum_{n=0}^{N-1} p_{i\ell}^{(n)} > 0 \tag{4.3.1}$$

That is, the (i, ℓ)-element of the sum matrix $\Sigma_{n=0}^{N-1}\mathbf{P}^{(n)}$ is nonzero.

For the rat-maze example,

$$\sum_{n=0}^{5} \mathbf{P}^{(n)} = \begin{array}{c} \\ F \\ 2 \\ 3 \\ 4 \\ 5 \\ S \end{array} \begin{array}{c} \begin{array}{cccccc} F & 2 & 3 & 4 & 5 & S \end{array} \\ \begin{pmatrix} 6.000 & 0 & 0 & 0 & 0 & 0 \\ 2.964 & 1.213 & 0.241 & 0.687 & 0.080 & 0.815 \\ 2.168 & 0.160 & 1.373 & 0.538 & 0.458 & 1.302 \\ 1.302 & 0.458 & 0.538 & 1.373 & 0.160 & 2.168 \\ 0.815 & 0.080 & 0.687 & 0.241 & 1.213 & 2.964 \\ 0 & 0 & 0 & 0 & 0 & 6.000 \end{pmatrix} \end{array}$$

showing that states 2, 3, 4, or 5 are not reachable from state F or S, and all states are reachable from states 2, 3, 4, or 5.

Let f_{ij} be the probability that the chain ever visits state j, given it starts from state i. We call it the *reaching probability*[1] from state i to state j. (The return probability f_{ii} is a special reaching probability in which $i = j$.) The matrix of reaching probabilities would be denoted by $\mathbf{F} \equiv [f_{ij}]_{i,j \in S}$.

To define reachability differently, we say that state ℓ is reachable from state i if and only if $f_{i\ell} > 0$.

---

[1] Some authors call it the *first passage probability* from state i to state j.

4.4 COMMUNICATION OF TWO STATES

If $i \to j$ and $j \to i$, then we say that the two states i and j *communicate* with each other, and we write $i \leftrightarrow j$. If not, we write $i \nleftrightarrow j$.

Clearly,

1. $i \leftrightarrow i$ (reflexivity).
2. If $i \leftrightarrow j$, then $j \leftrightarrow i$ (symmetry).
3. If $i \leftrightarrow j$ and $j \leftrightarrow k$, then $i \leftrightarrow k$ (transitivity).

4.5 CLASSES

All states that communicate with each other form a *class*. A state cannot belong to two distinct classes. Suppose state i belongs to two classes A and B. Let state j be an arbitrary member of A (thus, $j \leftrightarrow i$) and state k an arbitrary member of B (thus, $i \leftrightarrow k$). From the transitivity property, we must have $j \leftrightarrow k$ or j and k must belong to the same class, or $A = B$.

The chain's state space S can therefore be divided into one or more *disjoint* classes. A member of one class can be reachable from the members of the other classes but cannot communicate with them.

Also, once the chain leaves a state in class A to go to a state in a different class B, it cannot return to any state in A. (Otherwise, the states in A and B communicate; hence, $A = B$, contrary to our original assumption.)

There are three disjoint classes in the rat-maze example: $C_F \equiv \{F\}, C_2 \equiv \{2, 3, 4, 5\}$, and $C_S \equiv \{S\}$. State S, a member of class C_S, is reachable from all members of class C_2 but cannot communicate with any of them.

4.6 TRANSIENT AND RECURRENT CLASSES

We will now show that *recurrence and transience are class properties:* If a state is recurrent, all states in its class must be recurrent; if a state is transient, all states in its class must be transient.

Suppose state i is recurrent, and state j is reachable from state i. Because there is a nonzero probability of reaching state j from state i and because the chain keeps returning to state i, it will eventually visit state j. From state j, the chain must return to state i because state i is recurrent. Hence, states i and j communicate and belong to the same class. Furthermore, after the chain returns to state i, the whole process now restarts, so the chain must eventually visit state j again. Hence, state j is recurrent. It follows that a transient state cannot be reachable from a recurrent state. Recurrence is therefore a class property.

We said earlier that the chain cannot leave a class and return to it later. So once a chain enters a recurrent class, it is *trapped* there and cannot go to any other class.

On the other hand, suppose a transient state i communicates with some state j. If state j is recurrent, then from symmetry and the preceding argument, state i must be recurrent, which is a contradiction. Hence, state j must be transient. Thus, transience is also a class property.

Consider a chain with

$$\mathbf{P} = \begin{array}{c} \\ 1 \\ 2 \\ 3 \\ 4 \end{array} \begin{array}{c} \begin{array}{cccc} 1 & 2 & 3 & 4 \end{array} \\ \left(\begin{array}{cc|cc} .2 & .8 & 0 & 0 \\ .1 & .9 & 0 & 0 \\ \hline 0 & .5 & 0 & .5 \\ .3 & 0 & .7 & 0 \end{array}\right) \end{array}$$

Once the chain visits state 1 or 2, it will circulate between these two states forever. States 1 and 2 therefore form a recurrent class. From state 3, there is a .5 probability of moving to state 2, thus never returning to state 3, so state 3 is transient. Because $p_{43} = .7$ and $p_{34} = .5$, states 3 and 4 communicate and thus belong to the same transient class.

Markov Chains in Action: A Model for Brand Switching. Harary and Lipstein (1962) discussed a hypothetical brand-switching model among three brands A, B, and C. This model has a transition diagram as given in Figure 4.6.1. This diagram

> shows a situation in which consumers shift from brand B to brands A and C with some interchange between A and C. But once consumers leave brand B, they do not return. Thus brand B corresponds to a transient state of a Markov chain. Strategically, it is very weak. If the same conditions persist, it will be ruined. The product manager for brand B would do well to test his product against competitors, since it appears that once consumers sample A or C they show no tendency to go back to B....
>
> Brands A and C exchange customers between themselves but never relinquish customers to any other brand. Thus brands A and C form... [a recurrent class]. A desirable strategy for a manufacturer would be to establish a set of brands within a market that form... [a recurrent class]. The manufacturer always retains the consumer within his company franchise by providing a variety of products. Such a company position is then impregnable, at least until the introduction of significant change into the competitive complex, i.e., as long as the Markov chain remains stationary. ■

FIGURE 4.6.1 **Transition diagram for the brand-switching model.**

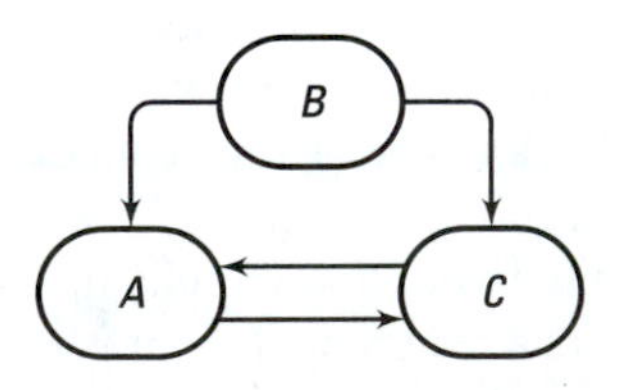

4.7 CANONICAL FORMS

Suppose there are r recurrent states and t transient states. If we group the states together in their classes and list the recurrent classes first, the transition matrix $\mathbf{P}$ can be written in the following *canonical* form:

$$\mathbf{P} = \begin{array}{c} \\ (r \text{ rows}) \\ (t \text{ rows}) \end{array} \begin{array}{c} \begin{array}{cc} (r \text{ columns}) & (t \text{ columns}) \end{array} \\ \left(\begin{array}{cc} \mathbf{R} & \mathbf{0} \\ \mathbf{S} & \mathbf{T} \end{array} \right) \end{array} \tag{4.7.1}$$

where

1. the elements of the $(r \times r)$ matrix $\mathbf{R}$ are the probabilities of the chain moving among the r recurrent states. Each row of this matrix adds up to 1. Furthermore, if there are ρ recurrent classes C_ℓ $(\ell = 1, 2, \ldots, \rho)$, then the states can also be organized so that

$$\mathbf{R} = \begin{pmatrix} \mathbf{R}_1 & \mathbf{0} & \cdots & \mathbf{0} \\ \mathbf{0} & \mathbf{R}_2 & \cdots & \mathbf{0} \\ \cdots & \cdots & \cdots & \cdots \\ \mathbf{0} & \mathbf{0} & \cdots & \mathbf{R}_\rho \end{pmatrix}$$

   The rows of each submatrix $\mathbf{R}_\ell$ $(\ell = 1, 2, \ldots, \rho)$ add to 1 and $\mathbf{R}_\ell$ contains the probabilities of the chain moving within the recurrent class C_ℓ. The elements of the $\mathbf{0}$ matrices are all zeros, representing the zero probabilities of the chain moving from one recurrent class to another.
2. the $(r \times t)$ matrix $\mathbf{0}$ contains the zero probabilities of the chain moving from a recurrent class to a transient class.
3. the $(t \times r)$ matrix $\mathbf{S}$ has the probabilities that the chain moves from the transient states to the recurrent states. With ρ recurrent classes, this matrix can take the form

$$\mathbf{S} \equiv (\mathbf{S}_1 \quad \mathbf{S}_2 \quad \cdots \quad \mathbf{S}_\rho) \tag{4.7.2}$$

   Its submatrix $\mathbf{S}_\ell$ $(\ell = 1, 2, \ldots, \rho)$ has the probabilities of the chain moving from the transient states to the states in class C_ℓ.
4. the $(t \times t)$ matrix $\mathbf{T}$ represents the probabilities of the chain moving among the t transient states.

The chain represented by the following transition matrix (with the states rearranged and renumbered to produce the appropriate canonical form) has two recurrent classes $A \equiv \{1, 2, 3\}$ and $B \equiv \{4, 5\}$ and one transient class $T \equiv \{6, 7, 8\}$:

$$\mathbf{P} = \begin{pmatrix} \mathbf{R}_1 & \mathbf{0} & \mathbf{0} \\ \mathbf{0} & \mathbf{R}_2 & \mathbf{0} \\ \mathbf{S}_1 & \mathbf{S}_2 & \mathbf{T} \end{pmatrix}$$

$$= \begin{array}{c} \\ 1 \\ 2 \\ 3 \\ 4 \\ 5 \\ 6 \\ 7 \\ 8 \end{array} \begin{array}{c} \begin{array}{cccccccc} 1 & 2 & 3 & 4 & 5 & 6 & 7 & 8 \end{array} \\ \left(\begin{array}{ccc|cc|ccc} .2 & .2 & .6 & 0 & 0 & 0 & 0 & 0 \\ 0 & .4 & .6 & 0 & 0 & 0 & 0 & 0 \\ .3 & .3 & .4 & 0 & 0 & 0 & 0 & 0 \\ \hline 0 & 0 & 0 & 0 & 1 & 0 & 0 & 0 \\ 0 & 0 & 0 & 1 & 0 & 0 & 0 & 0 \\ \hline .14 & .03 & .13 & .03 & .07 & .4 & .2 & 0 \\ .12 & .08 & 0 & .19 & .11 & .1 & .3 & .1 \\ .03 & .21 & .16 & .02 & .18 & .2 & 0 & .2 \end{array}\right) \end{array} \tag{4.7.3}$$

Not only is the canonical form more logical, revealing more of the chain's structure, but it is also helpful for calculations. For example, we now have

$$\mathbf{P}^n = \begin{pmatrix} \mathbf{R}^n & \mathbf{0} \\ ? & \mathbf{T}^n \end{pmatrix} \tag{4.7.4}$$

where

$$\mathbf{R}^n = \begin{pmatrix} \mathbf{R}_1^n & \mathbf{0} & \cdots & \mathbf{0} \\ \mathbf{0} & \mathbf{R}_2^n & \cdots & \mathbf{0} \\ \cdots & \cdots & \cdots & \cdots \\ \mathbf{0} & \mathbf{0} & \cdots & \mathbf{R}_\rho^n \end{pmatrix}$$

which is a more convenient form to use when calculating $\mathbf{P}^n$.

4.8 ABSORBING CHAINS

State i is said to be an *absorbing state* if and only if $p_{ii} = 1$. This means that, once the chain reaches an absorbing state, it stays there forever and cannot move to other states. An absorbing state is recurrent and forms a class by itself.

A chain is called an *absorbing chain* when all of its states are either transient or absorbing.

If there are a absorbing states and t transient states, the transition matrix $\mathbf{P}$ can be organized into the following canonical form:

$$\mathbf{P} = \begin{array}{c} (a \text{ rows}) \\ (t \text{ rows}) \end{array} \begin{array}{c} \begin{array}{cc} (a \text{ columns}) & (t \text{ columns}) \end{array} \\ \begin{pmatrix} \quad \mathbf{I} \quad & \quad \mathbf{0} \quad \\ \quad \mathbf{S} \quad & \quad \mathbf{T} \quad \end{pmatrix} \end{array}$$

where $\mathbf{I}$ is the $(a \times a)$ identity matrix and $\mathbf{S}$ is a $(t \times a)$ nonzero matrix.

The chain $\{R_n\}_{n=0,1,2,\ldots}$ in the rat-maze example is an absorbing chain with $a = 2$ absorbing states (F and S) and $t = 4$ transient states (2, 3, 4, and 5). The canonical form of its transition matrix is

$$\mathbf{P} = \begin{array}{c} \\ F \\ S \\ 2 \\ 3 \\ 4 \\ 5 \end{array} \begin{array}{c} \begin{array}{cccccc} F & S & 2 & 3 & 4 & 5 \end{array} \\ \left(\begin{array}{cc|cccc} 1 & 0 & 0 & 0 & 0 & 0 \\ 0 & 1 & 0 & 0 & 0 & 0 \\ \hline 1/2 & 0 & 0 & 0 & 1/2 & 0 \\ 1/3 & 0 & 0 & 0 & 1/3 & 1/3 \\ 0 & 1/3 & 1/3 & 1/3 & 0 & 0 \\ 0 & 1/2 & 0 & 1/2 & 0 & 0 \end{array}\right) \end{array} \tag{4.8.1}$$

4.9 IRREDUCIBLE CHAINS

A chain having only one class is said to be *irreducible*; otherwise, it is *reducible*. All states of an irreducible chain communicate with each other.

From the test of reachability (4.3.1), we can conclude that a finite chain having N states is irreducible if and only if all elements of the sum matrix $\mathbf{I}+\mathbf{P}+\mathbf{P}^2+\cdots+\mathbf{P}^{N-1}$ are strictly positive.

It is also true that all states in a finite irreducible chain must be recurrent. (Why?)

4.10 PERIODIC STATES

Let us now define the *periodicity* of a state. A state i is said to have a *period* d if, starting from state i, the chain can only revisit it d or a multiple of d steps later. In other words, state i has period d if and only if d is the smallest integer for which $p_{ii}^{(n)} = 0$ when n is *not* divisible by d. State i is said to be *periodic* if $d > 1$; otherwise, the state is *aperiodic*.

Consider a chain having the following transition matrix:

$$\mathbf{P} = \begin{array}{c} \\ A \\ B \\ C \\ D \\ E \\ F \\ G \end{array} \begin{array}{c} \begin{array}{ccccccc} A & B & C & D & E & F & G \end{array} \\ \left(\begin{array}{cc|ccc|cc} 0 & 0 & 1/4 & 1/4 & 1/2 & 0 & 0 \\ 0 & 0 & 2/5 & 2/5 & 1/5 & 0 & 0 \\ \hline 0 & 0 & 0 & 0 & 0 & 1/4 & 3/4 \\ 0 & 0 & 0 & 0 & 0 & 1/2 & 1/2 \\ 0 & 0 & 0 & 0 & 0 & 1/3 & 2/3 \\ \hline 2/3 & 1/3 & 0 & 0 & 0 & 0 & 0 \\ 1/4 & 3/4 & 0 & 0 & 0 & 0 & 0 \end{array}\right) \end{array} \tag{4.10.1}$$

Suppose this chain is in state A at step 0. At step 1, it must visit state C, D, or E and cannot yet return to state A. At step 2, it must visit state F or G and cannot yet return to state A. At step 3, it can return to state A (although it could go to state B

instead). Continuing, the chain cannot return to state A at steps 4, 5, 7, 8, ..., but only at steps 6, 9, 12, ... State A is therefore periodic with period $d = 3$.

Suppose that state i has period d. Consider all loops from state i, as defined in §3.1, that have a nonzero probability. Because the chain can only return to that state md steps later, the lengths of these loops must be a multiple of d.

Furthermore, a state of period $d > 0$ can be either transient or recurrent. It is periodic because, if the chain returns to it, it can only do so md steps later. But the chain does not have to return to it at these steps. In the following example, states 2 and 3 are both periodic and transient:

$$\mathbf{P} = \begin{array}{c} \\ 1 \\ 2 \\ 3 \end{array} \begin{array}{c} \begin{array}{ccc} 1 & 2 & 3 \end{array} \\ \left(\begin{array}{c|cc} 1 & 0 & 0 \\ \hline .3 & 0 & .7 \\ .4 & .6 & 0 \end{array}\right) \end{array}$$

4.11 PERIODIC CLASSES

Periodicity is also a class property; that is, if a state has period d, so do all other states in the same class.

To prove this, consider a chain $\{X_n\}_{n=0,1,2,\ldots}$ having two distinct states i and j that communicate with each other and hence belong to the same class. Without loss of generality, assume $X_0 = i$. There must be a path A, having a nonzero probability, starting with state i at step 0 and ending with state j at a certain step k. There must also be a path B, having a nonzero probability, starting with state j at step k and ending with state i at a certain step $\ell > k$.

The *combined path* $(A - B)$, which is path A immediately followed by path B, is a loop from state i. If state i has period d, then since loop $(A - B)$ has a nonzero probability, its length must be of the form $\ell = md$ for some integer m, and the length of path B must be of the form $md - k$ (Figure 4.11.1).

Consider now the combined path $(B - A)$, which is path B immediately followed by path A. This is a loop from state j that has a nonzero probability. It has a length of $(md - k) + k = md$. Thus, if state j is periodic with period e, then d is divisible by e.

From the symmetrical argument, we conclude that $d = e$, or the period of state j must be the same as that of state i.

FIGURE 4.11.1 State ℓ is reachable from state i.

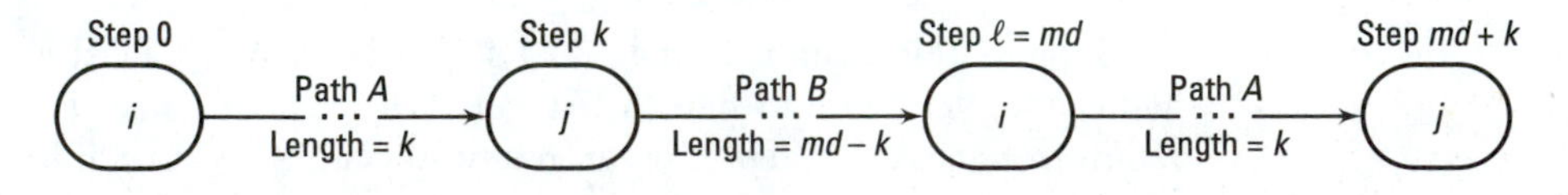

4.12 FINITE PERIODIC CHAINS

We call an irreducible chain having a state with period $d > 1$ a *periodic* chain. This implies that there is only one class, and all states of the chain have the same period.

The state space of a *finite* periodic chain with period d can be divided and reorganized into d disjoint nonnull *groups* $G_1, G_2, \ldots, G_d$. After visiting a state in group G_i $(1 \leq i < d)$, the chain must visit a state in group G_{i+1}; after visiting a state in group G_d, it must visit a state in group G_1. In other words, it has to follow the pattern: $G_1 - G_2 - \cdots - G_d - G_1 - G_2 - \cdots - G_d - G_1 - \cdots$. In this manner, once leaving group G_i, the chain can only return to it d steps later. For example, the state space of Chain (4.10.1), which has only one class, can be divided into three groups $G_1 \equiv \{A, B\}$, $G_2 \equiv \{C, D, E\}$, and $G_3 \equiv \{E, F\}$.

The transition matrix of a finite periodic chain, having period d, therefore can always be rearranged so that:

$$\mathbf{P} = \begin{array}{c} \\ G_1 \\ G_2 \\ \cdots \\ G_{d-1} \\ G_d \end{array} \begin{array}{c} \begin{array}{cccccc} G_1 & G_2 & G_3 & \cdots & G_{d-1} & G_d \end{array} \\ \begin{pmatrix} \mathbf{0} & \mathbf{D}_1 & \mathbf{0} & \cdots & \mathbf{0} & \mathbf{0} \\ \mathbf{0} & \mathbf{0} & \mathbf{D}_2 & \cdots & \mathbf{0} & \mathbf{0} \\ \cdots & \cdots & \cdots & \cdots & \cdots & \cdots \\ \mathbf{0} & \mathbf{0} & \mathbf{0} & \cdots & \mathbf{0} & \mathbf{D}_{d-1} \\ \mathbf{D}_d & \mathbf{0} & \mathbf{0} & \cdots & \mathbf{0} & \mathbf{0} \end{pmatrix} \end{array} \tag{4.12.1}$$

where $\mathbf{D}_i$ is the matrix giving the probability of the chain moving from a state in group G_i to a state in group G_{i+1}. The transition matrix (4.10.1) is in this form. Note that it is not neccessary that the matrices $\mathbf{D}_i$s be square.

Consider a seasonal chain, as introduced in §2.9, having N states. If we use a supplementary variable to study it, as in §2.14, then we have a special form of a periodic chain. For example, Chain (2.14.1) is periodic with $d = 2$. In this case, the matrices $\mathbf{D}_i$s are all square, having dimensions $(N \times N)$.

Of particular interest are the recurrent and periodic chains. Even after a large number of steps, the periodic pattern continues and the sequence of vectors $\{\boldsymbol{\pi}^{(n)}\}_{n=0,1,2,\ldots}$ therefore does not converge when $n \to \infty$. For example, consider the chain defined in Equation (4.10.1). Here, $\boldsymbol{\pi}^{(n)}$ keeps alternating among the following forms:

$$\begin{aligned} \pi^{(3m)} &= (* \quad * \quad 0 \quad 0 \quad 0 \quad 0 \quad 0) \\ \pi^{(3m+1)} &= (0 \quad 0 \quad * \quad * \quad * \quad 0 \quad 0) \\ \pi^{(3m+2)} &= (0 \quad 0 \quad 0 \quad 0 \quad 0 \quad * \quad *) \end{aligned}$$

where $*$ represents a nonnegative number.

Markov Chains in Action: Periodic Pattern in Counseling Interviews. Lichtenberg and Hummel (1976) examined the transcripts of six initial interviews conducted by two counselors. They defined ten states: introductory I, T_1, T_2, T_3, T_4, C_1, C_2, C_3, C_4, and terminal T_e. The chain is in state C_i $(1 \leq i \leq 4)$ when the counselor speaks and in state T_i

$(1 \leq i \leq 4)$ when the client speaks. The subscript indicates the kind of statement made: 1 if personal, affective, or self-disclosing; 2 if descriptive, impersonal, or nonaffective; 3 if cognitive or analytical; and 4 if directive, leading, structuring, or imperative. The following empirical transition matrix was obtained:

$$
\mathbf{P} = \begin{array}{c} \\ T_e \\ T_1 \\ T_2 \\ T_3 \\ T_4 \\ C_1 \\ C_2 \\ C_3 \\ C_4 \\ I \end{array}
\begin{array}{c|cccc|cccc|c}
T_e & T_1 & T_2 & T_3 & T_4 & C_1 & C_2 & C_3 & C_4 & I \\
1 & 0 & 0 & 0 & 0 & 0 & 0 & 0 & 0 & 0 \\ \hline
0 & 0 & 0 & 0 & 0 & 1 & 0 & 0 & 0 & 0 \\
.02 & 0 & 0 & 0 & 0 & .23 & .48 & .24 & .03 & 0 \\
0 & 0 & 0 & 0 & 0 & .19 & .62 & .15 & .04 & 0 \\
0 & 0 & 0 & 0 & 0 & .05 & .84 & .00 & .10 & 0 \\ \hline
0 & .00 & .50 & .42 & .08 & 0 & 0 & 0 & 0 & 0 \\
0 & .03 & .45 & .33 & .19 & 0 & 0 & 0 & 0 & 0 \\
0 & .00 & .50 & .50 & .00 & 0 & 0 & 0 & 0 & 0 \\
0 & .00 & .66 & .17 & .17 & 0 & 0 & 0 & 0 & 0 \\ \hline
0 & 0 & 0 & 0 & 1 & 0 & 0 & 0 & 0 & 0
\end{array}
$$

This chain has three classes: $C_T \equiv \{T_e\}$, recurrent, absorbing; $C_I \equiv \{I\}$, transient; and $C_2 \equiv \{T_1, T_2, T_3, T_4, C_1, C_2, C_3, C_4\}$, transient and periodic with period 2, since a step in the chain occurs when the conversation switches between the counselor and the client. ■

4.13 SUMMARY

In this chapter, we

1. defined the reaching probability f_{ij} as the probability that the chain ever visits state j, given it starts from state i.
2. distinguished between transient and recurrent states.
3. grouped states into disjoint classes, which are either transient or recurrent.
4. reorganized the transition matrix into a canonical form.
5. explained what is meant by irreducible chains, absorbing states, and absorbing chains.
6. distinguished between the periodic and aperiodic states and chains.
7. noted that chains having recurrent and periodic states do not have a limiting distribution.

PROBLEMS

4.1 Classify the states of the Markov chain defined in (4.10.1) by obtaining $\mathbf{P}+\mathbf{P}^2+\mathbf{P}^3$.

4.2 Classify the states of the Markov chains having the following transition matrices. Arrange them in canonical forms.

a.

b.

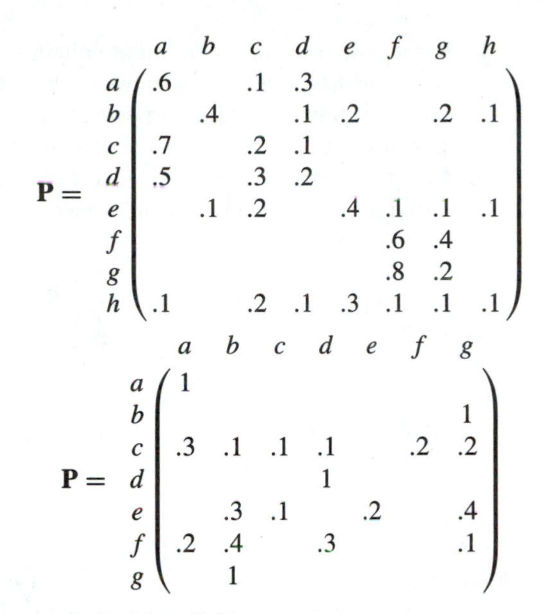

$$\mathbf{P} = \begin{array}{c} \\ a \\ b \\ c \\ d \\ e \\ f \\ g \\ h \end{array} \begin{array}{c} \begin{array}{cccccccc} a & b & c & d & e & f & g & h \end{array} \\ \begin{pmatrix} .6 & & .1 & .3 & & & & \\ & .4 & & .1 & .2 & & .2 & .1 \\ .7 & & .2 & .1 & & & & \\ .5 & & .3 & .2 & & & & \\ & .1 & .2 & & .4 & .1 & .1 & .1 \\ & & & & & .6 & .4 & \\ & & & & & .8 & .2 & \\ .1 & & .2 & .1 & .3 & .1 & .1 & .1 \end{pmatrix} \end{array}$$

$$\mathbf{P} = \begin{array}{c} \\ a \\ b \\ c \\ d \\ e \\ f \\ g \end{array} \begin{array}{c} \begin{array}{ccccccc} a & b & c & d & e & f & g \end{array} \\ \begin{pmatrix} 1 & & & & & & \\ & & & & & & 1 \\ .3 & .1 & .1 & .1 & & .2 & .2 \\ & & & 1 & & & \\ & .3 & .1 & & .2 & & .4 \\ .2 & .4 & & .3 & & & .1 \\ & 1 & & & & & \end{pmatrix} \end{array}$$

4.3 Classify the states in Problem 2.21.

4.4 Classify the states in Problem 2.3. What are their periods?

4.5 Classify the states of chain $\{Z_n\}_{n=0,1,2,\ldots}$ in Problem 2.27. What are their periods?

4.6 Classify the states of chain $\{Z_n\}_{n=0,1,2,\ldots}$ in Problem 2.28. What are their periods?

4.7 Let $\mathbf{P}$ be the transition matrix of a chain A.

a. Let $\mathbf{I}$ be an identity matrix having the same size as $\mathbf{P}$. Show that $\mathbf{Q} \equiv (\mathbf{P}+\mathbf{I})/2$ is also a transition matrix.

b. Call the chain having $\mathbf{Q}$ as transition matrix chain B. Show that two states in chain A communicate if and only if their equivalent states in chain B communicate.

4.8 Prove or give a counterexample to the following statement: If a finite Markov chain has m states, the period of each of its states divides m.

4.9 Prove that an irreducible finite Markov chain cannot be periodic if at least one of the diagonal elements of its transition matrix is strictly positive. Give an example to show that the converse is not true; that is, it is possible to have an irreducible finite aperiodic Markov chain in which $p_{ii} = 0$ for all $i \in S$.

4.10

a. For the reaching probability f_{ij}, use conditional arguments to show that

$$f_{ij} = p_{ij} + \sum_{m \neq j} p_{im} f_{mj} \quad \text{for all } m, i, j \in S$$

In matrix form,

$$\mathbf{F} \equiv [f_{ij}]_{i,j\in S} = \mathbf{P}(\mathbf{I} + \mathbf{F} - \mathbf{F}_d)$$

where $\mathbf{F}_d$ is the diagonal matrix, the (i, i)-element of which is f_{ii}.

b. For a particular state j in a finite chain with N states, this equation gives a set of N equations with N unknowns. Show that $[f_{ij} = 1]_{i \in S}$ is always one set of solutions. Since state i might be transient (that is, $f_{ii} < 1$), can you say that the preceding equations are always mutually independent?

c. Apply this equation to the taxicab problem in §3.9 with $j = A$. Show that the resulting equations are mutually independent and hence state A is recurrent.

d. For the following chain

$$\mathbf{P} = \begin{array}{c} \\ 1 \\ 2 \\ 3 \end{array} \begin{array}{c} \begin{array}{ccc} 1 & 2 & 3 \end{array} \\ \begin{pmatrix} 1 & 0 & 0 \\ .5 & 0 & .5 \\ .5 & .5 & 0 \end{pmatrix} \end{array}$$

show that, with $j = 2$, the preceding equations yield only two mutually independent equations for three unknowns. Is this consistent with the fact that state 2 is transient?

4.11 A Markov chain is said to be *doubly stochastic* when all columns of its transition matrix add up to 1; that is, $\sum_{i \in S} p_{ij} = 1$ for all $j \in S$ (besides the fact that all its rows must add up to 1). Show that all states in a finite doubly stochastic chain must be recurrent. *Hint:* Find $\sum_{i \in S} f_{ij}$ using Problem 4.10.

4.12 For a Markov chain $\{X_n\}_{n=0,1,2,\ldots}$, let $\ell_{ij}^{(n)}$ be the probability that the chain goes from state i to state j in n steps without ever visiting state i in between. We call it the *last visit probability*.

a. Use conditional arguments to show that

$$\ell_{ij}^{(n+1)} = \sum_{m \neq i} \ell_{im}^{(n)} p_{mj} \quad \text{for all } m, i, j \in S, \text{ and } n > 0$$

b. The probability that the chain goes from state i to state j without ever visiting state i in between is thus $\ell_{ij} \equiv \sum_{n=1}^{\infty} \ell_{ij}^{(n)}$. Show that

$$\ell_{ij} = \sum_{m \neq i} \ell_{im} p_{mj}$$

c. For a particular state i in a finite chain with N states, this equation gives a set of $N - 1$ equations with $N - 1$ unknowns. Show that $[\ell_{ij} = 0]_{i \in S}$ is always one set of solutions. Since state i might be transient (that is, $\sum_{j \neq i} \ell_{ij} > 0$), can you say that the preceding equations are always mutually independent?

d. Apply this equation to the taxicab problem in §3.9 with $i = A$. Show that the resulting equations are mutually independent and hence A is recurrent.

e. For the chain in Problem 4.10d, show that, with $i = 2$, the preceding equations yield only one independent equation for two unknowns. Is this consistent with the fact that state 2 is transient?

CHAPTER 5

Finite Absorbing Chains

5.1 INTRODUCTION

We have defined absorbing chains in general as those having only absorbing and transient states. We now turn our attention to the study of *finite* absorbing chains. Each such chain will have a set A of a absorbing states and a set T of t transient states. The transition matrices for these chains have the following canonical form:

$$\mathbf{P} = \begin{matrix} & (a \text{ columns}) & (t \text{ columns}) \\ (a \text{ rows}) & \mathbf{I} & \mathbf{0} \\ (t \text{ rows}) & \mathbf{S} & \mathbf{T} \end{matrix} \tag{5.1.1}$$

Every absorbing chain will eventually reach one of its absorbing states and stay there forever. The question that we wish to address is: *Which* absorbing state? or What is the probability that the chain will end up at a particular absorbing state rather than one of the others?

Before answering this question, we will first determine how many times an absorbing chain can be expected to visit a given transient state before being absorbed.

5.2 FIRST-STEP ANALYSIS

In this chapter, conditional arguments are often used to support our results. Its application to the study of Markov chains in the following manner is referred to as the *first-step analysis*. Suppose that we are interested in some quantity q_i connected with a Markov chain $\{X_n\}_{n=0,1,2,\ldots}$, which is dependent on the initial condition $X_0 = i$. First-step analysis suggests that, instead of trying to obtain q_i alone, we introduce other quantities q_js, which *exhaust all other possible initial conditions*. Conditional arguments can then be used to derive the necessary equations.

We now look at two important applications of first-step analysis:

1. The quantity q_i is the *conditional probability* α_i that a certain event E connected to chain $\{X_n\}_{n=0,1,2,\ldots}$ occurs, given that $X_0 = i$; that is,

$$q_i \equiv \alpha_i \equiv \Pr\{\text{Event } E \mid X_0 = i\}$$

Assume the chain's first transition is from state i to state j, or $X_1 = j$. Because of its memoryless property, the probability of event E now becomes independent of what state the chain was in at time 0 and only dependent on the fact that it starts anew from state j. In other words,

$$\Pr\{\text{Event } E \mid X_0 = i \text{ and } X_1 = j\} = \Pr\{\text{Event } E \mid X_1 = j\} = \alpha_j$$

With p_{ik} as the probability of the chain's first transition from state i to state k, the law of total probability (1.7.1) yields

$$\alpha_i = \sum_{j \in S} p_{ij} \Pr\{\text{Event } E \mid X_1 = j\} = \sum_{j \in S} p_{ij}\alpha_j$$

2. The quantity q_i is the *conditional expected number of steps* $n_i \equiv \mathcal{E}[n \mid X_0 = i]$ taken by the chain until a certain event E occurs.

With probability p_{ij}, the chain takes its first step to move from state i to state j. Given this first transition, because of the memoryless property,

$$\mathcal{E}[n \mid X_0 = i \text{ and } X_1 = j] = 1 + \mathcal{E}[n \mid X_1 = j] = 1 + n_j$$

which includes the first step already taken. The law of total probability for expectations (1.20.2) then gives

$$n_i = \sum_{k \in S} p_{ik}(1 + n_k) = 1 + \sum_{k \in S} p_{ik} n_k$$

If there are m possible initial conditions, first-step analysis would yield a set of m linear equations with m unknown q_js. If these equations are linearly independent, we can solve for the numerical value of q_i.

5.3 EXPECTED NUMBER OF VISITS TO A TRANSIENT STATE

Let v_{ij} be the number of visits a chain makes to state j throughout its life, having started from state i. We discussed $\mathcal{E}[v_{ij}]$ for a general Markov chain in §3.10. Suppose now state j is an absorbing state in an absorbing chain. In this case, the chain will either be absorbed into state j and stay there forever, or it will be absorbed into another state and thus not visit state j at all. In addition, if the chain starts from an absorbing state i, it will also stay there forever. Thus, the only case of real interest for us to study in this section is that of $\mathcal{E}[v_{ij}]$, where both i and j are transient states in an absorbing chain.

Returning to the chain $\{R_n\}_{n=0,1,2,\ldots}$ in the rat-maze example introduced in §2.10, for convenience its transition matrix in canonical form (4.8.1) is reproduced here:

$$\mathbf{P} = \begin{pmatrix} \mathbf{I} & \mathbf{0} \\ \mathbf{S} & \mathbf{T} \end{pmatrix} = \begin{array}{c} \\ F \\ S \\ 2 \\ 3 \\ 4 \\ 5 \end{array} \begin{array}{c} \begin{array}{cccccc} F & S & 2 & 3 & 4 & 5 \end{array} \\ \left(\begin{array}{cc|cccc} 1 & 0 & 0 & 0 & 0 & 0 \\ 0 & 1 & 0 & 0 & 0 & 0 \\ \hline 1/2 & 0 & 0 & 0 & 1/2 & 0 \\ 1/3 & 0 & 0 & 0 & 1/3 & 1/3 \\ 0 & 1/3 & 1/3 & 1/3 & 0 & 0 \\ 0 & 1/2 & 0 & 1/2 & 0 & 0 \end{array}\right) \end{array} \tag{5.3.1}$$

We want to find the expected number of times $\mathcal{E}[v_{23}]$ the rat visits room 3, starting from room 2, before it finds either food or shock.

To use first-step analysis, instead of studying $\mathcal{E}[v_{23}]$ alone, we consider the set of four quantities $\mathcal{E}[v_{i3}]$ (with $i = 2, 3, 4, 5$), which exhausts all possible initial conditions (excluding F and S, which we do not assume).

Suppose now that the rat starts from room 4, so the relevant quantity is $\mathcal{E}[v_{43}]$.

1. The probability that the rat reaches the absorbing room S after the first step is 1/3. If this occurs, the expected number of times the chain visits room 3 is $\mathcal{E}[v_{43} \mid R_1 = S] = 0$.
2. The probability that the rat reaches room 2 after the first step is 1/3. If this occurs, because of the memoryless property of the chain, the expected number of times it visits room 3 is the same as if it had started from room 2. Thus, $\mathcal{E}[v_{43} \mid R_1 = 2] = \mathcal{E}[v_{23}]$.
3. The probability that the rat reaches room 3 after the first step is 1/3. If this occurs, the expected number of times it visits room 3 is the same as if it had started from room 3, $\mathcal{E}[v_{43} \mid R_1 = 3] = \mathcal{E}[v_{33}]$.

Thus, as shown in the tree diagram in Figure 5.3.1(c),

$$\mathcal{E}[v_{43}] = (1/3)(0) + (1/3)\mathcal{E}[v_{23}] + (1/3)\mathcal{E}[v_{33}]$$

On the other hand, suppose the rat starts from room 3, thus *having already made one visit* to that room. The relevant quantity now is $\mathcal{E}[v_{33}]$.

1. The probability that the rat reaches the absorbing room F after the first step is 1/3. If this occurs, $\mathcal{E}[v_{33} \mid R_1 = F] = 1$.
2. The probability that the rat reaches room 4 after the first step is 1/3. If this occurs, $\mathcal{E}[v_{33} \mid R_1 = 4] = 1 + \mathcal{E}[v_{43}]$.
3. The probability that the rat reaches room 5 after the first step is 1/3. If this occurs, $\mathcal{E}[v_{33} \mid R_1 = 5] = 1 + \mathcal{E}[v_{53}]$.

Thus, as shown in the tree diagram in Figure 5.3.1(b),

$$\begin{aligned} \mathcal{E}[v_{33}] &= (1/3)(1) + (1/3)\,(1 + \mathcal{E}[v_{43}]) + (1/3)\,(1 + \mathcal{E}[v_{53}]) \\ &= 1 + (1/3)\mathcal{E}[v_{43}] + (1/3)\mathcal{E}[v_{53}] \end{aligned}$$

FIGURE 5.3.1 Tree diagram for the rat's number of visits to room 3.

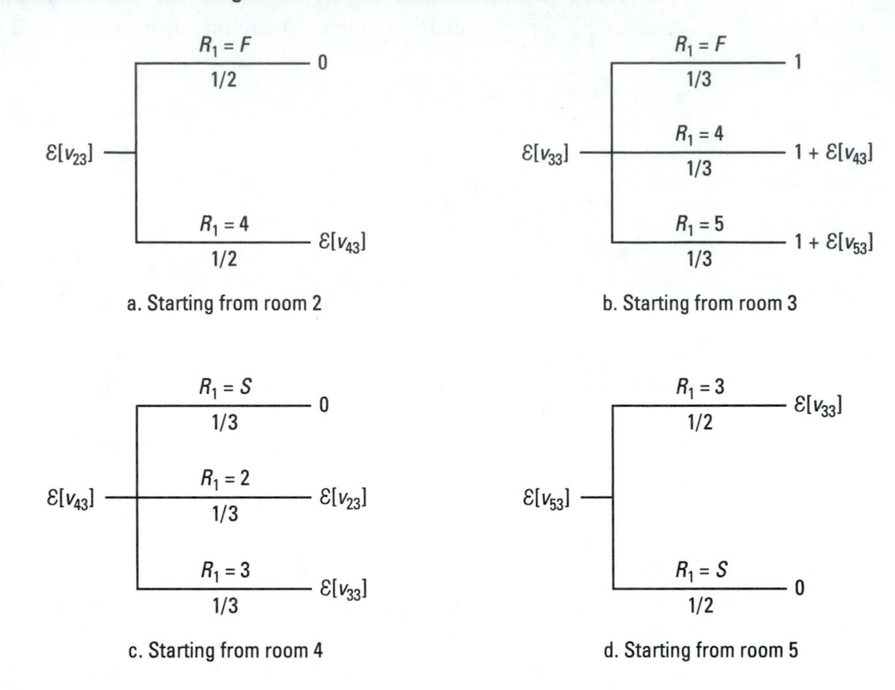

Similar arguments yield the following set of linear equations:

$$\begin{cases} \mathcal{E}[v_{23}] = & & + (1/2)\mathcal{E}[v_{43}] \\ \mathcal{E}[v_{33}] = 1 & & + (1/3)\mathcal{E}[v_{43}] + (1/3)\mathcal{E}[v_{53}] \\ \mathcal{E}[v_{43}] = & (1/3)\mathcal{E}[v_{23}] + (1/3)\mathcal{E}[v_{33}] & \\ \mathcal{E}[v_{53}] = & (1/2)\mathcal{E}[v_{33}] & \end{cases} \tag{5.3.2}$$

Solving them simultaneously, we obtain

$$\mathcal{E}[v_{23}] = 2/7 \qquad \mathcal{E}[v_{33}] = 10/7 \qquad \mathcal{E}[v_{43}] = 4/7 \qquad \mathcal{E}[v_{53}] = 5/7 \tag{5.3.3}$$

Suppose now that we also want to obtain $\mathcal{E}[v_{i4}]$, the expected number of times the rat visits room 4 before finding either food or shock, given it starts from room i (with $i = 2, 3, 4, 5$). Using similar first-step analysis, we should obtain the following set of equations:

$$\begin{cases} \mathcal{E}[v_{24}] = & & (1/2)\mathcal{E}[v_{44}] \\ \mathcal{E}[v_{34}] = & & + (1/3)\mathcal{E}[v_{44}] + (1/3)\mathcal{E}[v_{54}] \\ \mathcal{E}[v_{44}] = 1 + (1/3)\mathcal{E}[v_{24}] + (1/3)\mathcal{E}[v_{34}] & & \\ \mathcal{E}[v_{54}] = & + (1/2)\mathcal{E}[v_{34}] & \end{cases} \tag{5.3.4}$$

From this,

$$\mathcal{E}[v_{24}] = 5/7 \qquad \mathcal{E}[v_{34}] = 4/7 \qquad \mathcal{E}[v_{44}] = 10/7 \qquad \mathcal{E}[v_{54}] = 2/7 \tag{5.3.5}$$

The two sets of Equations (5.3.2) and (5.3.4) are very similar. Solving them independently as we did fails to take advantage of their common features. Combining them in matrix form, we obtain

$$\mathbf{U} = \mathbf{I} + \mathbf{T}\mathbf{U} \tag{5.3.6}$$

where

1. $\mathbf{U} \equiv [\mathcal{E}[v_{ij}]]_{i,j \in T}$ is the (4×4) transient-to-transient portion of the matrix $\mathbf{V} \equiv [\mathcal{E}[v_{ij}]]_{i,j \in S}$; that is,

$$\mathbf{U} = \begin{array}{c} \\ 2 \\ 3 \\ 4 \\ 5 \end{array} \begin{array}{c} \begin{array}{cccc} 2 & 3 & 4 & 5 \end{array} \\ \begin{pmatrix} \mathcal{E}[v_{22}] & \mathcal{E}[v_{23}] & \mathcal{E}[v_{24}] & \mathcal{E}[v_{25}] \\ \mathcal{E}[v_{32}] & \mathcal{E}[v_{33}] & \mathcal{E}[v_{34}] & \mathcal{E}[v_{35}] \\ \mathcal{E}[v_{42}] & \mathcal{E}[v_{43}] & \mathcal{E}[v_{44}] & \mathcal{E}[v_{45}] \\ \mathcal{E}[v_{52}] & \mathcal{E}[v_{53}] & \mathcal{E}[v_{54}] & \mathcal{E}[v_{55}] \end{pmatrix} \end{array}$$

2. $\mathbf{I}$ is the (4×4) identity matrix
3. $\mathbf{T}$ is the transient-to-transient portion of the transition matrix $\mathbf{P}$ as shown in (5.1.1) or (5.3.1)

We now have a simpler way to calculate all $\mathcal{E}[v_{ij}]$s simultaneously:

$$\mathbf{U} = (\mathbf{I} - \mathbf{T})^{-1} \tag{5.3.7}$$

For the rat-maze problem, we can calculate $\mathbf{U}$ from the canonical form (5.3.1) as

$$\mathbf{U} = \begin{pmatrix} 1 & 0 & -\frac{1}{2} & 0 \\ 0 & 1 & -\frac{1}{3} & -\frac{1}{3} \\ -\frac{1}{3} & -\frac{1}{3} & 1 & 0 \\ 0 & -\frac{1}{2} & 0 & 1 \end{pmatrix}^{-1} = \begin{array}{c} \\ 2 \\ 3 \\ 4 \\ 5 \end{array} \begin{array}{c} \begin{array}{cccc} 2 & 3 & 4 & 5 \end{array} \\ \begin{pmatrix} \frac{26}{21} & \frac{2}{7} & \frac{5}{7} & \frac{2}{21} \\ \frac{4}{21} & \frac{10}{7} & \frac{4}{7} & \frac{10}{21} \\ \frac{10}{21} & \frac{4}{7} & \frac{10}{7} & \frac{4}{21} \\ \frac{2}{21} & \frac{5}{7} & \frac{2}{7} & \frac{26}{21} \end{pmatrix} \end{array} \tag{5.3.8}$$

which is consistent with Equations (5.3.3) and (5.3.5).

It turns out that Equation (5.3.7) holds true for *all* finite absorbing chains. We can understand it for the general case by referring back to Equation (3.10.3), in which we proved that the expected number of visits to state j from state i throughout a chain's life is the (i, j)-element of the sum matrix $\sum_{n=0}^{\infty} \mathbf{P}^n$. Recall also from Equation (4.7.4) that the transient-to-transient portion of $\mathbf{P}^n$ is simply $\mathbf{T}^n$. Thus, the transient-to-transient portion of $\sum_{n=0}^{\infty} \mathbf{P}^n$ is equal to $\sum_{n=0}^{\infty} \mathbf{T}^n$. In Appendix (A.8), we say that the geometric series $\sum_{n=0}^{\infty} x^n$ converges to $(1 - x)^{-1}$ if and only if $|x| < 1$. Similarly, it can be shown that, as part of the transition matrix **P**, matrix **T** satisfies the necessary conditions for $\sum_{n=0}^{\infty} \mathbf{T}^n$ to converge elementwise to the matrix $(\mathbf{I} - \mathbf{T})^{-1}$.

This matrix $\mathbf{U} = (\mathbf{I} - \mathbf{T})^{-1}$ proves to be very important in the analysis of finite absorbing chains. It turns up in most results related to the chain. We refer to **U** as the *fundamental matrix.*

5.4 EXPECTED ABSORPTION TIME

Let u_i be the *number of steps* before the chain is absorbed into one of the absorbing states, starting from a transient state i. It is called the *absorption time* (*from state* i).

Note that the chain visits only the transient states before absorption. Thus, starting from a transient state i, the expected absorption time is simply the sum of the expected numbers of visits to all the transient states or, with T as the set of transient states,

$$\mathcal{E}[u_i] = \sum_{j \in T} \mathcal{E}[v_{ij}] \tag{5.4.1}$$

which is the *sum of the* i*-row* of the fundamental matrix **U**.

In matrix form,

$$\mathbf{u} \equiv [\mathcal{E}[u_i]]_{i \in T} = \mathbf{Ue}$$

where **e** is a column vector, all elements of which are equal to 1.

Returning to the rat-maze example, starting from room 2 the expected number of steps the rat makes before being either fed or shocked is thus the sum of the elements of the 2-row of the fundamental matrix (5.3.8):

$$\mathcal{E}[u_2] = \mathcal{E}[v_{22}] + \mathcal{E}[v_{23}] + \mathcal{E}[v_{24}] + \mathcal{E}[v_{25}] = \frac{26}{21} + \frac{2}{7} + \frac{5}{7} + \frac{2}{21} = \frac{49}{21}$$

The reader might want to obtain this quantity directly from the first-step analysis (Problem 5.44).

Absorbing Chains in Action: Individual's Criminal History. The incidents in an individual's criminal history can be classified as (I) injury, (T) theft, (D) damage, (C) combination of these elements, (N) none of these elements, or nonindex, and (F) desistance, or no further contact with the police. This last state is absorbing. From a reported cohort of offenders in Philadelphia, Figlio (1981) obtained the following transition matrix:

$$\mathbf{P} = \begin{array}{c} \\ F \\ N \\ I \\ T \\ D \\ C \end{array} \begin{array}{c} \begin{array}{cccccc} F & N & I & T & D & C \end{array} \\ \left(\begin{array}{c|ccccc} 1 & 0 & 0 & 0 & 0 & 0 \\ \hline .3068 & .4473 & .0685 & .1054 & .0228 & .0492 \\ .3314 & .4090 & .0920 & .0854 & .0222 & .0600 \\ .2126 & .4051 & .0530 & .2130 & .0235 & .0928 \\ .1770 & .5013 & .0882 & .1463 & .0529 & .0343 \\ .2478 & .3922 & .0703 & .1378 & .0169 & .1350 \end{array}\right) \end{array}$$

The following fundamental matrix **U** displays the expected number of offenses of each type before a person desists. Its row sums also give the expected numbers of crimes committed before desistance.

$$\mathbf{U} = \begin{matrix} & N & I & T & D & C \\ N & 2.51 & .24 & .41 & .08 & .21 \\ I & 1.43 & 1.25 & .38 & .08 & .21 \\ T & 1.62 & .24 & 1.58 & .09 & .28 \\ D & 1.77 & .29 & .51 & 1.12 & .22 \\ C & 1.54 & .25 & .48 & .08 & 1.32 \end{matrix} \rightarrow \mathbf{u} = \begin{matrix} N \\ I \\ T \\ D \\ C \end{matrix} \begin{pmatrix} 3.45 \\ 3.35 \\ 3.81 \\ 3.91 \\ 3.67 \end{pmatrix}$$

From the 10,214 offenses committed by the 3475 members of the birth cohort, Figlio computed the following *observed* mean number of offenses:

$$\begin{matrix} & N & I & T & D & C \\ N & 2.06 & .14 & .27 & .06 & .13 \\ I & 1.25 & 1.27 & .28 & .09 & .13 \\ T & 1.41 & .23 & 1.55 & .07 & .23 \\ D & 1.37 & .19 & .44 & 1.11 & .19 \\ C & 1.81 & .29 & .54 & .11 & 1.56 \end{matrix}$$

These observed values appear to be similar enough to the expected means generated by the fundamental matrix so that additional support is given to the model. ■

Absorbing Chains in Action: Length of a Newspaper Subscription. Deming and Glasser (1968) studied over 2000 subscriptions to the *Detroit Evening News* from March 1965 to March 1966.

The original motivation for the study was a tax-case that the *Detroit News* was involved in. The management of the *News* had, a few years earlier, purchased subscriptions from another Detroit newspaper when the later ceased publication. To provide a basis for the depreciation of such subscriptions, management required the average life of subscriptions to the *News*.

The subscribers were classified according to (a) whether the subscription was relatively recent and (b) whether the subscriber carried *News*-sponsored accident insurance, which was available only to subscribers. This classification resulted in seven categories of subscriber:

$U_0 \equiv$ Subscriber uninsured, subscription less than 1 year old
$U_1 \equiv$ Subscriber uninsured, subscription from 1 to less than 2 years old
$U_2 \equiv$ Subscriber uninsured, subscription 2 or more years old
$I_0 \equiv$ Subscriber insured, subscription less than 1 year old
$I_1 \equiv$ Subscriber insured, subscription from 1 to less than 2 years old
$I_2 \equiv$ Subscriber insured, subscription 2 or more years old
$C \equiv$ Subscription canceled

For example, a subscriber in state I_1 may shift to state I_2, U_2, or C. Note that insured subscribers would be more reluctant than uninsured ones to give up their subscriptions, lest they lose their insurance as well.

State C is absorbing; all other states are transient.

The following transition matrix was obtained:

$$\mathbf{P} = \begin{array}{c} \\ C \\ U_0 \\ U_1 \\ U_2 \\ I_0 \\ I_1 \\ I_2 \end{array} \begin{array}{c} \begin{array}{ccccccc} C & U_0 & U_1 & U_2 & I_0 & I_1 & I_2 \end{array} \\ \left(\begin{array}{c|cccccc} 1 & 0 & 0 & 0 & 0 & 0 & 0 \\ \hline .2451 & 0 & .7397 & 0 & 0 & .0152 & 0 \\ .1454 & 0 & 0 & .8404 & 0 & 0 & .0142 \\ .0588 & 0 & 0 & .9278 & 0 & 0 & .0134 \\ .1429 & 0 & 0 & 0 & 0 & .8571 & 0 \\ .1667 & 0 & 0 & 0 & 0 & 0 & .8333 \\ .0372 & 0 & 0 & .0433 & 0 & 0 & .9195 \end{array}\right) \end{array}$$

Hence,

$$\mathbf{U} = \begin{array}{c} \\ U_0 \\ U_1 \\ U_2 \\ I_0 \\ I_1 \\ I_2 \end{array} \begin{array}{c} \begin{array}{cccccc} U_0 & U_1 & U_2 & I_0 & I_1 & I_2 \end{array} \\ \left(\begin{array}{cccccc} 1 & .7397 & 9.7567 & 0 & .0152 & 1.9080 \\ 0 & 1 & 13.0484 & 0 & 0 & 2.3433 \\ 0 & 0 & 15.3861 & 0 & 0 & 2.5554 \\ 0 & 0 & 5.9177 & 1 & .8571 & 9.8565 \\ 0 & 0 & 6.9040 & 0 & 1 & 11.4992 \\ 0 & 0 & 8.2848 & 0 & 0 & 13.7990 \end{array}\right) \end{array} \rightarrow \mathbf{u} = \begin{array}{c} U_0 \\ U_1 \\ U_2 \\ I_0 \\ I_1 \\ I_2 \end{array} \left(\begin{array}{c} 13.42 \\ 16.39 \\ 17.94 \\ 17.63 \\ 19.40 \\ 22.08 \end{array}\right)$$

Thus, an uninsured subscriber whose subscription is less than 1 year old (state U_0) will keep subscribing, on average, for 13.42 years before canceling. Of these years, the subscriber will be insured, on average, for $.0152 + 1.9080 = 1.92$ years.

Given the number of subscribers in March 1965, the initial distribution among the transient states was obtained as:

$$\boldsymbol{\pi}_T^{(0)} = \begin{array}{c} \begin{array}{cccccc} U_0 & U_1 & U_2 & I_0 & I_1 & I_2 \end{array} \\ \left(\begin{array}{cccccc} .2083 & .1274 & .5070 & .0032 & .0081 & .1460 \end{array}\right) \end{array}$$

This gives the average life of a subscription as $\boldsymbol{\pi}_T^{(0)}\mathbf{u} = 17.42$ years. ■

Absorbing Chains in Action: Stay Times in a Mental Hospital. Meredith (1974) used Markov chains to evaluate various treatment programs available to retarded persons within California's state mental hospitals. Based primarily on self-help categories such as ambulation, arm-hand use, toileting, and expressive communication, the children were rated between 1 (total inability) and 4 (normal ability). Six groups were defined: Group 1 comprises children having scores from 1.0 to 1.5 (crib patients), group 2 those from 1.5 to 2, . . . , and group 6 those from 3.5 to 4 (normal ability).

The author wanted to compare the expected progress, long-range cost, and outcome of the Standard Care program with those of three other hospital programs identified as Intensive Training, School, and Behavior Modification.

From the cohort of children who were administered in 1970 under the Standard Care program, the following 1-year transition matrix was obtained:

$$
\mathbf{P} = \begin{array}{c} \\ 6 \\ \text{Death} \\ 1 \\ 2 \\ 3 \\ 4 \\ 5 \end{array} \begin{array}{c} \begin{array}{ccccccc} 6 & \text{Death} & 1 & 2 & 3 & 4 & 5 \end{array} \\ \left(\begin{array}{cc|ccccc} 1 & 0 & 0 & 0 & 0 & 0 & 0 \\ 0 & 1 & 0 & 0 & 0 & 0 & 0 \\ \hline .00 & .02 & .72 & .20 & .06 & .00 & .00 \\ .00 & .02 & .14 & .49 & .31 & .04 & .00 \\ .00 & .02 & .01 & .20 & .51 & .24 & .02 \\ .01 & .02 & .00 & .02 & .28 & .50 & .17 \\ .11 & .02 & .00 & .00 & .05 & .35 & .47 \end{array}\right) \end{array} \tag{5.4.2}
$$

Death and group 6 (representing placement in a foster home after having been judged normal) are absorbing. The other states are transient.

The following fundamental matrix gives the number of years a child is expected to be in each ability group under the Standard Care program:

$$
\mathbf{U} = \begin{array}{c} \\ 1 \\ 2 \\ 3 \\ 4 \\ 5 \end{array} \begin{array}{c} \begin{array}{ccccc} 1 & 2 & 3 & 4 & 5 \end{array} \\ \begin{pmatrix} 7.3 & 6.8 & 9.3 & 6.8 & 2.5 \\ 4.2 & 7.7 & 9.8 & 7.2 & 2.7 \\ 3.3 & 5.8 & 10.8 & 7.7 & 2.9 \\ 2.7 & 4.8 & 8.7 & 8.8 & 3.2 \\ 2.1 & 3.8 & 6.8 & 6.5 & 4.2 \end{pmatrix} \end{array} \to \mathbf{u} = \begin{array}{c} 1 \\ 2 \\ 3 \\ 4 \\ 5 \end{array} \begin{pmatrix} 32.7 \\ 31.6 \\ 30.5 \\ 28.2 \\ 23.4 \end{pmatrix}
$$

Analyzing this matrix, the author remarks: "[T]he stay times were somewhat similar, regardless of starting group, with Group 5 generally having the lowest stay times." ■

5.5 ABSORBING PROBABILITIES

The reaching probability f_{ij} for a general chain was defined in §4.3 as the probability that the chain will ever visit state j, starting from state i. Let us now study these reaching probabilities for an absorbing chain.

If the chain starts from an absorbing state $i \in A$, then $f_{ii} = 1$ and $f_{ij} = 0$ for all $j \neq i$. So let us henceforth assume that the chain starts from some transient state $i \in T$.

Consider now an absorbing state $j \in A$. The reaching probability f_{ij} would then be the probability that, starting from the transient state i, the chain visits state j and is absorbed into it instead of some other absorbing state. In this case, we call f_{ij} the *absorbing probability* from the transient state i to the absorbing state j.

Consider the rat-maze example. Suppose the rat starts from room 3. What is the probability that the rat is fed rather than shocked? Or what is the value of

$$
f_{3F} \equiv \Pr\{\text{rat finds food rather than shock} \mid R_0 = 3\}
$$

In using first-step analysis, instead of seeking the value of f_{3F} alone, we include it among the four quantities that exhaust all possible initial conditions (excluding the absorbing states F and S):

$$f_{iF} = \Pr\{\text{rat finds food rather than shock} \mid R_0 = i\} \quad \text{for all } i = 2, 3, 4, 5$$

Consider now the case where the rat starts from room 2 and hence f_{2F}:

1. The probability that the rat reaches room F after the first step is $1/2$. If this occurs, the probability of the rat finding food rather than shock is 1.
2. The probability that the rat reaches room 4 after the first step is $1/2$. If this occurs, the probability of it finding food rather than shock is the same as if it had started from room 4.

Hence,

$$f_{2F} = (1/2)(1) + (1/2) f_{4F}$$

Similar arguments for other initial conditions yield the following set of four linear equations with four unknowns:

$$\begin{cases} f_{2F} = (1/2)(1) & & + (1/2) f_{4F} \\ f_{3F} = (1/3)(1) & & + (1/3) f_{4F} + (1/3) f_{5F} \\ f_{4F} = & + (1/3) f_{2F} + (1/3) f_{3F} & \\ f_{5F} = & + (1/2) f_{3F} & \end{cases} \tag{5.5.1}$$

The solutions are

$$f_{2F} = 5/7 \qquad f_{3F} = 4/7 \qquad f_{4F} = 3/7 \qquad f_{5F} = 2/7$$

If we also want to obtain

$$f_{iS} = \Pr\{\text{rat finds shock rather than food} \mid R_0 = i\} \quad \text{for all } i = 2, 3, 4, 5$$

then using first-step analysis as before, we obtain the following set of four linear equations with four unknowns:

$$\begin{cases} f_{2S} = & & + (1/2) f_{4S} \\ f_{3S} = & & + (1/3) f_{4S} + (1/3) f_{5S} \\ f_{4S} = (1/3)(1) & + (1/3) f_{2S} + (1/3) f_{3S} & \\ f_{5S} = (1/2)(1) & + (1/2) f_{3S} & \end{cases} \tag{5.5.2}$$

which yields

$$f_{2S} = 2/7 \qquad f_{3S} = 3/7 \qquad f_{4S} = 4/7 \qquad f_{5S} = 5/7$$

agreeing with the fact that $f_{iS} = 1 - f_{iF}$.

Again, the two sets of Equations (5.5.1) and (5.5.2) are very similar, and solving them individually fails to take advantage of their common features. So combining them in matrix form gives

$$\mathbf{A} = \mathbf{S} + \mathbf{TA}$$

where

1. $\mathbf{A} \equiv [f_{ij}]_{i \in T, j \in A}$ is the transient-to-absorbing portion of the matrix $\mathbf{F} \equiv [f_{ij}]_{i,j \in S}$, or

$$\mathbf{A} \equiv \begin{pmatrix} f_{2F} & f_{2S} \\ f_{3F} & f_{3S} \\ f_{4F} & f_{4S} \\ f_{5F} & f_{5S} \end{pmatrix}$$

2. $\mathbf{S}$ and $\mathbf{T}$ are respectively the transient-to-absorbing and the transient-to-transient portions of the transition matrix $\mathbf{P}$ in canonical form (5.1.1).

From this,

$$\mathbf{A} = (\mathbf{I} - \mathbf{T})^{-1}\mathbf{S}$$

or

$$\mathbf{A} = \mathbf{US} \tag{5.5.3}$$

where $\mathbf{U}$ is the fundamental matrix as defined in (5.3.7).

For the rat-maze example, because we already have calculated $\mathbf{U}$ in Equation (5.3.8), the absorbing probabilities can be obtained as follows:

$$\mathbf{A} = \begin{pmatrix} \frac{26}{21} & \frac{2}{7} & \frac{5}{7} & \frac{2}{21} \\ \frac{4}{21} & \frac{10}{7} & \frac{4}{7} & \frac{10}{21} \\ \frac{10}{21} & \frac{4}{7} & \frac{10}{7} & \frac{4}{21} \\ \frac{2}{21} & \frac{5}{7} & \frac{2}{7} & \frac{26}{21} \end{pmatrix} \begin{pmatrix} \frac{1}{2} & 0 \\ \frac{1}{3} & 0 \\ 0 & \frac{1}{3} \\ 0 & \frac{1}{2} \end{pmatrix} = \begin{matrix} \\ 2 \\ 3 \\ 4 \\ 5 \end{matrix} \begin{matrix} \begin{matrix} F & S \end{matrix} \\ \begin{pmatrix} \frac{5}{7} & \frac{2}{7} \\ \frac{4}{7} & \frac{3}{7} \\ \frac{3}{7} & \frac{4}{7} \\ \frac{2}{7} & \frac{5}{7} \end{pmatrix} \end{matrix}$$

which agree with our previous results.

Equation (5.5.3) is true for all finite absorbing chains because, for such chains in general:

1. For all $n \geq 1$ and $i, j \in T$, $[\mathbf{T}^n]_{ij}$ is the probability that the chain starts from transient state i, still visits the transient state j at step n, and thus has not been absorbed by that time.
2. For all $n \geq 1$, $i \in T$, and $j \in A$, $[\mathbf{T}^n\mathbf{S}]_{ij}$ is the probability that the chain starts from transient state i, visits a certain transient state at step n, and then moves to the absorbing state j rather than any other absorbing state at step $n + 1$.

3. $\sum_{n=0}^{\infty}[\mathbf{T}^n\mathbf{S}]_{ij} = [(\sum_{n=0}^{\infty}\mathbf{T}^n)\mathbf{S}]_{i,j} = [(\mathbf{I}-\mathbf{T})^{-1}\mathbf{S}]_{i,j} = [\mathbf{US}]_{i,j}$ is the probability that the chain starts from transient state i and then moves to the absorbing state j rather than any other absorbing state regardless of the number of steps taken.

Note that, like the transition matrix $\mathbf{P}$, the absorbing matrix $\mathbf{A} \equiv [f_{ij}]_{i\in T, j\in A}$ gives only the conditional probabilities. The unconditional probability that the chain is eventually absorbed into a certain state can be found in the vector $\boldsymbol{\pi}_T^{(0)}\mathbf{A}$, where $\boldsymbol{\pi}_T^{(0)} \equiv [\Pr\{X_0\} = i]_{i\in T}$ is the initial distribution, on the condition that the chain does not start from an absorbing state.

Absorbing Chains in Action: Home or Death? In the study of the Standard Care program for retarded persons by Meredith (1974), the following absorbing probability matrix was also obtained from the transition matrix (5.4.2):

$$\mathbf{A} = \begin{array}{c} \\ 1 \\ 2 \\ 3 \\ 4 \\ 5 \end{array} \begin{array}{c} \begin{array}{ccccc} 1 & 2 & 3 & 4 & 5 \end{array} \\ \begin{pmatrix} 7.3 & 6.8 & 9.3 & 6.8 & 2.5 \\ 4.2 & 7.7 & 9.8 & 7.2 & 2.7 \\ 3.3 & 5.8 & 10.8 & 7.7 & 2.9 \\ 2.7 & 4.8 & 8.7 & 8.8 & 3.2 \\ 2.1 & 3.8 & 6.8 & 6.5 & 4.2 \end{pmatrix} \end{array} \begin{array}{c} \begin{array}{cc} 6 & \text{Death} \end{array} \\ \begin{pmatrix} .00 & .02 \\ .00 & .02 \\ .00 & .02 \\ .01 & .02 \\ .11 & .02 \end{pmatrix} \end{array} = \begin{array}{c} \\ 1 \\ 2 \\ 3 \\ 4 \\ 5 \end{array} \begin{array}{c} \begin{array}{cc} 6 & \text{Death} \end{array} \\ \begin{pmatrix} .35 & .65 \\ .37 & .63 \\ .39 & .61 \\ .44 & .56 \\ .53 & .47 \end{pmatrix} \end{array}$$

Note that the higher the ability group, the better the chance of ending in a foster home. Also, "[t]he prognosis appears to be that most children, under Standard Care, will never see a foster home before they die." ■

Absorbing Chains in Action: Allowance for Doubtful Accounts. In retail establishments, it is important to be able to estimate the allowance for doubtful accounts at fiscal year end. To use Markov chains, an account receivable is classified into age categories that reflect its stage of delinquency: current accounts, accounts 1 month past due, accounts 2 months past due, ... Furthermore, if an account is overdue for more than m months, it is classified as bad debt; if it is paid within m months, it is classified as paid. States bad debt and paid are absorbing, although a delinquent account may be repaid eventually. For example, from a random sample of approximately 1000 department store accounts, Cyert et al. (1962) obtained the following transition matrix for $m = 5$:

$$\mathbf{P} = \begin{array}{r} \\ \text{Paid} \\ \text{Bad debt} \\ \text{Current} \\ 1 \\ 2 \\ 3 \\ 4 \\ 5 \end{array} \begin{array}{c} \begin{array}{cc|cccccc} \text{Paid} & \text{Bad debt} & \text{Current} & 1 & 2 & 3 & 4 & 5 \end{array} \\ \left(\begin{array}{cc|cccccc} 1 & 0 & 0 & 0 & 0 & 0 & 0 & 0 \\ 0 & 1 & 0 & 0 & 0 & 0 & 0 & 0 \\ \hline .21 & 0 & .67 & .12 & 0 & 0 & 0 & 0 \\ .13 & 0 & .19 & .44 & .24 & 0 & 0 & 0 \\ .13 & 0 & .08 & .20 & .36 & .23 & 0 & 0 \\ .10 & 0 & .01 & .04 & .17 & .29 & .39 & 0 \\ .14 & 0 & .02 & 0 & .09 & .20 & .41 & .14 \\ .09 & .18 & .01 & .02 & .01 & .10 & .12 & .47 \end{array}\right) \end{array}$$

Note that the total balance method was used here: All dollars in the account balance are put in the age category corresponding to the oldest dollar. Thus, because it is possible to pay one 4-month overdue bill and still have another 2-month overdue bill, it is possible to go from state 4 to state 2 in one step.

If the process is permitted to continue indefinitely, all the dollars end up in either the paid state or the bad debt state. The entries in the first column of the matrix $\mathbf{A} = \mathbf{US}$ give the probabilities of dollars in each of the age categories being paid; those in the second column give the probabilities of becoming bad debts, which correspond to the *loss expectancy rates* developed on a judgment basis by most retail establishments.

At time 0, let the m-component vector

$$\mathbf{b} = (b_0 \quad b_1 \quad b_2 \quad \ldots \quad b_{m-1})$$

gives the dollar amount in each age category, except the bad debt and paid states. (b_0 is the amount in the current state.) The sum b of all these amounts is the total value of accounts receivable. The vector $(1/b)\mathbf{b}$ can be regarded as the conditional initial distribution $\boldsymbol{\pi}_T^{(0)}$ of the chain. The first entry in the two-component vector $\mathbf{bA}$ gives the amount expected to be paid, and the second entry gives the amount expected to become bad debt. ■

Absorbing Chains in Action: A Voting Technique. Mackinnon (1966) introduced a rather interesting voting technique called SPAN (*S*uccessive *P*roportional *A*dditive *N*umeration): A group of voters has to decide between a finite number of alternatives. Each voter begins with the same number of votes, which must be reallocated *round by round*. In each round, a voter may allocate some of his or her votes among the alternatives under consideration or transfer some to the other voters.

For example, suppose there are two alternatives, X and Y, and three voters, A, B, and C. Each voter initially has 100 votes.

1. Voter A gives one-half of his votes to voter B, one-fourth to alternative X, and one-fourth to alternative Y.
2. Voter B gives one-fourth of her votes to voter A, three-sixteenths to alternative X, and nine-sixteenths to alternative Y.
3. Voter C gives one-third of her votes to voter A, one-third to voter B, and one-third to alternative Y.

After the first round of votes, the 300 votes are distributed as follows:

1. Alternative X has $(1/4)100$ votes from A and $(3/16)100$ votes from B.
2. Alternative Y has a total of $(1/4 + 9/16 + 1/3)100$ votes.
3. Voter A has a total of $(1/4 + 1/3)100 = 700/12$ votes.
4. Voter B has a total of $(1/2 + 1/3)100 = 500/6$ votes.
5. Voter C has no votes.

In the second round, each voter must reallocate the number of votes he or she now has to the other voters or alternatives *in the same proportions as in the first round*. Thus, in the second round, the votes held by A and B are distributed as follows:

1. Alternative X receives an additional $(1/4)(700/12)$ votes from voter A and an additional $(3/16)(500/6)$ votes from voter B.
2. Alternative Y receives an additional $(1/4)(700/12)$ votes from voter A and an additional $(9/16)(500/6)$ votes from voter B.
3. Voter A now has $(1/4)(500/6)$ votes from B.
4. Voter B now has $(1/2)(700/12)$ votes from A.

The process is now repeated with another round. As the total number of votes allocated to the alternatives increases after each round, the process stops when a large proportion of the votes has been allocated to the alternatives. The alternative with the largest number of votes at the end is chosen.

McConway (1982) pointed out that this voting procedure can be represented as an absorbing Markov chain having the following transition probability matrix:

$$\mathbf{P} = \begin{array}{c} \\ X \\ Y \\ A \\ B \\ C \end{array} \begin{array}{c} \begin{array}{ccccc} X & Y & A & B & C \end{array} \\ \left(\begin{array}{cc|ccc} 1 & 0 & 0 & 0 & 0 \\ 0 & 1 & 0 & 0 & 0 \\ \hline 1/4 & 1/4 & 0 & 1/2 & 0 \\ 3/16 & 9/16 & 1/4 & 0 & 0 \\ 0 & 1/3 & 1/3 & 1/3 & 0 \end{array}\right) \end{array}$$

McConway then showed how the selected decision can be identified by first-step analysis. However, we can do better by calculating the fundamental matrix first:

$$\mathbf{U} = \begin{pmatrix} 1 & -1/2 & 0 \\ -1/4 & 1 & 0 \\ -1/3 & -1/3 & 1 \end{pmatrix}^{-1} = \begin{array}{c} \\ A \\ B \\ C \end{array} \begin{array}{c} \begin{array}{ccc} A & B & C \end{array} \\ \begin{pmatrix} 8/7 & 4/7 & 0 \\ 2/7 & 8/7 & 0 \\ 10/21 & 4/7 & 1 \end{pmatrix} \end{array}$$

The absorbing probabilities to states X and Y (or the probability of alternative X or Y being selected) can now be calculated as:

$$\mathbf{A} = \begin{array}{c} \\ A \\ B \\ C \end{array} \begin{array}{c} \begin{array}{ccc} A & B & C \end{array} \\ \begin{pmatrix} 8/7 & 4/7 & 0 \\ 2/7 & 8/7 & 0 \\ 10/21 & 4/7 & 1 \end{pmatrix} \end{array} \begin{array}{c} \begin{array}{cc} X & Y \end{array} \\ \begin{pmatrix} 1/4 & 1/4 \\ 3/16 & 9/16 \\ 0 & 1/3 \end{pmatrix} \end{array} = \begin{array}{c} \\ A \\ B \\ C \end{array} \begin{array}{c} \begin{array}{cc} X & Y \end{array} \\ \begin{pmatrix} 11/28 & 17/28 \\ 2/7 & 5/7 \\ 19/84 & 65/84 \end{pmatrix} \end{array}$$

Note that the technique can take the relative importance of the voters into consideration, as reflected in the question: Who has the right to make a decision first? All voters do not have to start in the first round, and the opinion of the person who has the right to start first would carry more weight than those of the others. For example:

1. If A is the leader of the group and has the right to make a decision first, there is a probability of 1/2 that he makes the decision himself and a probability of 1/2 that he delegates the decision making to B. If he chooses to make the decision, the initial distribution is

$$\pi_T^{(0)} = \begin{array}{c} \begin{array}{ccc} A & B & C \end{array} \\ \begin{pmatrix} 1 & 0 & 0 \end{pmatrix} \end{array}$$

and the proportion of votes which eventually goes to alternative X is $11/28$.

2. If A, B, and C are equal members of a committee and all cast the votes at the same time, the initial distribution is

$$\pi_T^{(0)} = \begin{matrix} A & B & C \\ (1/3 & 1/3 & 1/3) \end{matrix}$$

the proportion of votes that eventually goes to alternative X is

$$(1/3)(11/28) + (1/3)(2/7) + (1/3)(19/84) = 19/63$$

and the proportion of votes that eventually goes to alternative Y is

$$(1/3)(17/28) + (1/3)(5/7) + (1/3)(65/84) = 44/63 = 1 - 19/63 \quad \blacksquare$$

5.6 LIMITING DISTRIBUTIONS

In §3.9, we wondered whether the n-step transition matrix $\mathbf{P}^{(n)}$ converges. We now can address this question for absorbing chains and can also observe the consistency of the answer with what we have obtained so far in this chapter.

First, from the canonical form (5.1.1), we can prove (Problem 5.60) that

$$\mathbf{P}^{(k)} = \mathbf{P}^k = \begin{pmatrix} \mathbf{I} & \mathbf{0} \\ (\mathbf{I} + \mathbf{T} + \cdots + \mathbf{T}^k)\mathbf{S} & \mathbf{T}^k \end{pmatrix} \tag{5.6.1}$$

Since the matrix $\mathbf{T}^k$ converges elementwise to matrix $\mathbf{0}$ when $k \to \infty$, we have

$$\mathbf{P}^{(\infty)} \equiv \lim_{k\to\infty} \mathbf{P}^k = \begin{pmatrix} \mathbf{I} & \mathbf{0} \\ (\mathbf{I} + \mathbf{T} + \mathbf{T}^2 + \cdots)\mathbf{S} & \mathbf{0} \end{pmatrix} = \begin{pmatrix} \mathbf{I} & \mathbf{0} \\ \mathbf{A} & \mathbf{0} \end{pmatrix}$$

This confirms that:

1. The limiting probabilities are dependent on the initial conditions. (As discussed in §3.9, this happens when the rows of $\mathbf{P}^{(\infty)}$ are not identical.)
2. If the chain starts from an absorbing state, it will stay there forever (submatrix $\mathbf{I}$).
3. After an infinite number of steps, the probability of visiting a transient state is zero (submatrices $\mathbf{0}$).
4. Starting from a transient state, the limiting probability of visiting an absorbing state a is the probability f_{ia} in submatrix $\mathbf{A}$.

5.7 SUMMARY

We studied absorbing chains in this chapter and have shown that:

1. The fundamental matrix $\mathbf{U} = (\mathbf{I} - \mathbf{T})^{-1}$ gives the expected number of visits by the chain to the transient states.
2. The sums along the rows of the fundamental matrix give the expected numbers of steps before the chain is absorbed.
3. The matrix $\mathbf{A} = \mathbf{US}$ gives the probability of absorption to a particular absorbing state.

PROBLEMS

5.1 Consider the Markov chain in Problem 3.1. Find the probability that, starting from state C, the chain visits state A before state B

a. by using first-step analysis

b. by making the chain absorbing

5.2 Radioactive Rose Bengal is used as a tracer element in the study of biliary function in patients suffering from hepatic disease. After being injected into the blood, Rose Bengal moves between the blood (B) and liver (L) before leaving the body along with either urine (U) or feces (F). For patient D. W., having neonatal hepatitis at the Children's Medical Center, Dallas, Texas, the following 1-hour transition matrix was obtained by Saffer et al. (1976):

$$\mathbf{P} = \begin{array}{c} \\ U \\ F \\ B \\ L \end{array} \begin{array}{c} \begin{array}{cccc} U & F & B & L \end{array} \\ \left(\begin{array}{cc|cc} 1 & 0 & 0 & 0 \\ 0 & 1 & 0 & 0 \\ \hline .0204 & 0 & .6373 & .3423 \\ 0 & .0101 & .0377 & .9522 \end{array}\right) \end{array}$$

a. Find the expected number of hours Rose Bengal is in the blood or in the liver before it is found in the urine or feces.

b. Find the probability that Rose Bengal will eventually end up in the urine.

5.3 (Thompson and McNeal, 1967) A salesperson observes that the transition matrix for a customer's interest from one call to the next is as follows:

$$\mathbf{P} = \begin{array}{c} \\ S \\ L \\ 0 \\ 1 \\ 2 \end{array} \begin{array}{c} \begin{array}{ccccc} S & L & 0 & 1 & 2 \end{array} \\ \left(\begin{array}{cc|ccc} 1 & 0 & 0 & 0 & 0 \\ 0 & 1 & 0 & 0 & 0 \\ \hline .10 & .30 & 0 & .45 & .15 \\ .15 & .10 & 0 & .25 & .50 \\ .20 & .05 & 0 & .35 & .40 \end{array}\right) \end{array}$$

where the states are: (S) sale completed, (L) sale lost, (0) new customer, no history, (1) customer indicated a low degree of interest, (2) customer indicated a high degree of interest.

a. Find the expected number of calls for each type of customer.

b. Find the probability of making a sale to each type of customer.

5.4 Refer to Problem 2.6. Find the expected number of pairs of balls drawn before the urn contains only blue balls.

5.5 Refer to Problem 2.22 in which three tanks A, B, and C fight against each other and a surviving tank randomly picks one of the other surviving tanks as a target. Find

a. the expected duration of the entire fight

b. the probability that none survives after the fight

c. the probabilities that only tank A survives after the fight, only tank B survives, and only tank C survives.

5.6 Refer to Problem 2.23 in which three tanks A, B, and C fight against each other and a surviving tank picks its strongest opponent as a target. Find

a. the expected duration of the entire fight

b. the probability that none survives after the fight

c. the probabilities that only tank A survives after the fight, only tank B survives, and only tank C survives.

5.7 In Problem 2.8, *sib mating* unavoidably leads to the elimination of one of the genes a and b and hence the genotype ab.

a. Find the mean time until the disappearance of one of the genes.

b. Find the probability that gene a will disappear, starting from the pair (aa, ab).

5.8 Refer to Problem 2.26 in which Tim, Sam, Bob, and Jim play catch. Suppose Tim starts first.

a. Find the expected number of throws before Jim catches the ball for the first time.

b. If Jim has the ball, he will throw it to Tim with probability 2/3, or run away with the ball with probability 1/3. Find the expected number of throws in a game.

c. If Jim has the ball for the first time, he will throw it to Tim; if he has the ball for the second time, he will run away with it. Find the expected number of throws in a game.

d. Explain the relationships between the results.

5.9 A detective is chasing a fugitive around three cities A, B, and C. Assume that they move independently of each other. (Admittedly, this is an unrealistic assumption.) The detective's movements follow the transition matrix $\mathbf{P}_D$ and the fugitive's movements follow the transition matrix $\mathbf{P}_F$:

$$\mathbf{P}_D = \begin{array}{c} \\ A \\ B \\ C \end{array}\begin{array}{c} \begin{array}{ccc} A & B & C \end{array} \\ \begin{pmatrix} .2 & .5 & .3 \\ .3 & .3 & .4 \\ .1 & .4 & .5 \end{pmatrix} \end{array} \qquad \mathbf{P}_F = \begin{array}{c} \\ A \\ B \\ C \end{array}\begin{array}{c} \begin{array}{ccc} A & B & C \end{array} \\ \begin{pmatrix} .3 & .6 & .1 \\ .2 & .4 & .4 \\ .3 & .3 & .4 \end{pmatrix} \end{array}$$

The detective will catch the fugitive if they are in the same city.

a. Find the number of steps until this happens if the detective starts from city A and the fugitive from city B.

b. What is the probability that this occurs at city A?

5.10 N people, named $1, 2, \ldots, N$, play the following game: They perform a sequence of *multinomial* independent trials in which each trial has N outcomes $1, 2, \ldots, N$. In each trial, let p_i be the probability that the outcome is i $(\sum_{i=1}^{N} p_i = 1)$. Let k be a constant. If outcome i occurs k times in a row, then player i wins. Let $N = 3$ and $k = 3$.

a. Construct a Markov chain $\{X_n\}_{n=0,1,2,\ldots}$ in which the state space contains the similar outcomes already obtained in a row—that is, $S \equiv \{1, 11, 111, 2, 22, 222, 3, 33, 333\}$.

b. Let $p_1 = .6$, $p_2 = .3$, and $p_3 = .1$. Find the expected number of games before someone wins and the probability that player i wins.

5.11 Verify the results calculated by Figlio (1981) for an individual criminal history in §5.4.

5.12 Verify the results calculated by Deming and Glasser (1968) for the average length of a newspaper subscription in §5.4.

5.13 Verify the results calculated by Meredith (1974) for the stay times and the absorbing probabilities in a mental hospital in §5.4 and §5.5.

5.14 Consider the transition matrix for an accounts receivable chain obtained by Cyert et al. (1962) in §5.5.

a. Find the expected duration before a dollar in each delinquency state is paid or becomes bad debt.

b. Find the loss expectancy rate for each age category.

c. Suppose the initial accounts receivable vector is

$$\mathbf{b} = (\$300 \quad \$200 \quad \$300 \quad \$400 \quad \$250 \quad \$150)$$

What is the amount expected to be paid?

5.15 Genital herpes simplex virus infections are common sexually transmitted diseases. The lesions that characterize an episode of genital herpes may be classified into seven generally accepted stages: (0) macule, (1) papule, (2) vesicle, (3) pustule, (4) ulcer, (5) crust, and (6) healed. Of the 166 patients participating in a daily study by Badger and Vacek (1987), 84 were randomized to receive an antiviral drug and 82 to receive a placebo. For those receiving the placebo, the following 1-day transition matrix was obtained:

$$\mathbf{P} = \begin{array}{c} \\ 0 \\ 1 \\ 2 \\ 3 \\ 4 \\ 5 \\ 6 \end{array} \begin{array}{c} \begin{array}{ccccccc} 0 & 1 & 2 & 3 & 4 & 5 & 6 \end{array} \\ \begin{pmatrix} .43 & .23 & 0 & 0 & 0 & .13 & .20 \\ & .27 & .18 & .07 & .18 & .14 & .16 \\ & & .29 & .10 & .14 & .45 & .02 \\ & & & .35 & .37 & .24 & .04 \\ & & & & .59 & .26 & .14 \\ & & & & & .71 & .29 \\ & & & & & & 1 \end{pmatrix} \end{array}$$

Find the expected healing time, starting with each lesion stage.

5.16 A doctoral student at a university can go through the following states: ($S1$) enrolled but not advanced to candidacy, ($S3$) withdrawn prior to advancement to candidacy, ($S2F$) enrolled for the first semester of advancement to candidacy, ($S20$) enrolled for the second and subsequent semesters of advancement to candidacy, ($S4$) withdrawn after advancement to candidacy, and ($S5$) graduated. From the records of students at the Department of Educational Administration at the University of Texas at Austin from the fall semester 1969 to the fall semester 1978, most students in states $S3$ and $S4$ eventually returned to the university. Thus, only state $S5$ is absorbing. Bessent and Bessent (1980) obtained the following transition matrix:

$$\mathbf{P} = \begin{array}{c} \\ S5 \\ S1 \\ S3 \\ S2F \\ S20 \\ S4 \end{array} \begin{array}{c} \begin{array}{cccccc} S5 & S1 & S3 & S2F & S20 & S4 \end{array} \\ \begin{pmatrix} 1 & & & & & \\ & .25 & .48 & .27 & & \\ & .25 & .62 & .13 & & \\ & & & & .99 & .01 \\ .18 & & & & .7 & .12 \\ & & & & .74 & .26 \end{pmatrix} \end{array}$$

Verify that

> Calculated stay times showed: (1) a student may be expected to enroll for 2.3 semesters (or summer sessions) and will be unenrolled for 3 semesters before advancing to candidacy, (2) a student will be enrolled for dissertation about 6.6 semesters before graduating and the stay time is the same whether or not he has withdrawn previously.

5.17 Refer to the game of *Snakes and Ladders* in Problem 2.14. For simplicity, assume that the board has only 20 squares labeled from 1 to 20.

a. Without any ladder or snake, find the probability that square i is occupied during a game starting from square 1. *Hint:* Explain why $f_{ij} = \mathcal{E}[v_{ij}]$.

b. Without any ladder or snake, what is the expected number of tosses of a die a solo player has to make before the game ends?

c. With ladder (7, 10), what is the expected number of tosses of a die a solo player has to make before the game ends?

d. With ladder (14, 17), what is the expected number of tosses of a die a solo player has to make before the game ends?

e. Do ladders (7, 10) and (14, 17) have the same effect in shortening the game? Explain.

5.18 Two players A and B take turns playing on a gaming machine. All plays are independent. A plays first and can score a success against the machine with probability a; B with probability b. The first to score a success wins. Design an absorbing chain to find

a. the probability that A wins the game

b. the expected number of games A plays

c. the expected number of games B plays

Compare with Problem 1.31. *Hint:* Use Appendix (A.18).

5.19 Two players A and B are playing a sequence of games. They agreed that whoever wins a total of five games first wins \$1. Unfortunately, after seven games, in which A won four games and B won three, they have to stop. Let p be the probability that A wins in one game. How would the \$1 price be divided between them? *Hint:* Use Appendix (A.18).

5.20 A coin is biased with $\Pr\{H\} = p$ and $\Pr\{T\} = 1 - p$. Construct an absorbing chain to find

a. the expected number of tosses until two consecutive Hs or two consecutive Ts are obtained. Compare with Problem 1.38. *Hint:* Use Appendix (A.18).

b. the probability that two consecutive Hs are obtained before two consecutive Ts.

5.21 Three people A, B, and C are playing an infinite set of games as follows. At game n, two players play against each other while one watches. Whoever loses at game n will watch the other two playing at game $n+1$. Let p_{ij} be the probability that player i beats player j in a game ($i, j = A, B, C$; $p_{ij} + p_{ji} = 1$).

a. Let X_n be the player who is watching at game n. Find the transition matrix for $\{X_n\}_{n=0,1,2,\ldots}$.

b. Hence, show that the expected number of consecutive games that A plays is

$$\frac{p_{CB}(1+p_{AC}) + p_{BC}(1+p_{AB})}{1 - p_{AB}p_{AC}}$$

5.22 Refer to the woman trapped in a mine having two doors (left and right) in Problem 1.37. In the first attempt, she picks one of the two doors at random. Let D_n be the door she takes in the nth attempt. Assume now that the chain $\{D_n\}_{n=0,1,2,\ldots}$ has the following transition matrix:

$$\mathbf{P} = \begin{array}{c} \\ \text{Left} \\ \text{Right} \end{array} \begin{array}{c} \begin{array}{cc} \text{Left} & \text{Right} \end{array} \\ \begin{pmatrix} 1-\alpha & \alpha \\ \beta & 1-\beta \end{pmatrix} \end{array}$$

a. What is the expected number of doors she tries before reaching freedom?

b. What is the expected duration before she reaches freedom?

c. Verify the results in Problem 1.37.

5.23 A certain asset is offered for sale in an auction. Let x_i be the value of the ith bid. Assume all x_is are independent and identically distributed integer-valued random variables having $\Pr\{x_i = j\} \equiv a_j$ for all $j > 0$. Suppose the auction stops immediately after a bid that equals or exceeds a predetermined amount $M = \$3$. All bids arriving after this successful bid are considered to have zero value.

a. Let $X_n \equiv \max\{x_1, x_2, \ldots, x_n\}$. Find the transition matrix for $\{X_n\}_{n=0,1,2,\ldots}$. *Hint:* Use Problem 2.9.

b. Show that, independent of the maximum bid obtained so far, the expected number of further bids before the asset is sold is $1/\sum_{i=3}^{\infty} a_i$. Explain. *Hint:*

$$\begin{pmatrix} b_0 & -a_1 & -a_2 \\ 0 & b_1 & -a_2 \\ 0 & 0 & b_2 \end{pmatrix}^{-1} = \begin{pmatrix} 1/b_0 & a_1/(b_0 b_1) & a_2(a_1+b_1)/(b_0 b_1 b_2) \\ 0 & 1/b_1 & a_2/(b_1 b_2) \\ 0 & 0 & 1/b_2 \end{pmatrix}$$

5.24 An urn contains five balls labeled 1, 2, 3, 4, 5. Set up an absorbing chain to find the expected number of balls drawn with replacement until the same ball is drawn twice. Compare with Problem 1.50. *Hint:* Use Appendix (A.19).

5.25 An urn contains four balls, either black or white. A ball is drawn at random. If it is black, there is a probability $1-p$ that it will be returned to the urn; there is a probability p that a white ball will be returned to the urn instead. If it is white, it will be returned to the urn. This procedure is repeated indefinitely. Given i white balls initially in the urn, use the transition matrices obtained in Problem 2.5 to find

a. the expected number of further black balls drawn before the urn contains all white balls

b. the expected number of further balls, either black or white, drawn before the urn contains all white balls

Compare with Problem 1.14. *Hint:* Use Appendix (A.19) and

$$\begin{pmatrix} 3 & -3 & 0 \\ 0 & 2 & -2 \\ 0 & 0 & 1 \end{pmatrix}^{-1} = \begin{pmatrix} 1/3 & 1/2 & 1 \\ 0 & 1/2 & 1 \\ 0 & 0 & 1 \end{pmatrix}$$

5.26 Refer to Problem 2.21 in which the gambler bets all he has to get \$5. Suppose he starts with \$2.

a. With states 0 and 5 absorbing, verify that the fundamental matrix is:

$$\mathbf{U} = \frac{1}{1 - p^2q^2} \begin{array}{c} \begin{matrix} 1 & 2 & 3 & 4 \end{matrix} \\ \begin{pmatrix} 1 & p & p^2q & p^2 \\ pq^2 & 1 & pq & p \\ q & pq & 1 & p^2q \\ q^2 & pq^2 & q & 1 \end{pmatrix} \end{array}$$

b. Find the expected number of bets before he has to quit.

c. Find his probability of winning when

i. $p = 0$

ii. $p = .25$

iii. $p = .50$

iv. $p = .75$

v. $p = 1$

5.27 The gambler in Problem 5.26 is bold. Consider now another gambler who is timid: He only bets \$1 each time. This timid gambler also starts with \$2. Other assumptions remain the same as in Problem 2.21.

a. Verify that the new fundamental matrix is:

$$\mathbf{U} = \frac{1}{1 - 3pq + p^2q^2} \begin{array}{c} \begin{matrix} 1 & 2 & 3 & 4 \end{matrix} \\ \begin{pmatrix} 1-2pq & p(1-pq) & p^2 & p^3 \\ q(1-pq) & (1-pq) & p & p^2 \\ q^2 & q & (1-pq) & p(1-pq) \\ q^3 & q^2 & q(1-pq) & 1-2pq \end{pmatrix} \end{array}$$

b. Find the expected number of bets before he has to quit.

c. Find his probability of winning when

i. $p = 0$

ii. $p = .25$

iii. $p = .50$

iv. $p = .75$

v. $p = 1$

d. Compare the results with those in Problem 5.26.

5.28 In Problem 1.2, we studied the preliminary stage of a game of tennis between two players A and B. We now treat its final stage separately as a Markov chain $\{X_n\}_{n=0,1,2,\ldots}$ having absorbing states A and B. Show that its fundamental matrix is:

$$\mathbf{U} = \begin{array}{c} \\ AB \\ \text{Deuce} \\ AA \end{array} \begin{array}{c} \begin{array}{ccc} AB & \text{Deuce} & AA \end{array} \\ \begin{pmatrix} 1-pq & p & p^2 \\ q & 1 & p \\ q^2 & q & 1-pq \end{pmatrix} \end{array} \frac{1}{1-2pq}$$

Coupled with results obtained in Problem 1.2, find

a. the expected number of services in the entire game

b. the probability of A's winning the game when

i. $p = 0$

ii. $p = .25$

iii. $p = .50$

iv. $p = .75$

v. $p = 1$

5.29 A student in a college can be in one of the following states: (D) dropout, (G) graduation, (1) freshman, (2) sophomore, (3) junior, and (4) senior. Each year, let q be the probability of dropping out, r the probability of repeating the year, and p the probability of passing ($q + r + p = 1$). Show that the conditional probabilities of dropout or graduation are

$$\mathbf{A} = \begin{array}{c} \\ 1 \\ 2 \\ 3 \\ 4 \end{array} \begin{array}{c} \begin{array}{cc} D & G \end{array} \\ \begin{pmatrix} 1-t^4 & t^4 \\ 1-t^3 & t^3 \\ 1-t^2 & t^2 \\ 1-t & t \end{pmatrix} \end{array}$$

where $t \equiv p/(q+p)$. Thus, a person in a higher class has a better chance of graduating. *Hint:* Use Appendix (A.19).

5.30 Refer to the planned replacement policy for an instrument in Problem 2.29.

a. Find the average useful life of the instrument from the transition matrix obtained in Problem 2.29. *Hint:* Use Appendix (A.19).

b. From the fundamental matrix, find the probability that the instrument is replaced before its failure.

5.31 There are $N + 1$ points; each is assigned a value. They are ordered according to their values, the point with the lowest value is labeled 0 and that with the highest value is labeled N. An algorithm starts with point i and moves to one of the lower points $0, 1, \ldots, i-1$ with equal probability. Construct a Markov chain to show that the expected number of transitions to move from point i to point 0 is $\sum_{j=1}^{i-1} 1/j$. Compare with Problem 1.41. *Hint:*

$$\begin{pmatrix} 1 & 0 & 0 \\ -\frac{1}{2} & 1 & 0 \\ -\frac{1}{3} & -\frac{1}{3} & 1 \end{pmatrix}^{-1} = \begin{pmatrix} 1 & 0 & 0 \\ \frac{1}{2} & 1 & 0 \\ \frac{1}{2} & \frac{1}{3} & 1 \end{pmatrix}$$

5.32 A fair coin is tossed repeatedly. Let X_n be the nth outcome.

a. Construct a Markov chain to represent the enlarged chain

$$\{X_{n-2}, X_{n-1}, X_n\}_{n=0,1,2,\dots}$$

b. If we are only interested in the number of tosses until the pattern HHT appears, can we lump some of the states of this chain? (§2.17) Find the expected number of tosses until this pattern appears by setting up an absorbing chain.

c. If we are interested in the number of tosses until the pattern HTH appears instead, how do we lump? Would the expected number of tosses until this occurs be the same as that for the HHT pattern?

5.33 In Problem 2.8, *sib mating* unavoidably leads to the elimination of one of the genes a and b and hence the genotype ab. Suppose we now only want the expected number of generations before either gene a or gene b disappears, but not the probability of which gene disappears first. How can we lump some of the states together to form a smaller state space? Obtain the transition matrix of this lumped chain and hence the expected number of generations before either gene disappears. Compare with Problem 5.7.

5.34 Consider an absorbing Markov chain $\{X_n\}_{n=0,1,2,\dots}$ having transition matrix $\mathbf{P}$ in which $p_{ii} \neq 0$ for some $i \in T$. Thus the chain may remain at the same state during a transition. Consider now a chain $\{Y_n\}_{n=0,1,2,\dots}$ which records the movements of $\{X_n\}_{n=0,1,2,\dots}$ only when it changes states. For example, if the latter follows the path $a - b - b - d - a - a - c$, then the former follows $a - b - d - a - c$.

a. Explain why the transience portion of the transition matrix of $\{Y_n\}_{n=0,1,2,\dots}$ is

$$\mathbf{T}_Y = (\mathbf{I} - \mathbf{T}_d)^{-1}(\mathbf{T} - \mathbf{T}_d)$$

where $\mathbf{T}_d$ is a diagonal matrix, the (i, i)-element of which is p_{ii}.

b. Hence show that the fundamental matrix $\mathbf{U}_Y$ of chain $\{Y_n\}_{n=0,1,2,\dots}$ is

$$\mathbf{U}_Y = \mathbf{U}_X(\mathbf{I} - \mathbf{T}_d)$$

where $\mathbf{U}_X$ is the fundamental matrix of chain $\{X_n\}_{n=0,1,2,\dots}$.

c. Given chain $\{X_n\}_{n=0,1,2,\dots}$ starts from state i, let ζ_{ij} be the number of times it moves into state j before absorption; that is, the number of sojourns to state j (§2.19). Find $\mathcal{E}[\zeta_{ij}]$ for all $i, j \in T$.

d. Illustrate the previous results when

$$\mathbf{P} = \begin{array}{c} \\ A \\ B \\ C \\ D \\ E \end{array} \begin{array}{c} \begin{array}{ccccc} A & B & C & D & E \end{array} \\ \left(\begin{array}{cc|ccc} 1 & 0 & 0 & 0 & 0 \\ 0 & 1 & 0 & 0 & 0 \\ \hline 4/9 & 1/9 & 1/9 & 1/9 & 2/9 \\ 3/7 & 1/7 & 1/7 & 0 & 2/7 \\ 3/8 & 1/8 & 1/8 & 2/8 & 1/8 \end{array}\right) \end{array}$$

5.35 Consider a Markov chain $\{X_n\}_{n=0,1,2,\ldots}$ having state space $S \equiv \{0, 1, 2, \ldots, N\}$. At each transition, it always moves to a different state with equal probability. In other words, for all $0 \leq i \leq N$,

$$p_{ij} = \begin{cases} 0 & \text{for } j = i \\ 1/N & \text{for all } j \neq i \end{cases}$$

Show that the expected number of steps to go from state $i \neq 0$ to state 0 is always N. *Hint:* If $\mathbf{E}$ is an $(N \times N)$ matrix in which all elements equal 1, $[\mathbf{I} - (N+1)^{-1}\mathbf{E}]^{-1} = \mathbf{I} + \mathbf{E}$.

5.36 Consider a Markov chain with state space $S \equiv \{0, 1, 2, \ldots, N\}$. If it is at state i, it may remain at state i or move to another state with equal probability p_i; that is,

$$p_{ij} = \begin{cases} 1 - Np_i & \text{for } j = i \\ p_i & \text{for all } j \neq i \end{cases}$$

where $0 \leq Np_i \leq 1$ for all $0 \leq i \leq N$.

a. Show that the expected number of steps to go from state i to state 0 is

$$\mathcal{E}[u_i] = \frac{1}{N+1}\left(\frac{1}{p_i} + \sum_{k=1}^{N} \frac{1}{p_k}\right)$$

Hint: Use Problem 5.35.

b. Explain the relationship between this chain and the chain in Problem 5.35. *Hint:* Use Problem 5.34.

c. Find $\mathcal{E}[u_i]$ when $p_i = p$ for all $i \in S$. Explain the result.

5.37 A *multinomial* trial is an experiment having $N + 1$ outcomes, conveniently labeled $0, 1, 2, \ldots, N$. We denote the probability that outcome i occurs in each trial by p_i $(\sum_{j=0}^{N} p_i = 1)$.

a. Let X_n be the outcome of the nth trial. Find the transition matrix for $\{X_n\}_{n=0,1,2,\ldots}$.

b. Make state 0 absorbing. Use the results obtained in Problem 1.35 to construct the fundamental matrix. Verify that this matrix is $(\mathbf{I} - \mathbf{T})^{-1}$.

c. Hence, find the expected number of trials until outcome 0 is obtained. Explain the result.

5.38 (Whitaker, 1978) There are $N+1$ competing brands in the market labeled $0, 1, 2, \ldots, N$. Let X_n be the brand bought at time n. It was proposed that chain $\{X_n\}_{n=0,1,2,\ldots}$ has the following transition probabilities:

$$p_{ij} = \begin{cases} p_i w_j & \text{for } i \neq j \\ 1 - p_i(1 - w_i) & \text{for } i = j \end{cases}$$

where $0 < p_i \leq 1$ for all $i \in S$ and $\sum_{j=0}^{N} w_i = 1$. (Here, w_i is the purchasing persuasion of brand i, which is the proportion of all nonloyal consumers who are persuaded to purchase brand i next time; p_i is the proportion of consumers who will not repurchase brand i without persuasion next time.) If a customer is purchasing brand $i \neq 0$ now, show that the expected time she will purchase brand 0 for the first time is

$$\mathcal{E}[u_i] = \frac{1}{p_i} + \frac{1}{w_0} \sum_{j=1}^{N} \frac{w_j}{p_j}$$

Hint: Use Problem 5.37.

5.39 A *martingale* is a discrete-time chain $\{X_n\}_{n=0,1,2,\ldots}$ having state space $S \equiv \{0, 1, \ldots, N\}$ in which

$$\mathcal{E}[X_{n+1} \mid X_0, X_1, \ldots, X_n] = X_n \quad \text{for all } n \geq 0$$

Chain $\{X_n\}_{n=0,1,2,\ldots}$ can be a model for fair games in the sense that, given that a player has an amount of X_n at time n, the average amount at time $n+1$ is equal to X_n.

a. Suppose a chain $\{X_n\}_{n=0,1,2,\ldots}$ is both a martingale and a Markov chain with transition matrix **P**. Show (by induction or some other method) that $\sum_{j \in S} j p_{ij}^{(n)} = i$ for all $i \in S$ and $n > 0$.

b. Hence, if states 0 and N are absorbing, show that

$$f_{ij} = \lim_{n \to \infty} p_{ij}^{(n)} = \begin{cases} 1 - i/N & \text{for } j = 0 \\ 0 & \text{for all } j \neq 0 \text{ and } j \neq N \\ i/N & \text{for } j = N \end{cases}$$

(Note that the probability that the player reaches state N is proportional to the initial capital.)

5.40 Suppose a population always consists of N genes in any generation. There are two types of genes: A and B. Let X_n be the number of type A genes in the nth generation. Assume X_{n+1} is binomially distributed with the probability of success proportional to X_n, or

$$p_{ij} = \binom{N}{j} \left(\frac{i}{N}\right)^j \left(1 - \frac{i}{N}\right)^{N-j} \quad \text{for all } 0 \leq i, j \leq N$$

If there are now i type A genes, what is the probability that the type A genes will eventually disappear? *Hint:* Show that this is a martingale.

5.41 In an absorbing Markov chain $\{X_n\}_{n=0,1,2,\ldots}$, assume a discounted rate of α ($0 \le \alpha \le 1$); that is, a cost of \$1 is incurred at step n is equivalent to a cost of $\$\alpha^n$ at step 0. $\$\alpha^n$ is called the *present value* of \$1 at step n. Given $X_0 = i$, let v_i be the expected present value of the total cost incurred over the entire life of the chain.

a. Assume that there is a cost c_i incurred at step n whenever the chain visits a transient state i. Show that

$$[v_i]_{i \in S} = (\mathbf{I} - \alpha \mathbf{T})^{-1} \mathbf{c}_T$$

where $\mathbf{c}_T \equiv [c_i]_{i \in T}$. Compare with Problem 3.28.

b. Assume also a one-time cost of $\$c_k$ whenever the chain moves into an absorbing state k. Show that

$$\mathbf{t} \equiv [t_i]_{i \in A} = (\mathbf{I} - \alpha \mathbf{T})^{-1} (\mathbf{c}_T + \alpha \mathbf{S} \mathbf{c}_A)$$

where $\mathbf{c}_A \equiv [c_i]_{i \in A}$.

c. In Problem 5.3, suppose it costs \$3 to visit a new customer, \$2 to visit a customer with a low degree of interest, and \$1 to visit a customer with a high degree of interest. It also costs \$4 for a lost sale. How much profit should be made for a completed sale if $\alpha = .9$?

5.42

a. For a general absorbing chain as defined in Equation (5.1.1), assume we know that it will end up in a particular absorbing state $a \in A$. How do we modify the transition matrix to reflect this knowledge?

b. Find the transition matrix for the rat-maze example, given the rat will eventually find food.

c. Given the rat will eventually find food from room i ($i = 2, 3, 4, 5$), what is the expected number of steps before this occurs?

5.43 In an absorbing chain, let ℓ_{ij} ($i, j \in T$) be the probability that, starting from state i, the chain visits state j immediately before absorption. In other words, state j is the last transient state visited.

a. Use first-step analysis to express all ℓ_{ij}s in terms of the fundamental matrix.

b. Apply the results to the rat-maze example.

5.44 Use first-step analysis to verify that

$$\mathcal{E}[u_i] = 1 + \sum_{j \in T} p_{ij} \mathcal{E}[u_j] \quad \text{for all } i \in T$$

Hence, verify that the expected number of steps before the rat finds either food or shock is 49/21, given that it starts from room 2.

5.45 Show that $\mathbf{TU} = \mathbf{UT} = \mathbf{U} - \mathbf{I}$.

5.46

a. Use first-step analysis to show that

$$\mathcal{E}\left[(u_i)^2\right] = 1 + 2 \sum_{j \in T} p_{ij} \mathcal{E}[u_j] + \sum_{j \in T} p_{ij} \mathcal{E}\left[(u_j)^2\right] \quad \text{for all } i \in T$$

b. Hence, use Problem 5.45 to show that

$$\left[\mathcal{E}\left[(u_i)^2\right]\right]_{i\in S} = (2\mathbf{U} - \mathbf{I})\mathbf{u}$$

c. Find all the variances of u_is in the rat-maze example.

5.47

a. Show that

$$\left[\mathcal{E}\left[(v_{ij})^2\right]\right]_{i,j\in T} = \mathbf{U}(2\mathbf{U}_d - \mathbf{I})$$

where $\mathbf{U}_d$ is the diagonal matrix, the (i, i)-element of which is equal to $\mathcal{E}[v_{ii}]$. *Hint:* Use Problem 5.45.

b. Hence, find the variance of all v_{ij}s in the rat-maze example.

5.48 The reaching probability f_{ij} defined in §4.3 is the probability that a Markov chain ever visits state j from state i. When j is an absorbing state in an absorbing chain, we have studied f_{ij} as the absorbing probability in §5.5. Let us now study the probability that an absorbing chain reaches a certain transient state, starting from a transient state.

a. Explain why

$$\mathbf{I} - \mathbf{B}_d = (\mathbf{U}_d)^{-1}$$

where $\mathbf{B}_d$ and $\mathbf{U}_d$ are the diagonal matrices, the (i, i)-element of which are f_{ii} and $\mathcal{E}[v_{ii}]$, respectively, for all $i \in T$.

b. From the results obtained in Problems 4.10 and 5.45, show that

$$\mathbf{B} \equiv \left[f_{ij}\right]_{i,j\in T} = (\mathbf{U} - \mathbf{I})\,(\mathbf{U}_d)^{-1}$$

c. Hence, find all f_{ij}s in the rat-maze example.

d. Find all f_{i3}s in the rat-maze example by first-step analysis.

5.49 Consider the following chain:

$$\mathbf{P} = \begin{array}{c} \\ 1 \\ 2 \\ 3 \\ 4 \end{array}\begin{array}{c} \begin{array}{cccc} 1 & 2 & 3 & 4 \end{array} \\ \left(\begin{array}{c|ccc} 1 & 0 & 0 & 0 \\ \hline .3 & .2 & .4 & .1 \\ 0 & .4 & .1 & .5 \\ 0 & .3 & .5 & .2 \end{array}\right) \end{array}$$

a. Find the values of f_{32} and f_{42}.

b. Hence, find the expected numbers of steps before the chain reaches state 2, starting from state 3 and starting from state 4.

5.50 Consider the following chain:

$$\mathbf{P} = \begin{array}{c} \\ A \\ B \\ C \\ D \\ E \end{array}\begin{array}{c} \begin{array}{ccccc} A & B & C & D & E \end{array} \\ \left(\begin{array}{cc|ccc} 1 & 0 & 0 & 0 & 0 \\ 0 & 1 & 0 & 0 & 0 \\ \hline 4/9 & 1/9 & 1/9 & 1/9 & 2/9 \\ 3/7 & 1/7 & 1/7 & 0 & 2/7 \\ 3/8 & 1/8 & 1/8 & 2/8 & 1/8 \end{array}\right) \end{array}$$

Use the results in Problem 5.48 to find the matrix $\mathbf{F} \equiv [f_{ij}]_{i,j\in S}$.

5.51 Given that an absorbing chain starts from transient state i, let δ be the number of distinct transient states it visits before absorption (including state i).

a. Show that

$$[\mathcal{E}[\delta_i]]_{i\in T} = \mathbf{U}(\mathbf{U}_d)^{-1}\mathbf{e}$$

where $\mathbf{e}$ is a column, all elements of which are 1, and $\mathbf{U}_d$ is the diagonal matrix, the (i, i)-element of which is equal to $\mathcal{E}[v_{ii}]$. *Hint:* Use Problem 5.48.

b. Hence, find $[\mathcal{E}[\delta_i]]_{i\in T}$ for the rat-maze example.

5.52 Consider a Markov chain $\{X_n\}_{n=0,1,2,\dots}$ having state space S. We partition S into two disjoint subsets A and B; that is, $S = A \cup B$ and $A \cap B = \phi$. The transition matrix of this chain can now be arranged into the form

$$\mathbf{P} = \begin{array}{c} \\ A \\ B \end{array}\!\!\begin{array}{c} \begin{array}{cc} A & B \end{array} \\ \begin{pmatrix} \mathbf{P}_{AA} & \mathbf{P}_{AB} \\ \mathbf{P}_{BA} & \mathbf{P}_{BB} \end{pmatrix} \end{array}$$

Consider now chain $\{Y_n\}_{n=0,1,2,\dots}$, which records the location of $\{X_n\}_{n=0,1,2,\dots}$ only when $\{X_n\}_{n=0,1,2,\dots}$ visits A. For example, if $X_0 = i \in A$, then $Y_0 = i$. If $X_1, X_2, \dots, X_{n-1} \in B$ and $X_n = j \in A$, then $\{Y_n\}_{n=0,1,2,\dots}$ ignores the movements of $\{X_n\}_{n=0,1,2,\dots}$ between steps 1 and $n-1$ and only records $Y_1 = X_n = j \in A$. Chain $\{Y_n\}_{n=0,1,2,\dots}$ is called a *censored* chain of $\{X_n\}_{n=0,1,2,\dots}$. Let $\mathbf{Q}$ be the transition matrix of $\{Y_n\}_{n=0,1,2,\dots}$. Explain why

$$\mathbf{Q} = \mathbf{P}_{AA} + \mathbf{P}_{AB}(\mathbf{I} - \mathbf{P}_{BB})^{-1}\mathbf{P}_{BA}$$

5.53 Refer to Problem 5.10. Let $N = 3$, $k = 3$, $p_1 = .6$, $p_2 = .3$, and $p_3 = .1$.

a. Use Problem 5.52 to find the transition matrix for $\{Y_n\}_{n=0,1,2,\dots}$, which only records the movements of $\{X_n\}_{n=0,1,2,\dots}$ among states 1, 111, 2, 222, 3, and 333 in terms of p_is. Explain its elements.

b. Hence, find the probability that player i wins.

5.54 Consider the following *finite random walk* having $N + 1$ states with

$$\mathbf{P} = \begin{array}{c} \\ 0 \\ 1 \\ 2 \\ \cdots \\ N-1 \\ N \end{array}\!\!\begin{array}{c} \begin{array}{ccccccccc} 0 & 1 & 2 & 3 & \cdots & N-2 & N-1 & N \end{array} \\ \begin{pmatrix} r_0 & s_0 & & & \cdots & & & \\ q_1 & r_1 & s_1 & & \cdots & & & \\ & q_2 & r_2 & s_2 & \cdots & & & \\ \cdots & \cdots & \cdots & \cdots & \cdots & \cdots & \cdots & \cdots \\ & & & & \cdots & q_{N-1} & r_{N-1} & s_{N-1} \\ & & & & \cdots & & q_N & r_N \end{pmatrix} \end{array}$$

where $r_N < 1$, $r_0 + s_0 = 1$, $q_N + r_N = 1$, and $q_i + r_i + s_i = 1$ for all $1 \le i \le N-1$. We say this random walk has an *absorbing barrier* at 0.

a. Let n_{ij} be the number of steps for the chain to go from state i to state j. Use first-step analysis to show that $\mathcal{E}[n_{N,N-1}] = 1/q_N$ and $\mathcal{E}[n_{i,i-1}] = 1/q_i + s_i\mathcal{E}[n_{i+1,i}]/q_i$ for all $1 \le i \le N-1$.

b. Show that the expected number of steps to go from state 1 to state 0 is

$$\mathcal{E}[n_{1,0}] = \frac{1}{q_1}\left(1 + \frac{s_1}{q_2} + \frac{s_1}{q_2}\frac{s_2}{q_3} + \cdots + \frac{s_1}{q_2} \cdots \frac{s_{N-2}}{q_{N-1}}\frac{s_{N-1}}{q_N}\right)$$

c. Show that the expected number of steps to go from state $N-1$ to state N is

$$\mathcal{E}[n_{N-1,N}] = \frac{1}{s_{N-1}}\left(1 + \frac{q_{N-1}}{s_{N-2}} + \frac{q_{N-1}}{s_{N-2}}\frac{q_{N-2}}{s_{N-3}} + \cdots + \frac{q_{N-1}}{s_{N-2}} \cdots \frac{q_2}{s_1}\frac{q_1}{s_0}\right)$$

5.55 Consider the finite random walk in Problem 5.54. Assume now that we have two absorbing barriers at 0 and N, or $r_0 = 1$ and $r_N = 1$.

Define

$$\rho^{(0)} \equiv 1 \qquad \Omega_0 \equiv 0 \qquad \Omega_1 \equiv 1/s_1$$

$$\rho^{(i)} \equiv \prod_{j=1}^{i} (q_j/s_j) \quad \text{for all } 1 \le i < N$$

$$A_i \equiv \sum_{j=0}^{i} \rho^{(j)} \quad \text{for all } 0 \le i < N$$

$$\Omega_i \equiv \frac{1}{s_i}\left(\frac{q_i \cdots q_2}{s_{i-1} \cdots s_1} + \frac{q_i \cdots q_3}{s_{i-1} \cdots s_2} + \cdots + \frac{q_i}{s_{i-1}} + 1\right) \quad \text{for all } 2 \le i \le N$$

and

$$B_i \equiv \sum_{j=0}^{i} \Omega_j \quad \text{for all } 0 \le i \le N$$

a. Given the chain starts from a transient state i, use first-step analysis and the facts that $f_{0,0} = 1$ and $f_{0,N} = 0$ to show that the probability of absorption into state 0 is

$$f_{i,0} = 1 - \frac{A_{i-1}}{A_{N-1}} \quad \text{for all } 1 \le i \le N$$

b. Hence, if $q_i = q$ and $s_i = s$, then

$$f_{i,0} = \begin{cases} 1 - i/N & \text{when } q = s \\ \dfrac{\rho^i - \rho^N}{1 - \rho^N} & \text{when } q \ne s \end{cases}$$

where $\rho \equiv q/s$.

c. Given the chain starts from a transient state i, use first-step analysis and the fact that $\mathcal{E}[u_0] = \mathcal{E}[u_N] = 0$ to show that the expected time before absorption is

$$\mathcal{E}[u_i] = \frac{A_{i-1}}{A_{N-1}} B_{N-1} - B_{i-1} \quad \text{for all } 1 \le i \le N$$

d. Hence, if $q_i = q$ and $s_i = s$, then

$$\mathcal{E}[u_i] = \begin{cases} \dfrac{i(N-i)}{2s} & \text{when } q = s \\ \dfrac{1}{s(1-\rho)}\left(\dfrac{N(1-\rho^i)}{1-\rho^N} - i\right) & \text{when } q \neq s \end{cases}$$

e. Use the foregoing results to verify the answers in Problem 5.27.

5.56 $N+1$ points labeled $0, 1, 2, \cdots, N$ are evenly distributed on the circumference of a circle. A particle is moving from one point to another. At the ith movement, let the probability of its moving in the counterclockwise direction to the adjacent point be p and in the clockwise direction be $q \equiv 1-p$. Starting from point 0, show that the probability that it visits all points on the circle before returning to point 0 is $(p^N + q^N)(p-q)/(p^N - q^N)$. *Hint:* Use Problem 5.55.

5.57 A biased coin is tossed repeatedly. Assume $\Pr\{H\} = p$ and $\Pr\{T\} = 1-p$. Define an absorbing chain to find the expected number of tosses until k consecutive Hs are obtained. Compare with the result obtained in Problem 1.40. *Hint:* Use Appendix (A.20).

5.58 Consider the following chain, in which state 0 is absorbing:

$$\mathbf{P} = \begin{array}{c} \\ 0 \\ 1 \\ 2 \\ \cdots \\ N-1 \\ N \end{array} \begin{array}{c} \begin{array}{ccccccc} 0 & 1 & 2 & 3 & \cdots & N-1 & N \end{array} \\ \left(\begin{array}{c|cccccc} 1 & & & & \cdots & & \\ \hline q_1 & r_1 & s_1 & & \cdots & & \\ q_2 & & r_2 & s_2 & \cdots & & \\ \cdots & \cdots & \cdots & \cdots & \cdots & \cdots & \\ q_{N-1} & & & & \cdots & r_{N-1} & s_{N-1} \\ q_N & & & & \cdots & & r_N \end{array}\right) \end{array}$$

a. Let $r_i = 0$ for all $1 \le i \le N$.
 i. Find the fundamental matrix $\mathbf{U}$. *Hint:* Use Appendix (A.19).
 ii. Hence, find $\mathcal{E}[u_i]$ for all $1 \le i \le N$.

b. Now let $0 < r_i < 1$ for all $1 \le i \le N$. From the result obtained in part a, use Problem 5.34 to explain why

$$\mathcal{E}[u_i] = \frac{1}{(1-r_i)\gamma^{(i)}}\left(\gamma^{(i)} + \gamma^{(i+1)} + \cdots + \gamma^{(N)}\right)$$

where $\gamma^{(1)} \equiv (1-r_1)^{-1}$ and

$$\gamma^{(i)} \equiv \frac{1}{(1-r_1)}\frac{s_1}{(1-r_2)}\cdots\frac{s_{i-1}}{(1-r_i)} \quad \text{for all } 2 \le i \le N$$

c. Use first-step analysis to show that $\mathcal{E}[u_N] = (1-r_N)^{-1}$ and

$$\mathcal{E}[u_{i-1}] = \frac{1}{1-r_{i-1}} + \frac{s_{i-1}}{1-r_{i-1}}\mathcal{E}[u_i] \quad \text{for all } 1 \le i \le N-1$$

Hence, obtain the result for $\mathcal{E}[u_i]$ as in part b.

5.59 Consider the following chain, in which state N is absorbing:

$$\mathbf{P} = \begin{array}{c} \\ 0 \\ 1 \\ 2 \\ \cdots \\ N-1 \\ N \end{array} \begin{array}{c} \begin{array}{cccccccc} 0 & 1 & 2 & 3 & \cdots & N-1 & N \end{array} \\ \left(\begin{array}{cccccc|c} q_0 & s_0 & & & \cdots & & \\ q_1 & r_1 & s_1 & & \cdots & & \\ q_2 & & r_2 & s_2 & \cdots & & \\ \cdots & \cdots & \cdots & \cdots & \cdots & \cdots & \cdots \\ q_{N-1} & & & & \cdots & r_{N-1} & s_{N-1} \\ \hline & & & & \cdots & & 1 \end{array} \right) \end{array}$$

a. Let $r_i = 0$ for all $i \geq 1$. Obtain $\mathcal{E}[u_0]$ from the fundamental matrix. *Hint:* Use Appendix (A.20).

b. Now let $0 < r_i < 1$ for all $1 \leq i \leq N$. From the result obtained in part a, explain why

$$\mathcal{E}[u_0] = \frac{1}{(1 - r_N)\gamma^{(N)}} \left(\frac{1}{s_0} + \gamma^{(1)} + \cdots + \gamma^{(N-2)} + \gamma^{(N-1)} \right)$$

where $\gamma^{(i)}$ are defined as in Problem 5.58.

c. Let $n_{i,j}$ be the number of steps to go from state i to state j for the first time. Use first-step analysis to show that $\mathcal{E}[n_{0,1}] = {s_0}^{-1}$, $s_i \mathcal{E}[n_{i,i+1}] = 1 + q_i \mathcal{E}[n_{0,i}]$ and

$$\mathcal{E}[n_{0,i+1}] = \frac{1 - r_i}{s_i} \mathcal{E}[n_{0,i}] + \frac{1}{s_i}$$

Hence, obtain the result for $\mathcal{E}[u_0]$ as in part b.

d. Find $\mathcal{E}[u_0]$ when $s_i = s$, $r_i = r$, and $q_i = q$ for all $i \geq 0$.

e. Find $\mathcal{E}[u_0]$ when $s_i = s$, $r_i = 0$, and $q_i = q$ for all $i \geq 0$. Compare with Problem 5.57.

5.60 Use induction to prove Equation (5.6.1).

CHAPTER 6

Finite Nonabsorbing Chains

6.1 INTRODUCTION

In the previous chapter, we studied finite absorbing chains—that is, chains having only transient and absorbing states. These chains will surely be absorbed into one of the absorbing states. In this chapter, we study other types of *finite* chains, having a finite number of states N, and one recurrent class, which has more than one recurrent state. This means that the chains may move forever among different states within this recurrent class. We call them *nonabsorbing* chains.

Recall that we previously defined *irreducible* chains as those having only one class (§4.9). Also, an irreducible chain is said to be *periodic* if all its states have the same period $d > 1$; if $d = 1$, it is *aperiodic* (§4.12).

We develop this chapter in the following order:

1. We first study the behavior of finite, irreducible, and aperiodic chains, which we refer to as the *regular* chains.[1]
2. We then look at the behavior of finite, irreducible, and periodic chains.
3. Finally, we study the behavior of finite *reducible* chains having more than one class.

[1]Some authors call them the *ergodic* chains.

6.2 THE TAXICAB EXAMPLE

We now start with the regular chains. To illustrate this type of chain, let us return to the taxicab example introduced in §3.9: Tex drives his cab around three cities A, B, and C. Let T_n be the city he is in at the beginning of the nth hour. We assume that $\{T_n\}_{n=0,1,2,\ldots}$ is a Markov chain with the one-step transition matrix (3.9.2) reproduced here:

$$\mathbf{P} = \begin{array}{c} \\ A \\ B \\ C \end{array} \begin{array}{c} \begin{array}{ccc} A & B & C \end{array} \\ \begin{pmatrix} .4 & .1 & .5 \\ .2 & .2 & .6 \\ .3 & .4 & .3 \end{pmatrix} \end{array} \tag{6.2.1}$$

This is an example of a regular chain with $N = 3$.

6.3 LIMITING PROBABILITIES

If you will remember, back in §3.9, we wondered what the conditions were for the sequence of distributions $\boldsymbol{\pi}^{(n)}$ to converge to a vector $\boldsymbol{\pi}$ as $n \to \infty$ and whether this limiting vector, when it exists, is dependent on the initial conditions. We then showed in §4.10 that these sequences do not converge for finite, irreducible, and *periodic* chains.

Since the proof is somewhat beyond our scope here, we just state the following results for regular chains (which are finite, irreducible, and *aperiodic*):

1. As $n \to \infty$, $\pi_i^{(n)}$ converges to a *limiting probability* π_i for all $i \in S$, independent of the chain's initial conditions. Thus, the vector $\boldsymbol{\pi}^{(n)}$ converges to a *limiting distribution* $\boldsymbol{\pi} \equiv [\pi_i]_{i \in S}$. In other words, as $n \to \infty$, the n-step transition matrix $\mathbf{P}^{(n)}$ converges to a *limiting transition matrix* $\mathbf{P}^{(\infty)}$, each row of which is identical to $\boldsymbol{\pi}$.
2. The elements of $\boldsymbol{\pi}$ are the elements of the unique solution of

$$\boldsymbol{\pi} = \boldsymbol{\pi}\mathbf{P} \tag{6.3.1}$$

and

$$\sum_{i \in S} \pi_i = 1 \tag{6.3.2}$$

Note that $\mathbf{P}$ is premultiplied by $\boldsymbol{\pi}$ in Equation (6.3.1), which is not the same as $\mathbf{P}\boldsymbol{\pi}$, where it is postmultiplied. Note also that Equation (6.3.1) is equivalent to the following set of N linear equations:

$$\pi_j = \sum_{k \in S} \pi_k p_{kj} \quad \text{for all } j \in S \tag{6.3.3}$$

For the taxicab example, Equation (6.3.3) yields:

$$\begin{cases} \pi_A = .4\pi_A + .2\pi_B + .3\pi_C \\ \pi_B = .1\pi_A + .2\pi_B + .4\pi_C \\ \pi_C = .5\pi_A + .6\pi_B + .3\pi_C \end{cases} \tag{6.3.4}$$

Corresponding to a particular state j, Equation (6.3.3) has π_j on its left-hand side and the product of $\boldsymbol{\pi}$ and the j-column of $\mathbf{P}$ on its right-hand side. Recall that each row of the stochastic matrix $\mathbf{P}$ sums to 1, and hence, any one of the columns of $\mathbf{P}$ can be calculated from the others. Thus, one equation in (6.3.3) can be constructed from the rest, and only $N - 1$ of its equations are linearly independent. Observe also that if $\mathbf{y}$ is a solution of (6.3.1), then so is $c\mathbf{y}$ for any scalar c. Thus, there are an infinite number of solutions satisfying (6.3.1), and the N equations obtained from (6.3.3) cannot all be independent.

To show that one of those equations for the taxicab example can be derived from the rest, we add the first two equations of (6.3.4), obtaining

$$\pi_A + \pi_B = .5\pi_A + .4\pi_B + .7\pi_C$$

which is equivalent to the last equation in (6.3.4).

To have a set of N linearly independent equations, we therefore must delete one equation obtained from (6.3.3) (any one, preferably the most complicated) and substitute (6.3.2) for it. This substitution enables us to solve for the N unique numerical values of the π_is. In this manner, we can find that the limiting probability that Tex is in city A is $\pi_A = 32/105$, in city B is $\pi_B = 27/105$, and in city C is $\pi_C = 46/105$.

The limiting transition matrix for the taxicab example is thus

$$\mathbf{P}^{(\infty)} = \begin{matrix} & A & B & C \\ A & 32/105 & 27/105 & 46/105 \\ B & 32/105 & 27/105 & 46/105 \\ C & 32/105 & 27/105 & 46/105 \end{matrix} = \begin{matrix} & A & B & C \\ A & .30476 & .25714 & .43810 \\ B & .30476 & .25714 & .43810 \\ C & .30476 & .25714 & .43810 \end{matrix}$$

which is consistent with the previous discussions in §3.9 and the solution to Problem 3.23.

Equation (6.3.2) can also be written as $\boldsymbol{\pi}\mathbf{E} = \mathbf{e}$, where $\mathbf{E}$ and $\mathbf{e}$ are respectively the $(N \times N)$ matrix and the $(1 \times N)$ row having all elements equal to 1. Adding this equation with (6.3.1), we obtain $\boldsymbol{\pi}(\mathbf{I} + \mathbf{E} - \mathbf{P}) = \mathbf{e}$. This yields the following formula (Resnick, 1992a):

$$\boldsymbol{\pi} = \mathbf{e}\,(\mathbf{I} + \mathbf{E} - \mathbf{P})^{-1} \tag{6.3.5}$$

which can be very useful for numerical calculations.

For the limiting distribution in the taxicab example, $\boldsymbol{\pi}$ can be calculated as

$$\boldsymbol{\pi} = (1 \quad 1 \quad 1)\begin{pmatrix} (2-.4) & (1-.1) & (1-.5) \\ (1-.2) & (2-.2) & (1-.6) \\ (1-.3) & (1-.4) & (2-.3) \end{pmatrix}^{-1}$$

Actually, we should anticipate (6.3.1) from (2.4.5), which shows $\boldsymbol{\pi}^{(n)} = \boldsymbol{\pi}^{(n-1)}\mathbf{P}$. In addition, we can get an intuitive interpretation for (6.3.3) if we subtract $\pi_j p_{jj}$ from both sides of it:

$$\pi_j(1 - p_{jj}) = \sum_{k \neq j} \pi_k p_{kj} \quad \text{for all } j \in S \tag{6.3.6}$$

1. π_j is the limiting probability that the chain is in state j. Given that it is in state j, $1 - p_{jj}$ is the conditional probability that the chain leaves state j. Thus, the left-hand side of (6.3.6) is the limiting unconditional probability that the chain leaves state j.

2. The right-hand side of (6.3.6) is the limiting unconditional probability that the chain arrives at state j from all other states.

3. The right- and left-hand sides of (6.3.6) must be the same because the *flow* into a state must be equal to the flow out. This balancing of in- and outflow is the reason we say the system has reached the *steady state*, or *equilibrium*, as in §3.9.

At this point, the concept of the steady state of a Markov chain could stand some clarification. We said that, as $n \to \infty$, the *probability* of finding Tex at city i approaches π_i for $i = A$, B, and C. Note that we only talk about the convergence of the probabilities here, not the actual movements of Tex. Even after a very long duration of time, Tex still moves among the cities in the same manner as when he started at step 0, and his movement is still governed by the transition matrix (6.2.1).

Regular Chains in Action: Land Use in Equilibrium. Robinson (1980) used Markov chains to model the changes of land use among various states such as water, forest, grassland, agriculture, residential, and commercial/residential. In this case, the model assumed that the use of land continues to fluctuate among these states and will not settle to any particular state: "This concept of land use equilibrium differs from that of conventional land use theory where, if exogenous conditions do not change, land will eventually reach and remain in a particular use." ■

Regular Chains in Action: Market Share in Equilibrium. A hundred households in Leeds, Yorkshire, England, were asked about their weekly purchases of laundry cleaning products between January and June 1957. Their purchasing habits were divided into four states depending on whether the family was buying (1) detergent only, (2) soap powder only, (3) both detergent and soap powder, or (4) no detergent or soap powder at all.

From this interview data, Styan and Smith (1964) obtained the following 1-week transition matrix:

$$\mathbf{P} = \begin{array}{c} \\ 1 \\ 2 \\ 3 \\ 4 \end{array} \begin{array}{c} \begin{array}{cccc} \quad 1 \quad & \quad 2 \quad & \quad 3 \quad & \quad 4 \quad \end{array} \\ \begin{pmatrix} .6724 & .0857 & .0229 & .2190 \\ .0367 & .7179 & .0444 & .2010 \\ .1235 & .2407 & .5123 & .1235 \\ .1465 & .2584 & .0219 & .5732 \end{pmatrix} \end{array}$$

> [I]n this case the elements of . . . [$\boldsymbol{\pi}$] will contain the shares of the market attained by the various states if the switching pattern defined by **P** were to persist for a long period of time. It thus gives an indication of where the market is heading, which is a useful piece

of information for the market strategist. Comparison of . . . [π] with the current market shares will indicate how far from stationary the current market distribution is.

The limiting distribution

$$\pi = (.2114 \quad .4085 \quad .0614 \quad .3187)$$

can be compared with the following observed distributions

Week 1:	(.22	.44	.09	.25)
Week 26:	(.22	.37	.05	.36)
Average:	(.2104	.4123	.0642	.3131)

The closeness of . . . [π] to the average market share indicates that the market distribution is approximately stationary throughout the 26 weeks considered. ■

Regular Chains in Action: Evaluating Urban Land Use Policies. Land in an urban area can be used for many purposes. Bourne (1976) divided the uses into ten states: (1) low-density residential, (2) high-density residential, (3) office, (4) general commercial, (5) automobile commercial, (6) parking, (7) warehousing, (8) industry, (9) transportation, and (10) vacant land. A conversion in land use was defined as a major structural modification to, or replacement of, the building or buildings on that property. The data giving the changes in land use after each conversion in the city of Toronto for the years 1952 through 1962 led to the following transition matrix:

$$\mathbf{P} = \begin{array}{c} \\ 1 \\ 2 \\ 3 \\ 4 \\ 5 \\ 6 \\ 7 \\ 8 \\ 9 \\ 10 \end{array} \begin{array}{c} \begin{array}{cccccccccc} 1 & 2 & 3 & 4 & 5 & 6 & 7 & 8 & 9 & 10 \end{array} \\ \left(\begin{array}{cccccccccc} .13 & .34 & .10 & .04 & .04 & .22 & .03 & .02 & & .08 \\ .02 & .41 & .05 & .04 & & .04 & & & & .44 \\ & .07 & .43 & .05 & .01 & .28 & .14 & & & .02 \\ .02 & .01 & .09 & .30 & .09 & .27 & .05 & .08 & .01 & .08 \\ & & .11 & .07 & .70 & .06 & & .01 & & .05 \\ .08 & .05 & .14 & .08 & .12 & .39 & .04 & & .01 & .09 \\ .01 & .03 & .02 & .12 & .03 & .11 & .38 & .21 & .01 & .08 \\ .01 & .02 & .02 & .03 & .03 & .08 & .18 & .61 & & .02 \\ .01 & .18 & .14 & .04 & .10 & .39 & .03 & .03 & .08 & \\ .25 & .08 & .03 & .03 & .05 & .15 & .22 & .13 & & .06 \end{array} \right) \end{array}$$

This transition matrix yields the following limiting distribution (in percentages) for land use:

$$\pi = \begin{array}{c} \begin{array}{cccccccccc} 1 & 2 & 3 & 4 & 5 & 6 & 7 & 8 & 9 & 10 \end{array} \\ (\begin{array}{cccccccccc} 5.0 & 8.4 & 11.9 & 8.2 & 15.1 & 19.0 & 11.4 & 11.5 & .4 & 9.1 \end{array}) \end{array}$$

Markov chain theory now can be used to *evaluate the impact of various possible conversion policies*. For example, suppose the city wants to restrict future parking use. Then all future applications to change from state i ($i \neq 6$) to state 6 will be refused. This policy can be studied by setting $p_{i6} = 0$ for all $i \neq 6$ and readjusting other transition probabilities accordingly. Recalculating the limiting percentage of land use with this parking restriction, we obtain the following limiting probabilities in percentage:

$$\pi = \begin{pmatrix} 1 & 2 & 3 & 4 & 5 & 6 & 7 & 8 & 9 & 10 \\ 5.0 & 8.5 & 15.5 & 11.9 & 12.7 & .0 & 16.8 & 19.3 & .5 & 9.8 \end{pmatrix}$$

With the restriction on parking, all other uses are shown to increase in area, except for automobile-related commercial use. ■

6.4 LONG-TERM VISITING RATES

Consider now the number of visits $v_{ij}^{(n)}$ a chain makes to state j from state i between steps 0 and $n-1$ inclusive (§3.10). For a finite absorbing chain, if both states i and j are transient, this number is finite even when $n \to \infty$ and can be read from the fundamental matrix as discussed in §5.3. On the contrary, for a finite irreducible chain, since all its states must be recurrent, the chain keeps returning to state j, and this number will become infinite as $n \to \infty$.

For all chains, however, the *visiting rate*, or the *proportion of visits*,[2] to state j during the first n steps, defined as

$$g_{ij}^{(n)} \equiv \frac{1}{n}\mathcal{E}\left[v_{ij}^{(n)}\right] \tag{6.4.1}$$

is always a number between 0 and 1. Its unit is the number of visits *per unit time*.

From Equation (3.10.1),

$$g_{ij}^{(n)} \equiv \frac{1}{n}\mathcal{E}\left[v_{ij}^{(n)}\right] = \frac{1}{n}\sum_{m=0}^{n-1} p_{ij}^{(m)} \quad \text{for all } i, j \in S, \text{ and } n \geq 1$$

We are often interested in determining $\lim_{n\to\infty} g_{ij}^{(n)}$, which is the visiting rate to state j during a chain's *entire life*. We call this the *(long-term) visiting rate to state j*. This rate is zero if state j is transient and positive if it is recurrent.

Consider a state i in a finite irreducible chain. Because it is recurrent, the chain will eventually visit it and then will forget its past and start anew. As n approaches infinity, the number of visits to state i becomes infinite, and whether the chain starts from it at time 0 or a finite number of steps later becomes insignificant in the calculation of $\lim_{n\to\infty} g_{ij}^{(n)}$. We can therefore ignore the initial conditions and denote the visiting rate to state j in an irreducible chain simply as g_j:

$$g_j \equiv g_{ij} \equiv \lim_{n\to\infty} g_{ij}^{(n)} = \lim_{n\to\infty} \frac{1}{n}\mathcal{E}\left[v_{ij}^{(n)}\right] = \lim_{n\to\infty} \frac{1}{n}\sum_{m=0}^{n-1} p_{ij}^{(m)} \tag{6.4.2}$$

[2]To be consistent with our later discussion in Chapter 9, we use the term *visiting rate* more often than *proportion of visits*. The former gives a sense of progress, whereas the latter gives a sense of history. But in some applications, the latter term seems more appropriate.

Recall that we defined $p_{ij}^{(0)} \equiv \delta_{ij}$, where $\delta_{ij} = 1$ if $i = j$ and $\delta_{ij} = 0$ otherwise. Also, from the Chapman-Kolmogorov Equation (3.5.1), switching the order of summation, we have

$$\begin{aligned} g_j &= \lim_{n\to\infty} g_{ij}^{(n+1)} = \lim_{n\to\infty} \frac{1}{n+1} \sum_{m=0}^{n} p_{ij}^{(m)} \\ &= \lim_{n\to\infty} \frac{1}{n} \left(\delta_{ij} + \sum_{m=1}^{n} p_{ij}^{(m)} \right) = \lim_{n\to\infty} \frac{1}{n} \sum_{m=0}^{n-1} p_{ij}^{(m+1)} \\ &= \lim_{n\to\infty} \frac{1}{n} \sum_{m=0}^{n-1} \left(\sum_{k\in S} p_{ik}^{(m)} p_{kj} \right) \\ &= \sum_{k\in S} p_{kj} \left(\lim_{n\to\infty} \frac{1}{n} \sum_{m=0}^{n-1} p_{ik}^{(m)} \right) = \sum_{k\in S} p_{kj} g_k \end{aligned} \tag{6.4.3}$$

This equation can be written in matrix form as

$$\mathbf{g} = \mathbf{gP} \tag{6.4.4}$$

where

$$\mathbf{g} \equiv [g_j]_{j\in S} \tag{6.4.5}$$

Similar to Equation (6.3.1), $\mathbf{gP}$ is not the same as $\mathbf{Pg}$, and Equation (6.4.4) only provides $N-1$ mutually independent equations. To get a set of N independent equations that can be solved for the N values of the visiting rates, we must delete one equation obtained from (6.4.4) and substitute:

$$\sum_{j\in S} g_j = 1 \tag{6.4.6}$$

Remember now that a regular chain also has a limiting distribution. That distribution can be calculated from (6.3.1) and (6.3.2), which are identical to (6.4.4) and (6.4.6), respectively. Since these equations yield unique solutions, a simple yet interesting result is that, in a regular chain,

$$\pi_i = g_i \quad \text{for all } i \in S \tag{6.4.7}$$

That is, the limiting probabilities and the long-term visiting rates to the states are the same.

For the taxicab problem described earlier, Tex's long-term visiting rate to city A is $g_A = 32/105$, to city B is $g_B = 27/105$, and to city C is $g_C = 46/105$.

Finite Irreducible Chains in Action: Proportion of Times a Share Moves Up. For the period from January 1963 to April 1967, the *Financial Times* reported the total number of shares moving up (U), down (D), and unchanged (N) for approximately

2700 companies each day. From this source, Dryden (1969) obtained the following empirical daily transition matrix:

$$\mathbf{P} = \begin{array}{c} \\ U \\ D \\ N \end{array} \begin{array}{c} \begin{array}{ccc} U & D & N \end{array} \\ \begin{pmatrix} .586 & .073 & .340 \\ .070 & .639 & .292 \\ .079 & .064 & .857 \end{pmatrix} \end{array} \tag{6.4.8}$$

which gives the chain's long-term proportions of time the shares move up, down, or remain unchanged as

$$\mathbf{g} = (.156 \quad .154 \quad .687) \qquad \blacksquare$$

6.5 EXPECTED FIRST REACHING TIMES

Let n_{ij} be the number of steps before the chain reaches state j *for the first time*, starting from state i. We call it the chain's *first reaching time*[3] from state i to state j, and we are interested in its expected value $\mathcal{E}[n_{ij}]$.

First-step analysis can be used to calculate these expected values as follows. Suppose the chain $\{X_n\}_{n=0,1,2,\ldots}$ starts from state i.

1. The chain can go to state j after the first step with probability p_{ij}. If this occurs, then $\mathcal{E}[n_{ij} \mid X_1 = j] = 1$.
2. The chain can go to state k ($k \neq j$) after the first step with probability p_{ik}. If this occurs, then $\mathcal{E}[n_{ij} \mid X_1 = k \neq j] = 1 + \mathcal{E}[n_{kj}]$.

Thus, for all $i, j \in S$,

$$\mathcal{E}\left[n_{ij}\right] = p_{ij} + \sum_{k \neq j} p_{ik}\left(1 + \mathcal{E}\left[n_{kj}\right]\right) = 1 + \sum_{k \neq j} p_{ik}\mathcal{E}\left[n_{kj}\right] \tag{6.5.1}$$

For a particular state j in a finite irreducible chain, all $\mathcal{E}[n_{ij}]$s must be finite, and (6.5.1) can be used to obtain their numerical values.

For city A in the taxicab example, we can write

$$\begin{cases} \mathcal{E}\left[n_{AA}\right] = 1 + (.1)\mathcal{E}\left[n_{BA}\right] + (.5)\mathcal{E}\left[n_{CA}\right] \\ \mathcal{E}\left[n_{BA}\right] = 1 + (.2)\mathcal{E}\left[n_{BA}\right] + (.6)\mathcal{E}\left[n_{CA}\right] \\ \mathcal{E}\left[n_{CA}\right] = 1 + (.4)\mathcal{E}\left[n_{BA}\right] + (.3)\mathcal{E}\left[n_{CA}\right] \end{cases} \tag{6.5.2}$$

This set of equations yields $\mathcal{E}[n_{AA}] = 105/32$, $\mathcal{E}[n_{BA}] = 65/16$, and $\mathcal{E}[n_{CA}] = 15/4$. If Tex leaves city A at 8:00 A.M., he can expect to return to city A for the first time at $(8 + 105/32)$ A.M., or 11:16 A.M. If he leaves city B at 12:00 P.M., he can expect to reach city A for the first time at $(12 + 65/16)$ P.M., or 4:03 P.M.

In matrix form, (6.5.1) can be written as

$$\mathbf{N} = \mathbf{E} + \mathbf{P}(\mathbf{N} - \mathbf{N}_d) \tag{6.5.3}$$

[3] Some authors call it the *first passage time* from state i to state j.

where

1. $\mathbf{N} \equiv [\mathcal{E}[n_{ij}]]_{i,j \in S}$.
2. $\mathbf{N}_d$ is the matrix $\mathbf{N}$ with all nondiagonal elements replaced by zeros.
3. $\mathbf{E}$ is an $(N \times N)$ matrix in which all elements are equal to 1.

Finite Irreducible Chains in Action: Social Mobility. The expected first reaching times of Markov chains were used to study social mobility, especially as reflected in father-child occupation transitions. Beshers and Laumann (1967) defined the occupational categories as: (1) top professional, top business; (2) semiprofessional, middle business; (3) clerical, sales; (4) skilled; and (5) semiskilled and unskilled. Analyzing some classical social mobility data for the British, they obtained the following expected first reaching time matrix:

$$\mathbf{N} = \begin{array}{c} \\ 1 \\ 2 \\ 3 \\ 4 \\ 5 \end{array} \begin{array}{c} \begin{array}{ccccc} 1 & 2 & 3 & 4 & 5 \end{array} \\ \begin{pmatrix} 44.00 & 26.39 & 4.57 & 3.89 & 6.01 \\ 63.26 & 24.18 & 4.24 & 3.47 & 5.68 \\ 71.34 & 31.61 & 4.66 & 2.69 & 4.84 \\ 73.57 & 33.45 & 5.65 & 2.44 & 4.08 \\ 74.76 & 34.24 & 6.20 & 2.67 & 3.21 \end{pmatrix} \end{array}$$

As noted by the authors,

1. Along each column, the further away from the diagonal element, the numbers tend to be larger. This confirms that there is an underlying order of the occupations, which is the order listed above, indicating state 1 as the highest and state 5 the lowest.
2. The values above and below the diagonal are quite different. This means that it requires a much higher number of generations for upward mobility than for downward mobility. For example, it takes an average of 74.76 generations to go up from state 5 to state 1, while it only takes an average of 6.01 generations to go down from state 1 to state 5.
3. The values in the 1-column and the 2-column are much larger than those in other columns. This suggests that a "bottleneck" exists between the top two categories and the lower ones. ■

Actually, for a particular state j, the set of equations in (6.5.1) can be divided into two parts:

1. The equation having $\mathcal{E}[n_{jj}]$ on the left-hand side, or

$$\mathcal{E}\left[n_{jj}\right] = 1 + \sum_{k \neq j} p_{jk} \mathcal{E}\left[n_{kj}\right] \tag{6.5.4}$$

For the taxicab example, this is the first equation in (6.5.2).

2. The remaining $N-1$ equations, having $N-1$ unknowns $\mathcal{E}[n_{ij}]$s for all $i \neq j$. For the taxicab example, these are the second and third equations in (6.5.2), with two unknowns $\mathcal{E}[n_{BA}]$ and $\mathcal{E}[n_{CA}]$. Effectively then, we have to solve only these $N-1$ equations with $N-1$ unknowns.

The derivation of these $N-1$ equations is a disguised form of a method that we have already discussed in §5.4: It is equivalent to *converting state j into an absorbing state*, making all other states transient. Starting from the now transient state i, any visit to the now absorbing state j is the first visit. Hence, for $i \neq j$, $\mathcal{E}[n_{ij}]$ is equal to the expected time $\mathcal{E}[u_i]$ until absorption, which is the sum of the i-row of the fundamental matrix, as in Equation (5.4.1).

For the taxicab example, to calculate $\mathcal{E}[n_{iA}]$ for $i = B$ or $i = C$, we convert state A into an absorbing state. The transition matrix (6.2.1) now becomes:

$$\mathbf{P} = \begin{array}{c} \\ A \\ B \\ C \end{array} \begin{array}{c} \begin{array}{ccc} A & B & C \end{array} \\ \left(\begin{array}{c|cc} 1 & 0 & 0 \\ \hline .2 & .2 & .6 \\ .3 & .4 & .3 \end{array}\right) \end{array}$$

The fundamental matrix for this absorbing chain is

$$\mathbf{U} = (\mathbf{I}-\mathbf{T})^{-1} = \begin{pmatrix} (1-.2) & -.6 \\ -.4 & (1-.3) \end{pmatrix}^{-1} = \begin{array}{c} \\ B \\ C \end{array} \begin{array}{c} \begin{array}{cc} B & C \end{array} \\ \begin{pmatrix} \frac{35}{16} & \frac{30}{16} \\ \frac{5}{4} & \frac{10}{4} \end{pmatrix} \end{array} \tag{6.5.5}$$

giving

$$\mathcal{E}[n_{BA}] = \mathcal{E}[u_B] = \frac{35}{16} + \frac{30}{16} = \frac{65}{16} \qquad \mathcal{E}[n_{CA}] = \mathcal{E}[u_C] = \frac{5}{4} + \frac{10}{4} = \frac{15}{4}$$

Now (6.5.4), or the first equation of (6.5.2), yields $\mathcal{E}[n_{AA}] = 105/32$.

As we see later, this *absorbing method* provides more insight into the system than does Equation (6.5.1).

Finite Irreducible Chains in Action: Time until a Share Moves Up. In the study of share movements whose transition matrix is given in (6.4.8), Dryden (1969) also used the absorbing method to calculate the following mean first reaching time matrix:

$$\mathbf{N} = \begin{array}{c} \\ I \\ D \\ N \end{array} \begin{array}{c} \begin{array}{ccc} I & D & N \end{array} \\ \begin{pmatrix} * & 14.9 & 3.0 \\ 13.2 & * & 3.4 \\ 12.9 & 15.2 & * \end{pmatrix} \end{array}$$

> The mean first reaching times to any given state from each possible initial state are approximately equal. That is they do not depend strongly on the initial state. . . . If a share has either just increased or just decreased it will, on average, take about only three days before it experiences a day of no-change. ■

6.6 EXPECTED RETURN TIMES

Of the first reaching times n_{ij}s, we call n_{jj} the *return time* to state j and denote it by n_j.

The expected return time $\mathcal{E}[n_{jj}]$, or $\mathcal{E}[n_j]$, can be obtained from other $\mathcal{E}[n_{kj}]$s ($k \neq j$), as given in Equation (6.5.4). However, a more elegant method can be used.

First, we premultiply (6.5.3) by the visiting rate vector $\mathbf{g}$ of (6.4.5). Keeping in mind that $\mathbf{gP} = \mathbf{g}$, we have

$$\mathbf{gN} = \mathbf{gE} + \mathbf{gP}[\mathbf{N} - \mathbf{N}_d] = \mathbf{gE} + \mathbf{gN} - \mathbf{gN}_d$$

Hence,

$$\mathbf{gN}_d = \mathbf{gE}$$

The left-hand side, $\mathbf{gN}_d$, is a row, the elements of which are $g_i \mathcal{E}[n_i]$; the right-hand side, $\mathbf{gE}$, is a row with all elements equal to $\sum_{i=1}^{N} g_i = 1$. Thus, in a finite irreducible chain,

$$\mathcal{E}[n_i] = \frac{1}{g_i} \quad \text{for all } i \in S \tag{6.6.1}$$

which means that the expected return time to a state is the *reciprocal* of the visiting rate to that state.

Intuitively, this result should come as no surprise. If the visiting rate to a state is, say, one-third, then the chain should visit that state once every three steps on average.

For the taxicab, because $g_A = 32/105$, we have $\mathcal{E}[n_A] = 105/32$, which agrees with the value obtained earlier.

Finite Irreducible Chains in Action: Migration Patterns. Segal (1985) used expected return times to study population movements among three administrative regions of the African country of Malawi: (N) the Northern Region, (C) the Central Region, and (S) the Southern Region. The 1977 census gave the following 1-year transition matrix:

$$\mathbf{P} = \begin{matrix} & N & C & S \\ N & .970 & .019 & .012 \\ C & .005 & .983 & .012 \\ S & .004 & .014 & .982 \end{matrix}$$

(Thus, for example, approximately 97% of those resident in the Northern Region in 1976 were still there in 1977.)

As $\boldsymbol{\pi} = \mathbf{g} = (.117, .473, .409)$, the expected return times are $\mathcal{E}[n_N] = (1/.117) = 8.77$ years, $\mathcal{E}[n_C] = (1/.473) = 2.12$ years, and $\mathcal{E}[n_S] = (1/.409) = 2.43$ years. From these figures, Segal drew the following conclusions about the cultural implications of the migration patterns:

> If we use these figures as an ordinal measure of the relative permanence of migration, then we have an interesting result: Northern Region, 8.77; Central Region, 2.12; Southern Region, 2.43. The higher value for the North indicates that its out-migration is relatively more permanent than that from the other regions. . . . The relative permanence of emigration from the Northern Region along with the relatively small amount of movement in the reverse direction, also indicates that it will probably tend to retain its complex traditional cultural character for a longer period than either the Central or Southern Regions. ■

6.7 REGENERATIVE PROPERTY

In a regular chain, we have seen that the limiting probability π_i is equal to the visiting rate g_i for all $i \in S$. This equality is understandable because both π_i and g_i relate to the long-term behavior of the chain. But how about the expected number of steps before the chain returns to state i for the first time? Why does this quantity $\mathcal{E}[n_i]$ (which only relates to a finite number of steps at the beginning of the chain's life) have such a direct relationship, through (6.6.1), to the quantity π_i (which describes the behavior of the chain after a large number of steps)?

The answer lies in the memorylessness and time-homogeneity of the chain. Assume that at step 0 the chain is at a particular state, say, state i. Even after a large number of steps, whenever the chain visits state i, it *starts anew probabilistically*, and its probabilistic behavior from that time forward is the same as that starting from step 0.

Let us call all the steps at which the chain visits state i its *regenerative points*. The duration between any two consecutive regenerative points is the *regenerative cycle* from state i. A *cycle length* is the number of steps within a regenerative cycle—that is, the return time n_i. For finite irreducible chains, all cycle lengths must be finite (Figure 6.7.1).

The entire history of the chain is made up of such cycles. Even after a large number of steps, the chain must still be in one of these cycles. Furthermore, because the chain starts anew in the same manner at each regenerative point, the developments of the chain are *independent and probabilistically identical in different regenerative cycles*. These regenerative cycles therefore may be thought of as forming the units of the chain, or being the building blocks for the entire history of the chain. To understand the behavior of the chain throughout its history, even after a large number of steps, we need only understand *its behavior within one cycle* (which may as well be the first one, assumed to be probabilistically the same as all the others because the chain starts from state i at time 0).

Consider a simple example. Suppose the chain is a sequence of independent and identically distributed random variables. In that case, the chain starts anew at each step, the cycle lengths are always one, and the behavior of the chain at any time point is the same as that at step 1. The probability of getting, say, 4 on the 1472nd toss of a die is the same as getting 4 on the first toss.

FIGURE 6.7.1 The regenerative cycles from state 0.

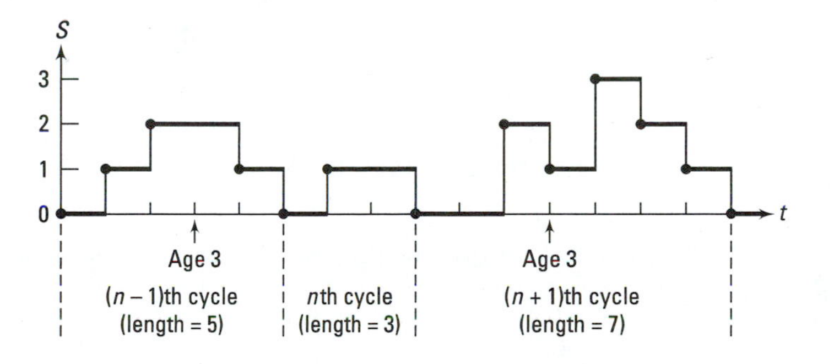

If the chain has gone through a steps since the last regenerative point, we call a the *age* of the chain. Regardless of whether the chain is in the first cycle or in the, say, 234th cycle, its behavior will be probabilistically identical if it has *the same age*—that is, if the distance from the last regenerative point is the same.

Consider a special periodic case in which the chain has to visit state i every fixed number of steps, say, five. Then the cycle lengths must always be five. At any step n, its age a within its current cycle is $a = n \bmod (5)$.[4] Thus, to obtain $\boldsymbol{\pi}^{(n)}$, we need not calculate $\mathbf{P}^n$, but only have to equate it with the corresponding distribution $\boldsymbol{\pi}^{(a)}$ within the first cycle. For example, at step 1472, the chain is at age 1472 mod (5) = 2, and hence $\boldsymbol{\pi}^{(1472)} = \boldsymbol{\pi}^{(2)}$.

The problem with aperiodic chains is that the cycle lengths are not constant; they are random variables. As an aperiodic chain develops, its age becomes more and more blurred. For illustration, suppose the cycle length can be either 4 or 5 days. At step 3, the chain is clearly at age 3 within the first cycle. However, at step 11, its age is no longer straightforward: If the first two cycles both have length 4, the chain is at age 3 of the third cycle; if the first two have lengths 4 and 5, it is at age 2; if the first two both have length 5, it is at age 1.

Referring to the mood problem studied in §2.3, there is a theory (actually, more of a fad) called biorhythm, which relies on the false assumption that cycle lengths in the human body are constant. Thus, someone's mood at a certain day, starting from his or her birthday, can be predicted depending on where that day is in a certain body cycle. Most scholars agree that our bodies have cycles; however, it is also rather evident that the cycle lengths are random variables, not constants. Even if these cycles vary only minimally (say, plus or minus 1 day in a cycle with an average length of 30 days), then it is not at all a simple task to determine where today is with respect to some body cycle of a 10-year-old child. The only thing we can say is that the probability that the child is happy today (π_h) is equal to the proportion of time that she is happy throughout her life (g_h). Since her life is made up of many probabilistically independent and identical cycles, this proportion in turn is equal to *the proportion of time that she is happy within an average cycle,* which is the ratio of the expected number of days she is happy in a cycle to the expected number of days in a cycle.

This discussion illustrates a general and very powerful result relating the long-term visiting rate g_k of a regular chain and, by the same token, its limiting probability π_k to the chain's behavior within the first cycle:

$$g_k = \pi_k = \frac{\mathcal{E}[\nu_{ik}]}{\mathcal{E}[n_i]} \quad \text{for all } k, i \in S \tag{6.7.1}$$

where ν_{ik} is the number of visits to state k within a regenerative cycle from state i and n_i is the length of the cycle from state i.

---

[4] $n \bmod (d)$ denotes the remainder when dividing n by d. For example, 27 mod (5) = 2.

Note that the chain visits state i only once within any cycle from state i. That is why $g_i = \pi_i = 1/\mathcal{E}[n_i]$, as in Equation (6.6.1).

The question now becomes: If the limiting results can be obtained from the chain's behavior within the first cycle, how can we study the first cycle from state i in isolation?

As in §6.5, the trick is to *convert state i into an absorbing state*, thus preventing the second cycle from occurring. The behavior of the irreducible chain in the first cycle from state i is the same as that of the resulting absorbing chain before its absorption into the only absorbing state i.

For the taxicab example, we used this method in §6.5 to obtain the expected cycle length from city A as $\mathcal{E}[n_A] = 105/32$. To obtain the expected number of visits Tex makes to city B within a cycle from city A, $\mathcal{E}[\nu_{AB}]$, we again use the fundamental matrix (6.5.5) and the first-step analysis.

1. Starting from city A, Tex remains at A after the first step with probability $p_{AA} = .4$. When this occurs, the cycle from state A only has one step, and the expected number of visits to B within this cycle is $\mathcal{E}[\nu_{AB} \mid T_1 = A] = 0$.
2. Starting from city A, Tex goes to city B after the first step with probability $p_{AB} = .1$. When this occurs, the fundamental matrix (6.5.5) gives the expected number of visits to B before Tex returns to A as $\mathcal{E}[\nu_{AB} \mid T_1 = B] = \mathcal{E}[\upsilon_{BB}] = 35/16$. (Note that the visit to B in the first step is included in $\mathcal{E}[\upsilon_{BB}]$.)
3. Starting from city A, Tex goes to city C after the first step with probability $p_{AC} = .5$. When this occurs, the fundamental matrix (6.5.5) gives the expected number of visits to B before Tex returns to A as $\mathcal{E}[\nu_{AB} \mid T_1 = C] = \mathcal{E}[\upsilon_{CB}] = 5/4$.

Thus, the expected number of visits Tex makes to city B within a cycle from city A is:

$$\mathcal{E}\,[\nu_{AB}] = (.4)(0) + (.1)(35/16) + (.5)(5/4) = 135/160$$

Equation (6.7.1) now gives

$$g_B = \pi_B = \frac{\mathcal{E}\,[\nu_{AB}]}{\mathcal{E}\,[n_A]} = \frac{135/160}{105/32} = \frac{27}{105}$$

A similar argument gives the expected number of times Tex visits city C within a cycle from city A as

$$\mathcal{E}\,[\nu_{AC}] = (.4)(0) + (.1)(30/16) + (.5)(10/4) = 23/16$$

giving

$$g_C = \pi_C = \frac{\mathcal{E}\,[\nu_{AC}]}{\mathcal{E}\,[n_A]} = \frac{23/16}{105/32} = \frac{46}{105}$$

These results are consistent with those we have previously obtained.

Regular Chains in Action: Reproductive Cycles of Female Rats. In toxicology, some useful information has been obtained from studies of the reproductive functioning of female rats, especially during their estrous cycles. Each cycle can be divided into five stages: states 1, 2, and 3 as days 1, 2, and 3 of diestrus, respectively; state 4 as proestrus; and state 5 as estrus. From vaginal smears of more than 500 animals taken over a period of 40 days, Girard and Sager (1987) obtained the following transition matrix:

$$\mathbf{P} = \begin{array}{c} \\ 1 \\ 2 \\ 3 \\ 4 \\ 5 \end{array} \begin{array}{c} \begin{array}{ccccc} 1 & 2 & 3 & 4 & 5 \end{array} \\ \begin{pmatrix} 0 & .937 & 0 & .016 & .011 \\ 0 & 0 & .729 & .229 & .042 \\ 0 & 0 & .245 & .426 & .329 \\ .014 & 0 & 0 & .005 & .981 \\ .742 & 0 & 0 & 0 & .258 \end{pmatrix} \end{array}$$

From Equation (6.7.1), because $\mathcal{E}[v_{ii}] = 1$ we have

$$\frac{\pi_j}{\pi_i} = \frac{\mathcal{E}\left[v_{ij}\right]}{\mathcal{E}\left[v_{ii}\right]} = \mathcal{E}\left[v_{ij}\right]$$

> [I]f the first day of diestrus is used to define the beginning of the cycle, then... $[1/\pi_1]$ provides the mean cycle length and... $[\pi_j/\pi_1]$, $j = 2, \ldots, 5$, can be interpreted as the mean number of days spent in the jth stage of the cycle. For example, in this particular choice of a model,... $[(\pi_1 + \pi_2 + \pi_3)/\pi_1]$ gives the average number of days of diestrus per cycle. ■

6.8 PERIODIC CHAINS

Having studied the aperiodic chains, let us now turn our attention to a periodic chain, which was defined as a finite irreducible chain in which all states have the same period $d > 1$.

In §4.12, we appropriately divided the state space of a periodic chain into d groups $G_1, G_2, \ldots, G_d$. We then showed that the chain's development always follows the pattern $G_1 - G_2 - \cdots - G_d - G_1 \cdots$ and that its transition matrix can be reorganized into the form (4.12.1):

$$\mathbf{P} = \begin{array}{c} \\ G_1 \\ G_2 \\ \cdots \\ G_{d-1} \\ G_d \end{array} \begin{array}{c} \begin{array}{cccccc} G_1 & G_2 & G_3 & \cdots & G_{d-1} & G_d \end{array} \\ \begin{pmatrix} \mathbf{0} & \mathbf{D}_1 & \mathbf{0} & \cdots & \mathbf{0} & \mathbf{0} \\ \mathbf{0} & \mathbf{0} & \mathbf{D}_2 & \cdots & \mathbf{0} & \mathbf{0} \\ \cdots & \cdots & \cdots & \cdots & \cdots & \cdots \\ \mathbf{0} & \mathbf{0} & \mathbf{0} & \cdots & \mathbf{0} & \mathbf{D}_{d-1} \\ \mathbf{D}_d & \mathbf{0} & \mathbf{0} & \cdots & \mathbf{0} & \mathbf{0} \end{pmatrix} \end{array} \qquad (6.8.1)$$

Real-Life Applications

continued on reverse side

An example of periodic chains was given in (4.10.1), which we reproduce here for convenience:

$$\mathbf{P} = \begin{array}{c} A \\ B \\ C \\ D \\ E \\ F \\ G \end{array} \overset{\begin{array}{ccccccc} A & B & C & D & E & F & G \end{array}}{\left(\begin{array}{cc|ccc|cc} 0 & 0 & 1/4 & 1/4 & 1/2 & 0 & 0 \\ 0 & 0 & 2/5 & 2/5 & 1/5 & 0 & 0 \\ \hline 0 & 0 & 0 & 0 & 0 & 1/4 & 3/4 \\ 0 & 0 & 0 & 0 & 0 & 1/2 & 1/2 \\ 0 & 0 & 0 & 0 & 0 & 1/3 & 2/3 \\ \hline 2/3 & 1/3 & 0 & 0 & 0 & 0 & 0 \\ 1/4 & 3/4 & 0 & 0 & 0 & 0 & 0 \end{array}\right)} \tag{6.8.2}$$

Here $d = 3$, $G_1 \equiv \{A, B\}$, $G_2 \equiv \{C, D, E\}$, $G_3 \equiv \{F, G\}$, and

$$\mathbf{D}_1 \equiv \begin{array}{c} A \\ B \end{array} \overset{\begin{array}{ccc} C & D & E \end{array}}{\begin{pmatrix} 1/4 & 1/4 & 1/2 \\ 2/5 & 2/5 & 1/5 \end{pmatrix}} \quad \mathbf{D}_2 \equiv \begin{array}{c} C \\ D \\ E \end{array} \overset{\begin{array}{cc} F & G \end{array}}{\begin{pmatrix} 1/4 & 3/4 \\ 1/2 & 1/2 \\ 1/3 & 2/3 \end{pmatrix}} \quad \mathbf{D}_3 \equiv \begin{array}{c} F \\ G \end{array} \overset{\begin{array}{cc} A & B \end{array}}{\begin{pmatrix} 2/3 & 1/3 \\ 1/4 & 3/4 \end{pmatrix}}$$

We observed in §4.12 that a periodic chain does not have limiting probabilities, but is there anything else we can say about its limiting behavior?

First, if we retrace the derivations of (6.4.4) and (6.4.6) relating to the long-term visiting rates, we can see that they hold true for all finite irreducible chains, periodic or not. Thus, although the limiting probabilities do not exist for a periodic chain, its visiting rates g_i do. For the preceding chain, g_is are the solutions of the following set of equations:

$$\begin{cases} g_1 = \frac{2}{3}g_6 + \frac{1}{4}g_7 \\ g_2 = \frac{1}{3}g_6 + \frac{3}{4}g_7 \\ g_3 = \frac{1}{4}g_1 + \frac{2}{5}g_2 \\ g_4 = \frac{1}{4}g_1 + \frac{2}{5}g_2 \\ g_5 = \frac{1}{2}g_1 + \frac{1}{5}g_2 \\ g_6 = \frac{1}{4}g_3 + \frac{1}{2}g_4 + \frac{1}{3}g_5 \\ 1 = g_1 + g_2 + g_3 + g_4 + g_5 + g_6 + g_7 \end{cases}$$

It also can be verified that the d-step transition matrix of a periodic chain having the one-step transition matrix $\mathbf{P}$, as in (6.8.1), is

$$\mathbf{P}^{(d)} = \begin{array}{c} \\ G_1 \\ G_2 \\ \cdots \\ G_{d-1} \\ G_d \end{array} \begin{array}{c} \begin{array}{cccccc} G_1 & G_2 & G_3 & \cdots & G_{d-1} & G_d \end{array} \\ \begin{pmatrix} \mathbf{P}_1 & \mathbf{0} & \mathbf{0} & \cdots & \mathbf{0} & \mathbf{0} \\ \mathbf{0} & \mathbf{P}_2 & \mathbf{0} & \cdots & \mathbf{0} & \mathbf{0} \\ \cdots & \cdots & \cdots & \cdots & \cdots & \cdots \\ \mathbf{0} & \mathbf{0} & \mathbf{0} & \cdots & \mathbf{P}_{d-1} & \mathbf{0} \\ \mathbf{0} & \mathbf{0} & \mathbf{0} & \cdots & \mathbf{0} & \mathbf{P}_d \end{pmatrix} \end{array} \tag{6.8.3}$$

where

$$\begin{cases} \mathbf{P}_1 \equiv \mathbf{D}_1 \cdots \mathbf{D}_d \\ \mathbf{P}_i \equiv \mathbf{D}_i \cdots \mathbf{D}_d \mathbf{D}_1 \cdots \mathbf{D}_{i-1} \quad \text{for all } 2 \le i \le d \end{cases} \tag{6.8.4}$$

For Chain (6.8.2), we can obtain

$$\mathbf{P}^{(3)} = \begin{array}{c} \\ G_1 \\ G_2 \\ G_3 \end{array} \begin{array}{c} \begin{array}{ccc} G_1 & G_2 & G_3 \end{array} \\ \begin{pmatrix} \mathbf{P}_1 & \mathbf{0} & \mathbf{0} \\ \mathbf{0} & \mathbf{P}_2 & \mathbf{0} \\ \mathbf{0} & \mathbf{0} & \mathbf{P}_3 \end{pmatrix} \end{array}$$

with

$$\mathbf{P}_1 \equiv \mathbf{D}_1\mathbf{D}_2\mathbf{D}_3 = \begin{array}{c} \\ 1 \\ 2 \end{array} \begin{array}{c} \begin{array}{cc} 1 & 2 \end{array} \\ \begin{pmatrix} 229/576 & 347/576 \\ 29/72 & 43/72 \end{pmatrix} \end{array}$$

$$\mathbf{P}_2 \equiv \mathbf{D}_2\mathbf{D}_3\mathbf{D}_1 = \begin{array}{c} \\ 3 \\ 4 \\ 5 \end{array} \begin{array}{c} \begin{array}{ccc} 3 & 4 & 5 \end{array} \\ \begin{pmatrix} 111/320 & 111/320 & 49/160 \\ 53/160 & 53/160 & 27/80 \\ 41/120 & 41/120 & 19/60 \end{pmatrix} \end{array}$$

$$\mathbf{P}_3 \equiv \mathbf{D}_3\mathbf{D}_1\mathbf{D}_2 = \begin{array}{c} \\ 6 \\ 7 \end{array} \begin{array}{c} \begin{array}{cc} 6 & 7 \end{array} \\ \begin{pmatrix} 43/120 & 77/120 \\ 349/960 & 611/960 \end{pmatrix} \end{array}$$

The form of matrix $\mathbf{P}^{(d)}$ in (6.8.3) confirms that, if we observe the chain only every d steps, ignoring what happens between, the chain behaves *as a recurrent chain* having d disjoint recurrent classes G_i $(i = 1, 2, \ldots, d)$. (The term *class* is only applicable when we ignore the steps between the two consecutive visits to G_i.) If it starts from G_i, it *always remains within* G_i every d steps.

Note that, if group G_i has k states, then $\mathbf{P}_i$ is a $(k \times k)$ matrix corresponding to those states and each row of which adds up to 1. The matrix $\mathbf{P}_i$ therefore can be considered a transition matrix giving the probabilities of the chain's movements within class G_i every d steps.

It follows that although $\lim_{n\to\infty} \mathbf{P}^{(n)}$ does not exist, $\lim_{m\to\infty} \mathbf{P}^{(md)}$ exists and

$$\lim_{m\to\infty} \mathbf{P}^{(md)} = \begin{array}{c} \\ G_1 \\ G_2 \\ \cdots \\ G_{d-1} \\ G_d \end{array} \begin{array}{c} \begin{array}{cccccc} G_1 & G_2 & G_3 & \cdots & G_{d-1} & G_d \end{array} \\ \begin{pmatrix} \mathbf{P}_1^{(\infty)} & \mathbf{0} & \mathbf{0} & \cdots & \mathbf{0} & \mathbf{0} \\ \mathbf{0} & \mathbf{P}_2^{(\infty)} & \mathbf{0} & \cdots & \mathbf{0} & \mathbf{0} \\ \cdots & \cdots & \cdots & \cdots & \cdots & \cdots \\ \mathbf{0} & \mathbf{0} & \mathbf{0} & \cdots & \mathbf{P}_{d-1}^{(\infty)} & \mathbf{0} \\ \mathbf{0} & \mathbf{0} & \mathbf{0} & \cdots & \mathbf{0} & \mathbf{P}_d^{(\infty)} \end{pmatrix} \end{array}$$

where

$$\mathbf{P}_i^{(\infty)} = \lim_{n\to\infty} \mathbf{P}_i^{(n)} = \lim_{n\to\infty} \mathbf{P}_i^n \quad \text{for all } 1 \le i \le d$$

The rows of $\mathbf{P}_i^{(\infty)}$ are identical to each other, which we shall denote by $\boldsymbol{\pi}_i$. Furthermore, $\boldsymbol{\pi}_i$ can be obtained from Equations (6.3.1) and (6.3.2); that is, $\boldsymbol{\pi}_i = \boldsymbol{\pi}_i \mathbf{P}_i$ and $\boldsymbol{\pi}_i \mathbf{E} = \mathbf{e}$.

For Chain (6.8.2), we obtain:

$$\lim_{m\to\infty} \mathbf{P}^{(3m)} = \begin{array}{c} \\ 1 \\ 2 \\ 3 \\ 4 \\ 5 \\ 6 \\ 7 \end{array} \begin{array}{c} \begin{array}{ccccccc} 1 & 2 & 3 & 4 & 5 & 6 & 7 \end{array} \\ \left(\begin{array}{cc|ccc|cc} \frac{232}{579} & \frac{347}{579} & 0 & 0 & 0 & 0 & 0 \\ \frac{232}{579} & \frac{347}{579} & 0 & 0 & 0 & 0 & 0 \\ \hline 0 & 0 & \frac{328}{965} & \frac{328}{965} & \frac{309}{965} & 0 & 0 \\ 0 & 0 & \frac{328}{965} & \frac{328}{965} & \frac{309}{965} & 0 & 0 \\ 0 & 0 & \frac{328}{965} & \frac{328}{965} & \frac{309}{965} & 0 & 0 \\ \hline 0 & 0 & 0 & 0 & 0 & \frac{349}{965} & \frac{616}{965} \\ 0 & 0 & 0 & 0 & 0 & \frac{349}{965} & \frac{616}{965} \end{array}\right) \end{array}$$

Actually, there is no need to use (6.3.1) and (6.3.2) to solve for all $\mathbf{P}_i^{(\infty)}$. It can be shown that (Problem 6.53):

$$\boldsymbol{\pi}_{i+1} = \boldsymbol{\pi}_i \mathbf{D}_i \tag{6.8.5}$$

For Chain (6.8.2), after obtaining $\boldsymbol{\pi}_1$, we can calculate

$$\boldsymbol{\pi}_2 = \boldsymbol{\pi}_1 \mathbf{D}_1 = \begin{pmatrix} \frac{232}{579} & \frac{347}{579} \end{pmatrix} \begin{pmatrix} 1/4 & 1/4 & 1/2 \\ 2/5 & 2/5 & 1/5 \end{pmatrix} = \begin{pmatrix} \frac{328}{965} & \frac{328}{965} & \frac{309}{965} \end{pmatrix}$$

and

$$\boldsymbol{\pi}_3 = \boldsymbol{\pi}_2 \mathbf{D}_2 = \begin{pmatrix} \frac{328}{965} & \frac{328}{965} & \frac{309}{965} \end{pmatrix} \begin{pmatrix} 1/4 & 3/4 \\ 1/2 & 1/2 \\ 1/3 & 2/3 \end{pmatrix} = \begin{pmatrix} \frac{349}{965} & \frac{616}{965} \end{pmatrix}$$

We also note here that the following limiting matrices exist for all $1 \le k \le d-1$:

$$\begin{aligned} \lim_{m\to\infty} \mathbf{P}^{(md+k)} &= \lim_{m\to\infty} \mathbf{P}^{(md)} \mathbf{P}^k \\ &= \mathbf{P}^k \lim_{m\to\infty} \mathbf{P}^{(md)} \end{aligned} \tag{6.8.6}$$

6.9 FINITE REDUCIBLE CHAINS

For the rest of the chapter, let us turn our attention to finite reducible chains—that is, finite chains in which not all states communicate with each other.

If there are no transient states, then because the chain cannot move from one recurrent class to the other, each recurrent class develops independently and can be studied separately as an irreducible chain. Thus, to avoid triviality, we assume that there is at least one transient state. Then there must be at least one recurrent class; otherwise, the chain will eventually have nowhere to go.

In general, let the chain's state space S comprise a set T of t transient states $t_1, t_2, \ldots, t_t$ and a set R of r recurrent states. The recurrent states can be grouped into ρ disjoint recurrent classes $C_1, C_2, \ldots, C_\rho$.

The chain's transition matrix $\mathbf{P}$ can take the canonical form (4.7.1),

$$\mathbf{P} = \begin{matrix} \\ (r \text{ rows}) \\ (t \text{ rows}) \end{matrix} \begin{matrix} (r \text{ columns}) & (t \text{ columns}) \\ \left(\begin{matrix} \mathbf{R} \\ \mathbf{S} \end{matrix} \right. & \left. \begin{matrix} \mathbf{0} \\ \mathbf{T} \end{matrix} \right) \end{matrix} \tag{6.9.1}$$

To illustrate, consider a chain having transition matrix (4.7.3) reproduced here:

$$\mathbf{P} = \begin{pmatrix} \mathbf{R}_1 & \mathbf{0} & \mathbf{0} \\ \mathbf{0} & \mathbf{R}_2 & \mathbf{0} \\ \mathbf{S}_1 & \mathbf{S}_2 & \mathbf{T} \end{pmatrix}$$

$$= \begin{matrix} \\ 1 \\ 2 \\ 3 \\ 4 \\ 5 \\ 6 \\ 7 \\ 8 \end{matrix} \begin{matrix} \begin{matrix} 1 & 2 & 3 & 4 & 5 & 6 & 7 & 8 \end{matrix} \\ \left(\begin{array}{ccc|cc|ccc} .2 & .2 & .6 & 0 & 0 & 0 & 0 & 0 \\ 0 & .4 & .6 & 0 & 0 & 0 & 0 & 0 \\ .3 & .3 & .4 & 0 & 0 & 0 & 0 & 0 \\ \hline 0 & 0 & 0 & 0 & 1 & 0 & 0 & 0 \\ 0 & 0 & 0 & 1 & 0 & 0 & 0 & 0 \\ \hline .14 & .03 & .13 & .03 & .07 & .4 & .2 & 0 \\ .12 & .08 & 0 & .19 & .11 & .1 & .3 & .1 \\ .03 & .21 & .16 & .02 & .18 & .2 & 0 & .2 \end{array} \right) \end{matrix} \tag{6.9.2}$$

In this chain, we have $\rho = 2$ recurrent classes: $C_1 \equiv \{1, 2, 3\}$ and $C_2 \equiv \{4, 5\}$.

It is easy to show that *all states in each recurrent class are lumpable* (§2.17). Thus, the chain is lumpable with respect to the partition $\tilde{S} \equiv \{C_1, C_2, \ldots, C_\rho, t_1, t_2, \ldots, t_t\}$. Furthermore, the transition matrix of the lumped chain takes the form:

$$\tilde{\mathbf{P}} = \begin{matrix} \\ (\rho \text{ rows}) \\ (t \text{ rows}) \end{matrix} \begin{matrix} (\rho \text{ columns}) & (t \text{ columns}) \\ \left(\begin{matrix} \mathbf{I} \\ \tilde{\mathbf{S}} \end{matrix} \right. & \left. \begin{matrix} \mathbf{0} \\ \mathbf{T} \end{matrix} \right) \end{matrix}$$

The (i, ℓ)-element of $\tilde{\mathbf{S}}$, corresponding to the transient state i and recurrent class C_ℓ ($\ell = 1, 2, \ldots, \rho$), is obtained by summing the elements in the i-row of matrix $\mathbf{S}_\ell$, as defined in Equation (4.7.2).

Chain (6.9.2) is lumpable with respect to the partition $\{C_1, C_2, 6, 7, 8\}$. Its lumped transition matrix is:

$$\tilde{\mathbf{P}} = \begin{array}{c} \\ C_1 \\ C_2 \\ 6 \\ 7 \\ 8 \end{array} \begin{array}{c} \begin{array}{ccccc} C_1 & C_2 & 6 & 7 & 8 \end{array} \\ \left(\begin{array}{cc|ccc} 1 & 0 & 0 & 0 & 0 \\ 0 & 1 & 0 & 0 & 0 \\ \hline .3 & .1 & .4 & .2 & 0 \\ .2 & .3 & .1 & .3 & .1 \\ .4 & .2 & .2 & 0 & .2 \end{array}\right) \end{array} \qquad (6.9.3)$$

Here, for example,

$$p_{6C_1} = .14 + .03 + .13 = .3$$

Note that $\tilde{\mathbf{P}}$ *has the form of an absorbing chain.*

6.10 TRANSIENT BEHAVIOR

Unlike regular chains, the behavior of a reducible chain is dependent on its initial conditions. For example, if the chain starts from a recurrent class, it will stay in that class forever. In this case, the chain can be studied as an irreducible one, as before. The only case we need to study is when the chain starts from a transient state. In this case, before the chain reaches a recurrent class, there is no need to distinguish the states within the recurrent classes. Lumping the states in each recurrent class as we did earlier allows us to study the behavior of the chain before its arrival at a recurrent class as that of an *absorbing chain.*

As an illustration, for Chain (6.9.2), to find the expected number of visits to the transient state 8, starting from the transient state 7, we simply obtain the fundamental matrix for $\tilde{\mathbf{P}}$ in (6.9.3):

$$\tilde{\mathbf{U}} = \begin{pmatrix} (1-.4) & -.2 & 0 \\ -.1 & (1-.3) & -.1 \\ -.2 & 0 & (1-.2) \end{pmatrix}^{-1} = \begin{array}{c} \\ 6 \\ 7 \\ 8 \end{array} \begin{array}{c} \begin{array}{ccc} 6 & 7 & 8 \end{array} \\ \begin{pmatrix} 140/79 & 40/79 & 05/79 \\ 25/79 & 120/79 & 15/79 \\ 35/79 & 10/79 & 100/79 \end{pmatrix} \end{array}$$

giving $\mathcal{E}[v_{78}] = 15/79$.

6.11 CLASS-ABSORPTION PROBABILITIES

After the chain reaches one of the recurrent classes, it will stay there forever, and its limiting behavior is independent of which state of that class it visits first. The important questions related to the limiting behavior of the chain therefore are not those concerning which *state* of the recurrent class the chain visits first, but rather which recurrent *class* it visits first. In other words, we want to know f_{iC_ℓ}, the probability that, starting from a transient state i, the chain will reach a particular recurrent class C_ℓ (for $\ell = 1, 2, \ldots, \rho$) rather than any other recurrent class.

Treating each recurrent class as an absorbing state, the problem is essentially the same as that for absorption probabilities studied in §5.5. For our Chain (6.9.2), Equation (5.5.3) yields:

$$\begin{pmatrix} f_{6C_1} & f_{6C_2} \\ f_{7C_1} & f_{7C_2} \\ f_{8C_1} & f_{8C_2} \end{pmatrix} = \tilde{\mathbf{U}}\tilde{\mathbf{S}}$$

$$= \begin{pmatrix} 140/79 & 40/79 & 5/79 \\ 25/79 & 120/79 & 15/79 \\ 35/79 & 10/79 & 100/79 \end{pmatrix} \begin{pmatrix} .3 & .1 \\ .2 & .3 \\ .4 & .2 \end{pmatrix}$$

$$= \begin{pmatrix} 520/790 & 270/790 \\ 375/790 & 415/790 \\ 525/790 & 265/790 \end{pmatrix}$$

6.12 LONG-TERM VISITING RATE

Our definition of the long-term visiting rate g_{ij} remains the same as in (6.4.2). However, for finite reducible chains, this quantity is dependent on the initial conditions.

1. If states i and j belong to different recurrent classes, the chain will never visit state j from state i. Hence, $g_{ij} = 0$.

2. If states i and j belong to the same recurrent classes, we can treat this class alone as an irreducible chain. The visiting rate g_{ij} becomes g_j, which is independent of the starting state, and can be obtained by using (6.4.6) and (6.4.4).

With Chain (6.9.2), treating class $C_1 \equiv \{1, 2, 3\}$ alone as an irreducible chain yields

$$\begin{cases} g_{11} = g_{21} = g_{31} = g_1 = 3/16 \\ g_{12} = g_{22} = g_{32} = g_2 = 5/16 \\ g_{13} = g_{23} = g_{33} = g_3 = 8/16 \end{cases}$$

Although class $C_2 \equiv \{4, 5\}$ is periodic with period 2, the visiting rates still exist and are given by

$$\begin{cases} g_{44} = g_{54} = g_4 = 1/2 \\ g_{45} = g_{55} = g_5 = 1/2 \end{cases}$$

3. If state i is transient and state j is recurrent and belongs to class C_ℓ, then

- **a.** the chain will end up within C_ℓ with probability f_{iC_ℓ}. When this occurs, the visiting rate to state $j \in C_\ell$ is g_j. (Why?)
- **b.** the chain will never visit C_ℓ with probability $1 - f_{iC_\ell}$. When this occurs, the visiting rate to state $j \in C_\ell$ is 0.

Hence,

$$g_{ij} = f_{iC_\ell} g_j \quad \text{for all } i \in T \text{ and } j \in C_\ell$$

4. If both state i and state j are transient, then $g_{ij} = 0$.

For Chain (6.9.2), the visiting rates can now be presented in the following matrix:

$$[g_{i,j}]_{i,j\in S} =$$

$$\begin{array}{c} \\ 1\\2\\3\\4\\5\\6\\7\\8 \end{array}
\begin{array}{c}
\begin{array}{ccc|cc|ccc} 1 & 2 & 3 & 4 & 5 & 6 & 7 & 8 \end{array} \\
\left(\begin{array}{ccc|cc|ccc}
\frac{3}{16} & \frac{5}{16} & \frac{8}{16} & 0 & 0 & 0 & 0 & 0 \\
\frac{3}{16} & \frac{5}{16} & \frac{8}{16} & 0 & 0 & 0 & 0 & 0 \\
\frac{3}{16} & \frac{5}{16} & \frac{8}{16} & 0 & 0 & 0 & 0 & 0 \\
\hline
0 & 0 & 0 & \frac{1}{2} & \frac{1}{2} & 0 & 0 & 0 \\
0 & 0 & 0 & \frac{1}{2} & \frac{1}{2} & 0 & 0 & 0 \\
\hline
\left(\frac{520}{790}\right)\left(\frac{3}{16}\right) & \left(\frac{520}{790}\right)\left(\frac{5}{16}\right) & \left(\frac{520}{790}\right)\left(\frac{8}{16}\right) & \left(\frac{270}{790}\right)\left(\frac{1}{2}\right) & \left(\frac{270}{790}\right)\left(\frac{1}{2}\right) & 0 & 0 & 0 \\
\left(\frac{375}{790}\right)\left(\frac{3}{16}\right) & \left(\frac{375}{790}\right)\left(\frac{5}{16}\right) & \left(\frac{375}{790}\right)\left(\frac{8}{16}\right) & \left(\frac{415}{790}\right)\left(\frac{1}{2}\right) & \left(\frac{415}{790}\right)\left(\frac{1}{2}\right) & 0 & 0 & 0 \\
\left(\frac{525}{790}\right)\left(\frac{3}{16}\right) & \left(\frac{525}{790}\right)\left(\frac{5}{16}\right) & \left(\frac{525}{790}\right)\left(\frac{8}{16}\right) & \left(\frac{265}{790}\right)\left(\frac{1}{2}\right) & \left(\frac{265}{790}\right)\left(\frac{1}{2}\right) & 0 & 0 & 0
\end{array}\right)
\end{array}$$

6.13 LIMITING DISTRIBUTION

All reducible chains eventually have to end up within one of their recurrent classes. Thus, $\lim_{n\to\infty} p_{ij}^{(n)} = 0$ if state j is transient.

If state j belongs to an *aperiodic* recurrent class C_ℓ, then

1. the chain will end up within C_ℓ with probability f_{iC_ℓ}. When this occurs, the limiting probability $\lim_{n\to\infty} p_{ij}^{(n)}$ is the same as if the chain starts from within C_ℓ; that is, it is the same as the visiting rate g_j.
2. the chain will never visit C_ℓ with probability $1 - f_{iC_\ell}$. When this occurs, $\lim_{n\to\infty} p_{ij}^{(n)} = 0$.

Hence, for an aperiodic recurrent class C_ℓ,

$$\lim_{n\to\infty} p_{ij}^{(n)} = f_{iC_\ell} g_j = g_{ij} \quad \text{for all } i \in T \text{ and } j \in C_\ell$$

In our present example, since state 1 is recurrent and aperiodic,

$$\lim_{n\to\infty} p_{61}^{(n)} = g_{61} = \left(\frac{520}{790}\right)\left(\frac{3}{16}\right)$$

On the other hand, if state j belongs to a recurrent and *periodic* class C_ℓ, $\lim_{n\to\infty} p_{ij}^{(n)}$ does not exist. In our example, since state 5 is recurrent and periodic with period 2, $\lim_{n\to\infty} p_{45}^{(n)}$ and $\lim_{n\to\infty} p_{85}^{(n)}$ do not exist, although $\lim_{n\to\infty} p_{45}^{(2n)}$ and $\lim_{n\to\infty} p_{85}^{(2n)}$ do.

We summarize the results for our current example in the following matrix:

$$\mathbf{P}^{(\infty)} = \begin{array}{c} \\ 1 \\ 2 \\ 3 \\ 4 \\ 5 \\ 6 \\ 7 \\ 8 \end{array} \begin{array}{c} \begin{array}{cccccccc} 1 & 2 & 3 & 4 & 5 & 6 & 7 & 8 \end{array} \\ \left(\begin{array}{ccc|cc|ccc} \frac{3}{16} & \frac{5}{16} & \frac{8}{16} & 0 & 0 & 0 & 0 & 0 \\ \frac{3}{16} & \frac{5}{16} & \frac{8}{16} & 0 & 0 & 0 & 0 & 0 \\ \frac{3}{16} & \frac{5}{16} & \frac{8}{16} & 0 & 0 & 0 & 0 & 0 \\ \hline 0 & 0 & 0 & * & * & 0 & 0 & 0 \\ 0 & 0 & 0 & * & * & 0 & 0 & 0 \\ \hline \left(\frac{520}{790}\right)\left(\frac{3}{16}\right) & \left(\frac{520}{790}\right)\left(\frac{5}{16}\right) & \left(\frac{520}{790}\right)\left(\frac{8}{16}\right) & * & * & 0 & 0 & 0 \\ \left(\frac{375}{790}\right)\left(\frac{3}{16}\right) & \left(\frac{375}{790}\right)\left(\frac{5}{16}\right) & \left(\frac{375}{790}\right)\left(\frac{8}{16}\right) & * & * & 0 & 0 & 0 \\ \left(\frac{525}{790}\right)\left(\frac{3}{16}\right) & \left(\frac{525}{790}\right)\left(\frac{5}{16}\right) & \left(\frac{525}{790}\right)\left(\frac{8}{16}\right) & * & * & 0 & 0 & 0 \end{array}\right) \end{array}$$

where * means that the quantity does not exist.

6.14 SUMMARY

1. In a regular (that is, finite, irreducible, and aperiodic) chain, the limiting probabilities π_i are equal to the long-term visiting rates g_i and are the solutions of (6.3.1) and (6.3.2).
2. In a finite, irreducible, and periodic chain, the limiting probabilities π_i do not exist, but the long-term visiting rates g_i do, and they are the solutions of (6.4.6) and (6.4.4).
3. In a finite irreducible chain, the expected first reaching times $\mathcal{E}[n_{ij}]$ are the solutions of (6.5.1); the expected return times $\mathcal{E}[n_i]$ are the reciprocal of the long-term visiting rate g_i.
4. In a regular chain, the limiting probability π_k and the long-term visiting rates g_i are the ratio of the expected number of visits to state k during a regenerative cycle from state i and the expected cycle length from state i.
5. A periodic chain with period $d > 1$ has a limiting distribution only if we study it every d steps.
6. The proportions of visits and the limiting probabilities of a finite reducible chain can be obtained by lumping together all states in each recurrent class.

PROBLEMS

6.1 If π is the limiting distribution of a finite chain having transition matrix $\mathbf{P}$, explain why it is also the limiting distribution of the chain having transition matrix $\mathbf{P}^n$ for all $n < \infty$.

6.2 Refer to Problem 2.13 in which John is married to Mary and the probability that he or she is happy at day n is $(1 + H_{n-1})/4$, where H_n is the number of happy persons among John and Mary during day n. What is the fraction of days that both are happy?

6.3 Suppose a particle is moving among five positions -2, -1, 0, 1, and 2, according to the following transition matrix:

$$\mathbf{P} = \begin{array}{c} \\ -2 \\ -1 \\ 0 \\ 1 \\ 2 \end{array} \begin{array}{c} \begin{array}{ccccc} -2 & -1 & 0 & 1 & 2 \end{array} \\ \left(\begin{array}{ccccc} & 1/3 & 1/3 & 1/3 & \\ 1/2 & & 1/2 & & \\ & 1/2 & & 1/2 & \\ & & 1/2 & & 1/2 \\ & 1/3 & 1/3 & 1/3 & \end{array} \right) \end{array}$$

Use a symmetrical property of the chain to find its limiting distribution.

6.4 Suppose the rat-maze experiment introduced in §2.10 is not terminated when the rat visits room S or F. Instead, it is allowed to run from these rooms into one of the two adjacent rooms. Assume the probability that the rat takes a particular door is $1/2$. Use a symmetrical property of the chain to find the long-term proportion of times that it visits each room.

6.5 Equation (6.3.1) alone yields $N - 1$ mutually independent equations with N unknowns and thus has an infinite number of solutions. One solution vector $\mathbf{y} \equiv [y_i]_{i \in S}$ can be obtained by giving an *arbitrary* value to one unknown and then solving for the others. Since the limiting distribution $\boldsymbol{\pi}$ is also a solution, we must have $\boldsymbol{\pi} = c\mathbf{y}$, where $c = 1/\sum y_i$ is a constant which normalizes $\mathbf{y}$ so that $\sum \pi_i = 1$. By doing this, we effectively reduce the number of unknowns from N to $N - 1$. (Let us call this the *normalization method*.)

a. Consider the taxicab example. Eliminating the first equation of (6.3.4) gives two mutually independent equations having three unknowns. Now assign to π_A an arbitrary value, say, 10. This results in two unknowns with two equations. Solve these two equations for the values of π_B and π_C. Finally, normalize the results to obtain the correct values of π_A, π_B, and π_C.

b. Use the normalization method to find the limiting distribution of a chain with

$$\mathbf{P} = \begin{array}{c} \\ 1 \\ 2 \\ 3 \end{array} \begin{array}{c} \begin{array}{ccc} 1 & 2 & 3 \end{array} \\ \left(\begin{array}{ccc} 1/3 & 2/3 & 0 \\ 0 & 3/4 & 1/4 \\ 1/5 & 2/5 & 2/5 \end{array} \right) \end{array}$$

c. Use the normalization method to find the limiting distribution of a chain with

$$\mathbf{P} = \begin{array}{c} \\ A \\ B \\ C \\ D \end{array} \begin{array}{c} \begin{array}{cccc} A & B & C & D \end{array} \\ \left(\begin{array}{cccc} .1 & .3 & .2 & .4 \\ .3 & .1 & .3 & .3 \\ .3 & .1 & .1 & .5 \\ .4 & .1 & .2 & .3 \end{array} \right) \end{array}$$

6.6 There are three brands X, Y, and Z in a hypothetical market. A company buys one of these brands on a regular basis. Let X_n be the brand it buys at the nth order. We assume the discrete-time chain $\{X_n\}_{n=0,1,2,\ldots}$ is Markovian with the following transition matrix:

$$
\mathbf{P} = \begin{array}{c} \\ X \\ Y \\ Z \end{array} \begin{array}{c} \begin{array}{ccc} X & Y & Z \end{array} \\ \begin{pmatrix} .5 & .2 & .3 \\ .1 & .6 & .3 \\ .2 & .4 & .4 \end{pmatrix} \end{array}
$$

Find the proportion of times that the company buys brand X, Y, or Z.

6.7 In a sequence of dependent trials, the probability of a success at the nth trial is $(k+1)/(k+2)$, where k is the number of successes in the previous two trials. Find the limiting probability of a success.

6.8 The transition matrix for the duration of eruptions of the Old Faithful geyser was given in (2.13.2). Find the limiting probability of a high eruption.

6.9 Verify the results calculated by Styan and Smith (1964) for the equilibrium market share in §6.3.

6.10 Verify the results calculated by Bourne (1976) for evaluating land use policies in §6.3.

6.11 Verify the results calculated by Dryden (1969) for the stock price movements in §6.4 and §6.5.

6.12 Verify the results calculated by Segal (1985) for migration patterns in §6.6.

6.13 Taylor and Reid (1987) recorded the status of 258 teachers in a school district in October for each of three academic years 1983–1986. The status can be: (N) new, (C) continuing, (R) resigned, (T) retired, (L) on leave, (S) sabbatical, (I) ill, and (D) deceased. The transition matrix obtained is as follows:

$$
\mathbf{P} = \begin{array}{c} \\ N \\ C \\ R \\ T \\ L \\ S \\ I \\ D \end{array} \begin{array}{c} \begin{array}{cccccccc} N & C & R & T & L & S & I & D \end{array} \\ \begin{pmatrix} & .773 & .194 & & & & .033 & \\ & .896 & .040 & .016 & .022 & .007 & .017 & .002 \\ 1 & & & & & & & \\ 1 & & & & & & & \\ & .178 & .533 & & .289 & & & \\ & 1 & & & & & & \\ & .806 & & .139 & & & .055 & \\ 1 & & & & & & & \end{pmatrix} \end{array}
$$

Note that a resigned, retired, or deceased teacher is replaced by a new teacher. Assume the average cost for a new teacher is \$13,640; a continuing teacher is \$19,840; a resigned teacher is −\$4562; a retired teacher is −\$11,160; a teacher on leave is −\$6200; a teacher on sabbatical is \$23,560; an ill teacher is \$17,301; and a deceased teacher is −\$11,160. Assume no inflation. What is the limiting expected cost for a teacher?

6.14 Land uses can change to and from the following states: (1) water, (2) forest, (3) grassland, (4) agriculture, (5) residential, and (6) commercial/residential. Find the limiting distribution from the following transition matrix for San Juan Island (Washington), as studied by Robinson (1980),

$$\mathbf{P} = \begin{array}{c} \\ 1 \\ 2 \\ 3 \\ 4 \\ 5 \\ 6 \end{array} \begin{array}{c} \begin{array}{cccccc} 1 & 2 & 3 & 4 & 5 & 6 \end{array} \\ \begin{pmatrix} .915 & .010 & .040 & .010 & .025 & 0 \\ .006 & .946 & .014 & .015 & .015 & .004 \\ .013 & .050 & .845 & .057 & .032 & .003 \\ .002 & .017 & .014 & .948 & .014 & .005 \\ .015 & 0 & 0 & .045 & .932 & .008 \\ 0 & 0 & 0 & .030 & 0 & .970 \end{pmatrix} \end{array}$$

Give an interpretation of the limiting distribution.

6.15 For the taxicab example, find $\mathcal{E}[n_{ji}]$ for all $i, j = A, B, C$.

6.16 For the chain defined in Problem 6.5b,

a. use (6.5.1) to find the values of $\mathcal{E}[n_{i1}]$ for all $i = 1, 2, 3$
b. make state 1 absorbing to find the values of $\mathcal{E}[n_{i1}]$ for all $i = 1, 2, 3$
c. use the regenerative method to find the limiting distribution

6.17 For the chain defined in Problem 6.5c,

a. use (6.5.1) to find the values of $\mathcal{E}[n_{iA}]$ for all $i = A, B, C, D$
b. make state A absorbing to find the values of $\mathcal{E}[n_{iA}]$ for all $i = A, B, C, D$
c. use the regenerative method to find the limiting distribution

6.18 For the chain with three brands X, Y, and Z defined in Problem 6.6,

a. use (6.5.1) to find the values of $\mathcal{E}[n_{iX}]$ for all $i = X, Y, Z$
b. make state X absorbing to find the values of $\mathcal{E}[n_{iX}]$ for all $i = X, Y, Z$
c. use the regenerative method to find the limiting distribution

6.19 From the transition matrix obtained by Girard and Sager (1987) for reproductive cycles of female rats in §6.7, find

a. the limiting distribution
b. the cycle length
c. the expected duration of diestrus in each cycle
d. the expected duration of proestrus in each cycle
e. the expected duration of estrus in each cycle

6.20 By using (6.3.1) and (6.3.2), find the limiting distribution $\boldsymbol{\pi}$ for a two-state Markov chain introduced in Problem 3.24. Do the results agree with those obtained there?

6.21 A transition matrix is called *doubly stochastic* if all its columns add up to 1, besides the fact that all its rows add up to 1. In a doubly stochastic chain having N states, verify that $\pi_i = 1/N$ for all $i \in S$ (see also Problem 4.11).

6.22 Consider a drunkard who is walking on a circle. He can only stand on one of five positions, equally spaced on the circle, labeled 1, 2, 3, 4, 5. At any step, the probability that he makes one step in the counterclockwise direction is p and one step in the clockwise direction is $1 - p$. Find the limiting distribution of his location.

6.23 Let Y_n be the sum of n independent rolls of a fair die. Find the limiting distribution for $\{X_n\}_{n=0,1,2,\ldots}$, where $X_n = Y_n \bmod (8)$. *Hint:* Use Problem 2.11.

6.24 Let $\{X_n\}_{n=0,1,2,\ldots}$ be a sequence of independent and identically distributed random variables taking values in $\{0, 1, 2, 3, 4\}$, with the common distribution $p_i \equiv \Pr\{X_i = i\}$ for all $0 \le i \le 4$. Let $Y_0 \equiv 0$ and $Y_{n+1} \equiv Y_n + X_n$. What is the expected number of steps until Y_n is divisible by 5? *Hint:* Use Problem 2.10.

6.25 A card in a stack of n cards is randomly selected and then put on top of the stack. If we continue in this manner for a long time, show that the stack is well shuffled; that is, show that any one of the $n!$ possible orders is equally likely to occur.

6.26 A man has r umbrellas, which he leaves either at home or at work. At the beginning of each day, if it rains, he takes one, if available at home, to work. At the end of each day, if it rains, he takes one, if available at work, home. If it is not raining, he does not take any. Assume that, independent of the past, it rains with probability p at the beginning and at the end of each day. Define an appropriate Markov chain to find the long-term proportion of times he gets wet.

6.27 N men, named $1, 2, \ldots, N$ take turns playing a game. As long as a player wins the game, he continues to play. When player i loses, he lets player $i + 1$ play $(i = 1, 2, \ldots, N - 1)$. When player N loses, he lets player 1 play. Let p_i be the probability that player i wins a game $(1 \le i \le N)$. Let g_i be the long-term proportion of times that player i plays the games. Express g_i in terms of p_is by constructing a Markov chain and then

a. using Equation (6.4.4)

b. using the regenerative arguments

6.28 Consider a Markov chain $\{X_n\}_{n=0,1,2,\ldots}$ having state space $S \equiv \{0, 1, 2, \ldots, N\}$. At each transition, it always moves to a different state with equal probability. In other words, for all $0 \le i \le N$,

$$p_{ij} = \begin{cases} 0 & \text{for } j = i \\ 1/N & \text{for all } j \ne i \end{cases}$$

a. Show that the expected return time to state i is $\mathcal{E}[n_i] = N + 1$. *Hint:* Use Problem 5.35.

b. Thus, show that $\pi_i = 1/(N + 1)$ for all i. Is this consistent with the fact that the chain is double stochastic? (Problem 6.21)

6.29 Consider a Markov chain with state space $S \equiv \{0, 1, 2, \ldots, N\}$. If it is at state i, it may remain at state i or move to another state with equal probability p_i; that is,

$$p_{ij} = \begin{cases} 1 - Np_i & \text{for } j = i \\ p_i & \text{for all } j \ne i \end{cases}$$

where $0 \le Np_i \le 1$ for all $0 \le i \le N$.

a. Show that the expected return time to state i is

$$\mathcal{E}[n_i] = p_i \sum_{k=0}^{N} \frac{1}{p_k}$$

Hint: Use Problem 5.36.

b. Hence, find the limiting distribution. Compare with the results in Problem 3.26.
c. Also obtain the limiting distribution from Equation (6.3.1).

6.30 (Whitaker, 1978) There are $N + 1$ competing brands in the market labeled $0, 1, 2, \ldots, N$. Let X_n be the brand bought at time n. It was proposed that chain $\{X_n\}_{n=0,1,2,\ldots}$ has the following transition probabilities:

$$p_{ij} = \begin{cases} p_i w_j & \text{for } i \neq j \\ 1 - p_i(1 - w_i) & \text{for } i = j \end{cases}$$

where $0 < p_i \leq 1$ for all $i \in S$ and $\sum_{j=0}^{N} w_i = 1$.

a. Find the expected return time to brand 0. Hence, show that

$$\pi_i = \frac{w_i/p_i}{\sum_{j=0}^{N}(w_j/p_j)} \qquad \text{for all } i \in S$$

which is proportional to both w_i and $1/p_i$. *Hint:* Use Problem 5.38.

b. Also obtain the limiting distribution from Equation (6.3.1).

c. Assume $p_1 = .30$, $p_2 = .60$, $p_3 = .90$, $w_1 = .30$, $w_2 = .50$, $w_3 = .20$. Find the limiting brand shares.

6.31 Three people A, B and C are playing an infinite set of games as follows. At game n, two players play against each other while one watches. Whoever loses at game n will watch the other two playing at game $n + 1$. Let p_{ij} be the probability that player i beats player j in a game ($i, j = A, B, C$; $p_{ij} + p_{ji} = 1$).

a. Find the expected number of games between player i's two consecutive turns to watch. *Hint:* Use Problem 5.21.

b. Hence, show that the long-term proportion of games that A is watching is

$$g_A = \frac{1 - p_{AB}p_{AC}}{3 - p_{AB}p_{AC} - p_{BA}p_{BC} - p_{CA}p_{CB}}$$

c. Also obtain g_A from Equation (6.4.4).

6.32 Items are produced by a machine one at a time. Let the probability that an item is defective be p. Consider the following sampling plan. In stage A, each item is inspected. If s consecutive nondefective items are found, the plan changes to stage B, in which each item is inspected with probability α. As soon as a defective item is found, the sampling plan reverts back to stage A. This pattern repeats indefinitely.

If the nth item is found defective, let $X_n = 0$; otherwise, let X_n be the number of previously inspected items found to be nondefective. If $X_n > s$, let $X_n = s$.

a. Use Equation (6.4.4) to find g_0 and hence find the long-term proportion of items being inspected. *Hint:* Use Problem 2.12.

b. Let the regenerative points be at the start of plan A. Use the regenerative argument to find the long-term proportion of items being inspected. *Hint:* Use Problem 1.40.

6.33 N people, named $1, 2, \ldots, N$, play the following game. They perform a sequence of *multinomial* independent trials in which each trial has N outcomes $1, 2, \ldots, N$. In each trial, let p_i be the probability that the outcome is i ($\sum_{i=1}^{N} p_i = 1$). Let k be a constant. If outcome i occurs k times in a row, then player i wins.

a. What is the long-term rate that player i wins? *Hint:* Use Problem 1.40.

b. What is the long-term rate that someone wins?

c. Hence, find the expected number of tosses between two consecutive wins.

d. What is the probability that player i is the first one to win the game? Compare with the answer in Problem 5.20b.

e. When $N = 2$, draw a graph showing the limiting probability that player i wins as a function of k and p.

f. When $N = 3$; $k = 3$; and $p_1 = .2$, $p_2 = .6$, and $p_3 = .1$, verify the answer in Problem 5.10.

6.34 Two competing treatments A and B are being compared by testing on patients. It is desirable that the number of patients exposed to the inferior treatment be kept minimum. In the *two-armed bandit rule* (Robbins, 1952), we first choose treatment A or B at random. If A is tested in the nth trial and shows a success (S), we will test A again in the $(n + 1)$th trial; if it shows a failure (F), we will switch to B. A similar rule is applied to B. Let α and β be the probability of success in each patient by treatments A and B, respectively.

a. Consider the chain having state space $\{(A, S), (A, F), (B, S), (B, F)\}$. Find its transition matrix.

b. Show that the long-term proportion of success is

$$g^{(1)} = \frac{\alpha + \beta - 2\alpha\beta}{2 - (\beta + \alpha)}$$

i. by using Equation (6.4.4)

ii. by using the regenerative arguments

c. Had we known that treatment A is better (that is, $\alpha > \beta$), we would use A alone. The long-term proportion of success in this case would be $g^{(2)} = \alpha$. The value of $g^{(2)} - g^{(1)}$ is the price we have to pay for our ignorance of which treatment is better, and hence we must use the two-armed bandit rule. Is this value positive? What is its maximum value?

d. On the other hand, if we randomly pick a treatment and stick to it regardless of the result of the previous trial, what is the long-term proportion of success $g^{(3)}$ in this use? Is $g^{(1)} - g^{(3)}$ positive? What is its maximum value?

6.35 Consider a Markov chain $\{X_n\}_{n=0,1,2,...}$ having transition matrix $\mathbf{P}$ in which $p_{ii} \neq 0$ for some $i \in S$. Thus, the chain may remain at the same state during a transition. Consider now a chain $\{Y_n\}_{n=0,1,2,...}$, which records the movements of $\{X_n\}_{n=0,1,2,...}$ only when it changes states. For example, if the latter follows the path $a - b - b - d - a - a - c$, then the former follows $a - b - d - a - c$. (See also Problem 5.34.)

a. Explain why the transition matrix for $\{Y_n\}_{n=0,1,2,...}$ is

$$\mathbf{Q} = (\mathbf{I} - \mathbf{P}_d)^{-1}(\mathbf{P} - \mathbf{P}_d)$$

where $\mathbf{P}_d$ is a diagonal matrix, the (i, i)-element of which is p_{ii}.

b. Hence, verify that the limiting distribution of $\{Y_n\}_{n=0,1,2,\ldots}$ is *proportional* to

$$\boldsymbol{\rho} = \boldsymbol{\pi}_X(\mathbf{I} - \mathbf{P}_d)$$

where $\boldsymbol{\pi}_X$ is the limiting distribution of $\{X_n\}_{n=0,1,2,\ldots}$.

c. Use the taxicab problem to illustrate the result.

6.36 Consider a Markov chain with:

$$\mathbf{P} = \begin{array}{c} \\ 0 \\ 1 \\ 2 \\ \cdots \\ N-1 \\ N \end{array} \begin{array}{c} \begin{array}{cccccc} 0 & 1 & 2 & \cdots & N-1 & N \end{array} \\ \left(\begin{array}{cccccc} q_0 & s_0 & & \cdots & & \\ q_1 & r_1 & s_1 & \cdots & & \\ q_2 & & r_2 & \cdots & & \\ \cdots & \cdots & \cdots & \cdots & \cdots & \cdots \\ q_{N-1} & & & \cdots & r_{N-1} & s_{N-1} \\ q_N & & & \cdots & & r_N \end{array} \right) \end{array}$$

a. Let $r_i = 0$ for all $1 \le i \le N$. Use Equation (6.4.4) to show that its limiting distribution is

$$\pi_i = \frac{s_0 s_1 \cdots s_{i-1}}{1 + s_0 + s_0 s_1 + \cdots + s_0 s_1 \cdots s_{N-1}} \qquad \text{for all } 1 \le i \le N$$

b. Now let $0 < r_i < 1$ for all $1 \le i \le N$. Show that its limiting distribution is

$$\pi_i = \frac{\delta^{(i)}}{1 + \delta^{(1)} + \delta^{(2)} + \cdots + \delta^{(N)}} \qquad \text{for all } 0 \le i \le N$$

where $\delta^{(0)} \equiv 1$ and

$$\delta^{(i)} \equiv \frac{s_0}{(1 - r_1)} \frac{s_1}{(1 - r_2)} \cdots \frac{s_{i-1}}{(1 - r_i)} \qquad \text{for all } i \ge 1$$

i. by using Equation (6.4.4).
ii. by using the result obtained in part a and Problem 6.35.
iii. by using a regenerative cycle from state 0. *Hint:* Use Problem 5.58.
iv. by using a regenerative cycle from state i ($1 \le i \le N$). *Hint:* Use Problem 5.59.

c. Verify the results in Problem 6.32.

d. Verify the results in Problem 6.27.

6.37 Consider an N-state Markov chain with:

$$\mathbf{P} = \begin{array}{c} \\ 1 \\ 2 \\ 3 \\ \cdots \\ N-1 \\ N \end{array} \begin{array}{c} \begin{array}{ccccc} 1 & 2 & \cdots & N-1 & N \end{array} \\ \left(\begin{array}{ccccc} \alpha_1 & \alpha_2 & \cdots & \alpha_{N-1} & \alpha_N \\ 1 & & \cdots & & \\ & 1 & \cdots & & \\ \cdots & \cdots & \cdots & \cdots & \cdots \\ & & \cdots & & \\ & & \cdots & 1 & \end{array} \right) \end{array}$$

This chain can be used to represent the *remaining life* r_n of an instrument having finite life $L \le N$ at time n. If its remaining life at time n is $r_n > 1$, its remaining life is reduced by 1 at time $n+1$. If its remaining life at time n is 1, it will expire at time $n+1$ and be replaced by an equivalent instrument, having the same life distribution.

a. Find its limiting distribution

i. by using Equation (6.4.4)

ii. by using the regenerative arguments

b. Thus, show that the limiting expected remaining life is

$$\lim_{n\to\infty} \mathcal{E}[r_n] = \frac{\mathcal{E}[L^2]}{2\mathcal{E}[L]} + \frac{1}{2}$$

6.38 Consider a *finite random walk* with

$$\mathbf{P} = \begin{array}{c} \\ 0 \\ 1 \\ 2 \\ \cdots \\ N-1 \\ N \end{array} \begin{array}{c} \begin{array}{cccccccc} 0 & 1 & 2 & 3 & \cdots & N-2 & N-1 & N \end{array} \\ \left(\begin{array}{cccccccc} r_0 & s_0 & & & \cdots & & & \\ q_1 & r_1 & s_1 & & \cdots & & & \\ & q_2 & r_2 & s_2 & \cdots & & & \\ \cdots & \cdots & \cdots & \cdots & \cdots & \cdots & \cdots & \cdots \\ & & & & \cdots & q_{N-1} & r_{N-1} & s_{N-1} \\ & & & & \cdots & & q_N & r_N \end{array} \right) \end{array}$$

where $r_0 + s_0 = 1$, $q_N + r_N = 1$, and $q_i + r_i + s_i = 1$ for all $0 < i < N$. Assume we have the *reflecting barriers* at 0 and N, or $r_0, r_N < 1$.

a. Explain why the chain is aperiodic if at least one r_i is greater than 0.

b. Let $\delta^{(0)} \equiv 1$ and

$$\delta^{(i)} \equiv \frac{s_0}{q_1}\frac{s_1}{q_2}\cdots\frac{s_{i-1}}{q_i} \quad \text{for all } 1 \le i$$

Show that its limiting distribution is

$$\pi_i = \frac{\delta^{(i)}}{1 + \delta^{(1)} + \delta^{(2)} + \cdots + \delta^{(N)}} \quad \text{for all } 0 \le i \le N$$

i. by using Equation (6.4.4)

ii. by using Problem 5.54 and the regenerative arguments

6.39 Consider the Ehrenfest model studied in Problem 2.3. Use Problem 6.38 to show that the limiting distribution of the number of balls in urn A is

$$\pi_j = \binom{2N}{j} 2^{-2N} \quad \text{for all } 0 \le j \le 2N$$

Compare with the result obtained in Problem 3.13. *Hint:* $\sum_{i=0}^{N} \binom{N}{i} = 2^N$.

6.40 Use Problem 6.38 to find the limiting distributions for the chains in Problem 2.4. *Hint:* $\sum_{i=0}^{N} \binom{N}{i}^2 = \binom{2N}{N}$.

6.41 Consider a Markov chain having the following transition matrix:

$$\mathbf{P} = \begin{array}{c} 0 \\ 1 \\ 2 \\ 3 \\ 4 \end{array} \begin{pmatrix} & 1 & & & \\ \frac{\mu}{3\lambda+\mu} & & \frac{3\lambda}{3\lambda+\mu} & & \\ & \frac{2\mu}{2\lambda+2\mu} & & \frac{2\lambda}{2\lambda+2\mu} & \\ & & \frac{3\mu}{\lambda+3\mu} & & \frac{\lambda}{\lambda+3\mu} \\ & & & 1 & \end{pmatrix}$$

(columns labeled 0, 1, 2, 3, 4)

Use Problem 6.38 to prove that

$$g_0 = \frac{1}{2(1+\rho)^3} \qquad g_1 = \frac{1+3\rho}{2(1+\rho)^3} \qquad g_2 = \frac{3\rho(1+\rho)}{2(1+\rho)^3}$$

$$g_3 = \frac{\rho^3+3\rho^2}{2(1+\rho)^3} \qquad g_4 = \frac{\rho^3}{2(1+\rho)^3}$$

where $\rho \equiv \lambda/\mu$.

6.42 Let $a_1, a_2, a_3, \ldots$ be a sequence of positive numbers and let $A_i \equiv 1 + a_1 + a_2 + \cdots + a_i$. Consider the following chain:

$$\mathbf{P} = \begin{array}{c} 0 \\ 1 \\ \cdots \\ N-2 \\ N-1 \\ N \end{array} \begin{pmatrix} \frac{1}{A_1} & \frac{a_1}{A_1} & & \cdots & & \\ \frac{1}{A_2} & \frac{a_1}{A_2} & \frac{a_2}{A_2} & \cdots & & \\ \cdots & \cdots & \cdots & \cdots & \cdots & \cdots \\ \frac{1}{A_{N-1}} & \frac{a_1}{A_{N-1}} & \frac{a_2}{A_{N-1}} & \cdots & \frac{a_{N-1}}{A_{N-1}} & \\ \frac{1}{A_N} & \frac{a_1}{A_N} & \frac{a_2}{A_N} & \cdots & \frac{a_{N-1}}{A_N} & \frac{a_N}{A_N} \\ \frac{1}{A_N} & \frac{a_1}{A_N} & \frac{a_2}{A_N} & \cdots & \frac{a_{N-1}}{A_N} & \frac{a_N}{A_N} \end{pmatrix}$$

(columns labeled 0, 1, 2, …, $N-1$, N)

Show that the limiting distribution of this chain is

$$\pi_i = \frac{\delta^{(i)}}{1 + \delta^{(1)} + \delta^{(2)} + \cdots + \delta^{(N)}} \qquad \text{for all } 0 \le i \le N$$

where $\delta^{(0)} \equiv 1$, $\delta^{(1)} \equiv a_1$, and

$$\delta^{(i)} \equiv \frac{a_1}{1}\frac{a_2}{A_1}\frac{a_3}{A_2}\cdots\frac{a_i}{A_{i-1}} \qquad \text{for all } 2 \le i \le N$$

6.43

a. Suppose that costs are discounted at rate α $(0 \le \alpha \le 1)$. Thus, a cost of \$1 incurred at step n is equivalent to a cost of $\$\alpha^n$ at step 0. $\$\alpha^n$ is called the *present* *value* of \$1 at step n. Assume that whenever an irreducible chain visits state j, it has to pay \$$c_j$. Let v_i be the expected present value of the total cost incurred over its entire life, starting from state i. Show that

$$[v_i]_{i\in S} = (\mathbf{I} - \alpha\mathbf{P})^{-1}\mathbf{c}$$

where $\mathbf{c} \equiv [c_i]_{i\in S}$. Compare with Problem 5.41.

b. The total cost incurred over the entire life of an irreducible chain is ∞. However, show that its expected cost per transition over its entire life, or its *long-term cost rate*, is $\boldsymbol{\pi}\mathbf{c}$.

c. In the taxicab, Tex has to pay a toll of \$3 to enter city A, \$2 to enter City B, and \$4 to enter city C. Starting from city A, what is the expected present value of his toll cost over his entire career when $\alpha = 0.8$? What is his long-term cost rate?

6.44 Consider an irreducible Markov chain $\{X_n\}_{n=0,1,2,\ldots}$ having state space S. We partition S into two disjoint subsets A and B; that is, $S = A \cup B$ and $A \cap B = \phi$. The transition matrix of this chain now can be arranged into the form

$$\mathbf{P} = \begin{matrix} & \begin{matrix} A & B \end{matrix} \\ \begin{matrix} A \\ B \end{matrix} & \begin{pmatrix} \mathbf{P}_{AA} & \mathbf{P}_{AB} \\ \mathbf{P}_{BA} & \mathbf{P}_{BB} \end{pmatrix} \end{matrix}$$

Consider now chain $\{Y_n\}_{n=0,1,2,\ldots}$, which records the location of $\{X_n\}_{n=0,1,2,\ldots}$ only when $\{X_n\}_{n=0,1,2,\ldots}$ visits A. Chain $\{Y_n\}_{n=0,1,2,\ldots}$ is called a *censored* chain of $\{X_n\}_{n=0,1,2,\ldots}$. Let $\boldsymbol{\pi}$ be the limiting distribution of $\{X_n\}_{n=0,1,2,\ldots}$. We partition $\boldsymbol{\pi}$ into $(\boldsymbol{\pi}_A, \boldsymbol{\pi}_B)$, corresponding to the subsets A and B. Use Problem 5.52 to show that the limiting distribution of $\{Y_n\}_{n=0,1,2,\ldots}$ is proportional to $\boldsymbol{\pi}_A$.

6.45 Consider a Markov chain $\{X_n\}_{n=0,1,2,\ldots}$ having state space S and transition matrix $\mathbf{P}$. Chain $\{Y_n\}_{n=0,1,2,\ldots}$ is called a *thinned* chain of $\{X_n\}_{n=0,1,2,\ldots}$ if it records the location of $\{X_n\}_{n=0,1,2,\ldots}$ at step n with probability ρ_{ij} and ignores it with probability $1 - \rho_{ij}$, assuming $\{X_n\}_{n=0,1,2,\ldots}$ moves from state i to state j at step n.

a. Let $\mathbf{Q}$ be the transition matrix of $\{Y_n\}_{n=0,1,2,\ldots}$. Show that

$$\mathbf{Q} = (\mathbf{I} - \mathbf{P} + \mathbf{R})^{-1}\mathbf{R}$$

where $\mathbf{R} \equiv [p_{ij}\rho_{ij}]_{i,j\in S}$. *Hint:* This is a censored chain.

b. Find the limiting distribution of $\{Y_n\}_{n=0,1,2,\ldots}$ in terms of the limiting distribution of $\{X_n\}_{n=0,1,2,\ldots}$.

c. In the taxicab example, Tex is supposed to record all his movements in a logbook, but sometimes he forgets. Let ρ_{ij} be the probability that he records a trip from city i to city j $(i, j = A, B, C)$. Suppose

$$[\rho_{ij}]_{i,j\in S} = \begin{matrix} & \begin{matrix} A & B & C \end{matrix} \\ \begin{matrix} A \\ B \\ C \end{matrix} & \begin{pmatrix} .95 & .97 & .94 \\ .96 & .91 & .98 \\ .95 & .95 & .99 \end{pmatrix} \end{matrix}$$

Find the proportions of his visits to the cities as recorded in his logbook.

6.46 Recall from §2.13 that a Markov chain of order 2 can be made a chain of order 1 by suitably enlarging its state space. For an irreducible Markov chain of order 1 having transition matrix $\mathbf{P} \equiv [p_{ij}]_{i,j \in S}$ and limiting distribution $\boldsymbol{\pi} \equiv [\pi_i]_{i \in S}$, let us enlarge its state space in a similar manner; that is, we want to study the vector chain $\{X_{n-1}, X_n\}_{n=0,1,2,\dots}$.

a. Find the transition matrix of the expanded chain in terms of p_{ij}s.

b. Verify that

$$\lim_{n\to\infty} \Pr\{(X_n, X_{n+1}) = (i, j)\} = \pi_i p_{ij} \quad \text{for all } i, j \in S$$

Explain using the regenerative property.

c. Illustrate with the taxicab example. What is the limiting probability that Tex drives to city B immediately after visiting city A?

6.47 Consider an irreducible Markov chain.

a. Use first-step analysis to show that

$$\mathcal{E}\left[\left(n_{ij}\right)^2\right] = 1 + 2\sum_{k\neq j} p_{ik}\mathcal{E}\left[n_{kj}\right] + \sum_{k\neq j} p_{ik}\mathcal{E}\left[\left(n_{kj}\right)^2\right] \quad \text{for all } i, j \in S$$

Compare with Problem 5.46.

b. Hence, find $\mathcal{E}[(n_{ij})^2]$ for all $i, j = A, B, C$ in the taxicab example.

6.48 Consider an irreducible Markov chain.

a. Use Problem 6.47 to show that

$$g_j\mathcal{E}\left[\left(n_{jj}\right)^2\right] = 1 + 2\sum_{i\in S}\sum_{k\neq j} g_i p_{ik}\mathcal{E}\left[n_{kj}\right] \quad \text{for all } j \in S$$

b. Hence, find $\mathcal{E}[(n_i)^2]$ for all $i = A, B, C$ in the taxicab example.

6.49 Refer to Problem 2.17 in which a *sequential search method* is used to find an item among three items A, B, and C.

a. It is now desirable to find the limiting probability for state ABC only. Show how the other states can be lumped together (§2.17).

b. Hence, find the limiting distribution for the chain $\{X_n\}_{n=0,1,2,\dots}$.

c. Show that, with this method of relocation, the most frequently searched item will likely be in location 1 and the least searched item in location 3.

6.50

a. Let X_n be the state of machine A at time n, which can be either up or down. If it is up at time n, let the probability that it will be down at time $n+1$ be d; if it is down at time n, let the probability that it will be up at time $n+1$ be u. Find the limiting distribution for $\{X_n\}_{n=0,1,2,\dots}$. *Hint:* Use Problem 3.24.

b. Let Y_n be the state of another machine B at time n, which can also be either up or down. It behaves independently and identically with machine A. Find the transition matrix for the vector chain $\{X_n, Y_n\}_{n=0,1,2,\dots}$. Guess the limiting distribution and verify that it satisfies Equation (6.3.1).

c. Suppose we only want to know the number of machines that are up at time n. How can we lump the states? From this, obtain the limiting distribution of the number of machines that are up.

6.51 Refer to the weather in Problems 2.27 and 3.20.

a. Let X_n be the kind of weather at 8:00 A.M. on day n. What is the limiting distribution of $\{X_n\}_{n=0,1,2,\ldots}$?

b. Let Y_n be the kind of weather at 4:00 P.M. on day n. What is the limiting distribution of $\{Y_n\}_{n=0,1,2,\ldots}$? Can we obtain this from the limiting distribution of $\{X_n\}_{n=0,1,2,\ldots}$?

c. Let T_n be the kind of weather at 12:00 A.M. on day n. What is the limiting distribution of $\{T_n\}_{n=0,1,2,\ldots}$? Can we obtain this from the limiting distribution of $\{X_n\}_{n=0,1,2,\ldots}$ or $\{Y_n\}_{n=0,1,2,\ldots}$?

d. What is $\lim_{m\to\infty} \mathbf{P}_Z^{(3m)}$ of the chain $\{Z_n\}_{n=0,1,2,\ldots}$?

6.52 Refer to the bombers in Problems 2.28 and 3.21. Let $p_D = .4$ and $p_N = .7$.

a. What is the limiting distribution of $\{X_n\}_{n=0,1,2,\ldots}$? Find the expected value and compare it with that obtained in Problem 3.21.

b. What is the limiting distribution of $\{Y_n\}_{n=0,1,2,\ldots}$? Can we obtain this from the limiting distribution of $\{X_n\}_{n=0,1,2,\ldots}$?

c. What is $\lim_{m\to\infty} \mathbf{P}_Z^{(2m)}$ of the chain $\{Z_n\}_{n=0,1,2,\ldots}$?

6.53 Prove Equation (6.8.5).

6.54 For Chain (6.8.2), use Equation (6.8.6) to find $\lim_{m\to\infty} \mathbf{P}^{(3m+1)}$ and $\lim_{m\to\infty} \mathbf{P}^{(3m+2)}$.

6.55 Consider a reducible chain with

$$\mathbf{P} = \begin{array}{c} \\ 1 \\ 2 \\ 3 \\ 4 \\ 5 \\ 6 \\ 7 \\ 8 \end{array} \begin{array}{c} \begin{array}{cccccccc} 1 & 2 & 3 & 4 & 5 & 6 & 7 & 8 \end{array} \\ \left(\begin{array}{ccc|cc|ccc} 1/3 & 2/3 & 0 & 0 & 0 & 0 & 0 & 0 \\ 0 & 3/4 & 1/4 & 0 & 0 & 0 & 0 & 0 \\ 1/5 & 2/5 & 2/5 & 0 & 0 & 0 & 0 & 0 \\ \hline 0 & 0 & 0 & 2/5 & 3/5 & 0 & 0 & 0 \\ 0 & 0 & 0 & 1/4 & 3/4 & 0 & 0 & 0 \\ \hline 1/9 & 2/9 & 1/9 & 1/9 & 0 & 1/9 & 1/9 & 2/9 \\ 1/7 & 0 & 2/7 & 0 & 1/7 & 1/7 & 0 & 2/7 \\ 1/8 & 2/8 & 0 & 1/8 & 0 & 1/8 & 2/8 & 1/8 \end{array}\right) \end{array}$$

a. Find the limiting probabilities and put the results in matrix form. *Hint:* Use Problem 6.5b.

b. Find the reaching probability f_{ij}s (for all $i, j \in S$) as defined in §4.3 and put the results in matrix form. *Hint:* Use Problem 5.50.

CHAPTER 7

Infinite Chains

7.1 INFINITE RANDOM WALKS

In this chapter, we study the *infinite* chains—that is, chains having an infinite number of states. A classic example of this type is a chain with state space $S \equiv \{0, 1, 2, 3, \ldots\}$ and transition matrix

$$\mathbf{P} = \begin{array}{c} \\ 0 \\ 1 \\ 2 \\ 3 \\ \vdots \end{array} \begin{array}{c} \begin{array}{ccccc} 0 & 1 & 2 & 3 & \cdots \end{array} \\ \begin{pmatrix} r_0 & s_0 & 0 & 0 & \cdots \\ q_1 & r_1 & s_1 & 0 & \cdots \\ 0 & q_2 & r_2 & s_2 & \cdots \\ 0 & 0 & q_3 & r_3 & \cdots \\ \vdots & \vdots & \vdots & \vdots & \ddots \end{pmatrix} \end{array} \tag{7.1.1}$$

where $q_k, r_k, s_k \geq 0$ for all $k \geq 0$, $r_0 + s_0 = 1$, and $q_k + r_k + s_k = 1$ for all $k \geq 1$. This chain is known as the *(simple) random walk (with barrier at 0)*.

Seeing this name, we should not conclude that some scientists, probably with nothing better to do, have been overconcerned about the whereabouts of a wanderer. In fact, random walk finds applications in many fields of interest. For example, let X_n be the position on the x axis of a diffusing particle subject to random collisions and impulses immediately after step n. Assume that, at each step, the particle makes a jump, the size of which is a random variable, taking value from the set $\{-1, 0, 1\}$. The chain $\{X_n\}_{n=0,1,2,\ldots}$ is a simple random walk without barrier (that is, with state space $\{\ldots, -2, -1, 0, 1, 2, \ldots\}$).

Biologists use random walks to study the meanderings of some foraging species and the migration of certain bacteria.

Random Walks in Action: Determining Molecular Structure. According to Weiss (1994),

> One of the most important applications of random walk theory, as attested to by the award of a Nobel Prize in chemistry to Herbert Hauptman and Jerome Karle in 1985, is to the determination of molecular structure from data collected in x-ray scattering experiments. A considerable amount of the analysis required to interpret scattering data relies on the theory of random walks. ■

We use random walks as an example in our discussion of infinite chains.

7.2 INFINITE CHAINS

While most results obtained previously for the finite chains are still applicable for infinite chains, the study of infinite chains encounters a number of difficulties that do not exist for finite chains.

The first difficulty is, in theory at least, that the number of parameters needed to describe the chain may be infinite. For the random walk in (7.1.1), the values of q_ks and r_ks may be different for different values of k, and therefore must be independently assigned. This is clearly impractical. One common solution is to *truncate* the sample space, lumping all the states in its *tail* together, thus converting the chain into a finite one. The other solution is to find some *functional relationships* between the parameters, thus reducing the number of parameters needed to specify the chain. In this chapter, we assume the following relationships: $s_i = s$, $q_i = 1 - s \equiv q$, and $r_i = 0$ for all $i \geq 1$. In this manner, the foregoing random walk can be described by only two parameters s_0 and s as follows:

$$\mathbf{P} = \begin{array}{c} \\ 0 \\ 1 \\ 2 \\ 3 \\ \vdots \end{array} \begin{array}{c} \begin{array}{ccccc} 0 & 1 & 2 & 3 & \dots \end{array} \\ \begin{pmatrix} r_0 & s_0 & 0 & 0 & \dots \\ q & 0 & s & 0 & \dots \\ 0 & q & 0 & s & \dots \\ 0 & 0 & q & 0 & \dots \\ \vdots & \vdots & \vdots & \vdots & \ddots \end{pmatrix} \end{array} \tag{7.2.1}$$

Multiplying the infinite-dimensional matrix $\mathbf{P}$ with itself is the second difficulty, especially when it has an infinite number of parameters. Even when some functional relationships exist between the parameters, the method derived for multiplying one type of infinite matrix might be so model dependent that it cannot be readily adapted for another. Thus, unless we want to study a specific type of chain, there is not much we can say on this matter.

The third difficulty is in the state classification. Recall that, if a finite chain is absorbing, it eventually must end up at an absorbing state; if irreducible, all states must be recurrent. These properties no longer are necessarily true for infinite chains, as we discuss next.

7.3 INFINITE ABSORBING CHAINS

Consider an infinite *absorbing* chain in which the absorbing states form a subset A of the sample space S and the transient states form another subset T.

Starting from a transient state, the number of transient states the chain refuses to revisit keeps increasing as it develops. If the number of transient states is finite, the chain must eventually end up at an absorbing state; otherwise, it has no place to go! Let us now assume in this section that the number of transient states is infinite, in which case the chain may *never* be absorbed and remain forever within the set of transient states.

Recall that we denoted the reaching probability from state i to state j by f_{ij} in §4.3. Let us now denote the probability that the chain will never be absorbed from a transient state i by

$$\bar{a}_i \equiv 1 - \sum_{j \in A} f_{ij} \quad \text{for all } i \in T$$

Because we can show by first-step analysis that

$$\bar{a}_i = \sum_{m \in T} p_{im} \bar{a}_m \quad \text{for all } i \in T$$

let us consider the following set of equations

$$y_i = \sum_{m \in T} p_{im} y_m \quad \text{for all } i \in T \tag{7.3.1}$$

Note that $\mathbf{0} \equiv [y_i = 0]_{i \in T}$ is a solution of this set of equations. So if this set admits a *unique* solution, then $\bar{a}_i = 0$ for all $i \in T$, and the chain will surely be absorbed regardless of its initial conditions.

On the other hand, if, besides the null solution $\mathbf{0}$, (7.3.1) also has a *bounded positive* solution $\mathbf{y} \equiv [y_i]_{i \in T}$ (that is, $0 < y_i$ for *some* $i \in T$ and $y_i < \infty$ for all $i \in T$), then we shall show that there is a nonzero probability $0 < \bar{a}_i$ that the chain will never be absorbed from state i for some $i \in T$.

To start, if $\mathbf{y}$ is bounded and positive, it is always possible to find a scalar γ such that $z_i \equiv \gamma y_i \leq 1$ for all $i \in T$. Clearly, $\mathbf{z} \equiv \gamma \mathbf{y}$ is also a solution of (7.3.1). In essence, the proof reduces to showing that $z_i \leq \bar{a}_i$ for all $i \in T$.

Let us denote the probability that the chain, starting from state i, has not been absorbed by the step n as

$$\bar{a}_i^{(n)} \equiv \Pr\{X_n \in T \mid X_0 = i \in T\}$$

Using first-step analysis, we obtain

$$\bar{a}_i^{(1)} = \sum_{j \in T} p_{ij} \qquad \bar{a}_i^{(n+1)} = \sum_{j \in T} p_{ij} \bar{a}_j^{(n)} \qquad \text{for all } n \geq 1$$

To prove that $z_i \leq \bar{a}_i$, we shall prove by induction that

$$z_i \leq \bar{a}_i^{(n)} \quad \text{for all } n \geq 1 \tag{7.3.2}$$

First, because all $z_i \leq 1$ satisfies Equation (7.3.1), we have

$$z_i = \sum_{j \in T} p_{ij} z_j \leq \sum_{j \in T} p_{ij} = \bar{a}_i^{(1)}$$

Hence, (7.3.2) is true for $n = 1$. Now we assume that (7.3.2) is true for n; that is, $z_j \leq \bar{a}_j^{(n)}$. This yields

$$z_i = \sum_{j \in T} p_{ij} z_j \leq \sum_{j \in T} p_{ij} \bar{a}_j^{(n)} = \bar{a}_i^{(n+1)}$$

Thus, (7.3.2) is also true for $n + 1$, and hence for all $n \geq 0$.

We have shown that, if (7.3.1) has a solution $\mathbf{y} \equiv [y_i]_{i \in T}$ such that $0 < y_i < \infty$ for some $i \in T$, then $0 < \bar{a}_i$ and the chain, starting from state i, may never be absorbed.

Assume now that (7.3.1) has a bounded and positive solution set $\mathbf{y} \equiv [y_i]_{i \in T}$. Let Y be the smallest number that is larger than all y_is. Then $\mathbf{y}/Y$ is the largest solution set that is less than 1. Thus, we must have $\bar{a}_i = y_i/Y$.

For illustration, consider the random walk (7.2.1). Let $r_0 = 1$; hence, $s_0 = 0$. State 0 is now absorbing, and all other states are transient.

This chain is normally known as the *gambler's ruin problem* because the states of the chain can represent the fortune of a gambler who is playing against an infinitely rich adversary. At each move (or step), the probability that the gambler wins \$1 from his adversary is s, and the probability that he loses is q. Is there any possibility that he will be ruined—that is, be absorbed into the absorbing ruin-state 0?

Fortunately, from our previous discussion, our gambler does not always end up at bankruptcy. Under favorable conditions, there is a nonzero probability that he remains solvent forever. This occurs when (7.3.1) admits a bounded positive solution.

For this chain, (7.3.1) takes the form:

$$\begin{cases} y_1 = s y_2 \\ y_i = q y_{i-1} + s y_{i+1} \quad \text{for all } i > 1 \end{cases} \tag{7.3.3}$$

Because $q + s = 1$ and $y_i = q y_i + s y_i$ for all $i \geq 1$, we can write (7.3.3) as

$$\begin{cases} (y_2 - y_1) = y_1 \, (q/s) \\ (y_{i+1} - y_i) = (y_i - y_{i-1}) \, (q/s) \quad \text{for all } i > 1 \end{cases}$$

Applying these equations recursively yields

$$(y_{i+1} - y_i) = y_1 \, (q/s)^i \quad \text{for all } i \geq 1$$

Thus, for all $i \geq 2$,

$$\begin{aligned} y_i &= (y_i - y_{i-1}) + (y_{i-1} - y_{i-2}) + \cdots + (y_2 - y_1) + y_1 \\ &= \left[(q/s)^{i-1} + (q/s)^{i-2} + \cdots + (q/s) + 1\right] y_1 \\ &= \frac{1 - (q/s)^i}{1 - (q/s)} y_1 \end{aligned} \tag{7.3.4}$$

Case $q \geq s$: Here $q/s \geq 1$, and hence $\lim_{i\to\infty} y_i = \infty$. The only bounded solution for (7.3.3) is the null solution $\mathbf{0}$. Thus, $\bar{a}_i = 0$ for all $i \in T$, and the chain will surely be absorbed into the absorbing state 0.

If the gambler's losing probability q is greater than *or even equal to* his winning probability s, he surely will be ruined if he keeps playing indefinitely. (Unfair!)

Case $q < s$: Here the right-hand side of (7.3.4) converges to a finite value, giving $\lim_{i\to\infty} y_i = y_1/(1 - q/s)$. Thus, if we give y_1 any bounded positive value, all y_is will also be bounded positive. The chain therefore may be able to avoid absorption.

Given a particular value of y_1, the smallest number that is larger than all y_is is $Y \equiv y_1/(1 - q/s)$. Dividing all y_is by Y, we obtain the probability of the chain not being absorbed into state 0, starting from state i, as:

$$\bar{a}_i = 1 - (q/s)^i \quad \text{for all } i \geq 1$$

Hence, the probabilities of the chain being absorbed into state 0 are:

$$f_{i0} = (q/s)^i \quad \text{for all } i \geq 1 \tag{7.3.5}$$

Since the chain's tendency to move up (s) is always higher than its tendency to move down (q), it tends to drift to the *infinite state* ∞. Once "reaching" state ∞, the chain tends to remain there and has no inclination to return to the finite states. We can therefore consider this chain as having *two* absorbing states: ∞ and 0. The chain will not surely be absorbed into state 0 because it may be absorbed into state ∞ instead.

The gambler may be able to avoid bankruptcy if his winning probability s is *strictly greater* than his losing probability q. In this case, there is still a nonzero probability $(q/s)^i$ that he will be ruined eventually. (Unfair!) The higher his initial capital i, the lower his ruin-probability.

Like all absorbing chains, all long-term visiting rates g_is to the transient states are zeros, and all their expected return times $\mathcal{E}[n_i]$ are infinite.

7.4 INFINITE IRREDUCIBLE CHAINS

We now turn our attention to the infinite *irreducible* chains, in which all states communicate with each other.

We argued before that every finite Markov chain must have at least one recurrent state. After some finite number of steps, one of its transient states becomes unreturnable; after another finite number of steps, another transient state becomes likewise; then another. . . . If all states in a finite chain are transient, the chain eventually runs out of states and has nowhere to go. This argument, however, is not applicable for an infinite chain. An infinite chain may refuse to revisit one transient state after the other, yet never exhausts the pool of its potential destinations. In other words, all of the states in an infinite irreducible chain may be transient.

Consider two arbitrary states i and j in a general irreducible chain. Their being in an irreducible chain requires that they communicate or that the reaching probability f_{ij} is nonzero. However, it is not necessary that $f_{ij} = 1$. Only when state j is recurrent does $f_{ij} = 1$. Consequently, state j is transient (and hence the chain) if we can find at least one state i such that $f_{ij} < 1$.

Consider now an arbitrary state in a general irreducible chain conveniently labeled 0. Let us convert state 0 into an absorbing state, thus making all other states transient. The transience condition $f_{i0} < 1$ is equivalent to the existence of a nonzero probability $\bar{a}_i \equiv 1 - f_{i0} > 0$ that the chain will never be absorbed into the now absorbing state 0 from a state i. From our discussion in the previous section about the absorbing chains, we immediately have the following result: An infinite irreducible chain is transient if the set of equations

$$y_i = \sum p_{im} y_m \quad \text{for all } m, i \in S, \text{ and } m, i \neq 0 \tag{7.4.1}$$

has a bounded positive solution.

In our random walk (7.2.1), let us now assume that $s_0 > 0$ and $s > 0$. This makes the chain irreducible. If $s_0 = 1$ (or $r_0 = 0$), we say that the chain has a *reflecting barrier* at state 0 because it immediately bounces back to state 1 once it hits state 0. On the other hand, if $s_0 < 1$, the chain may sojourn at state 0 for some time before being reflected back; we say that it has a *partial reflecting barrier* at state 0.

In either case, when we make state 0 an absorbing state, (7.4.1) becomes (7.3.3).

Case $q \geq s$**:** (7.3.3) only admits the null solution $\mathbf{0}$. The chain will surely reach state 0 and is therefore recurrent.

When $q > s$, the chain has the tendency to drift toward state 0 because the probability q of its moving toward state 0 is strictly greater than the probability s of its moving away from it. After hitting state 0, it is reflected back but keeps drifting to that state again. State 0 is therefore recurrent and hence the chain.

When $q = s$, starting from any state i, the chain has no inclination to drift away in either direction but keeps revisiting it, making it recurrent.

Case $q < s$**:** The chain is transient because (7.3.3) has a bounded positive solution. This should come as no surprise: When the probability s of the chain's moving toward the higher states is strictly greater than its probability q of moving toward the lower states, in the long run, it will drift toward state ∞, not returning to any finite state.

7.5 POSITIVE-RECURRENCE

In the previous section, we derived a method to test whether an irreducible chain is transient or recurrent. If transient, the chain does not have a limiting distribution. So let us now concentrate on the limiting behavior of infinite irreducible recurrent chains.

In §6.4, we showed that the visiting rates g_is in a finite recurrent chain are independent of the initial condition and satisfy both (6.4.6) and (6.4.4) simultaneously. It is easy to retrace the arguments to see that these equations are also applicable for infinite recurrent chains; that is, the visiting rate vector $\mathbf{g} \equiv [g_i]_{i \in S}$ is the solution of both

$$\mathbf{g} = \mathbf{g}\mathbf{P} \tag{7.5.1}$$

and

$$\sum_{i \in S} g_i = 1 \tag{7.5.2}$$

The problem is that we now have an *infinite number of equations* with an infinite number of unknowns. Here is one common method of solution.[1]

Step 1: Consider Equation (7.5.1) alone. Note that, if vector $\mathbf{g} \equiv [g_i]_{i \in S}$ is a solution of (7.5.1), then so is $c\mathbf{g} \equiv [cg_i]_{i \in S}$ for any scalar c. Thus, there is an infinite number of vectors satisfying (7.5.1), one of which can be obtained by giving an arbitrary value to one unknown and then solving for the others.

In this step, we first select a particular state and conveniently label it 0. We then try to express all g_i in terms of g_0.

Step 2: We substitute the expressions obtained in step 1 to the left-hand side of Equation (7.5.2). This provides one function for one unknown g_0, enabling us to obtain its value. This value in turn yields a set of values g_is that satisfies both (7.5.1) and (7.5.2).

For illustration, in the random walk with a reflective barrier at state 0 (that is, $r_0 = 0$, $s_0 = 1$), let $q \geq s$. The chain is therefore recurrent, and (7.5.1) takes the form:

$$\begin{cases} g_0 = \quad q g_1 \\ g_1 = g_0 \quad + q g_2 \\ g_2 = \quad s g_1 \quad + q g_3 \\ g_3 = \quad s g_2 + \quad \cdots \\ \vdots \end{cases}$$

Step 1: We express all g_is in terms of g_0:

$$g_1 = \left(\frac{1}{q}\right) g_0$$

$$g_2 = \left(\frac{1}{q}\right)\left(\frac{1}{q} g_0 - g_0\right) = \left(\frac{1}{q}\right)\left(\frac{s}{q}\right) g_0$$

$$g_3 = \left(\frac{1}{q}\right)\left(\frac{s}{q^2} - \frac{s}{q}\right) g_0 = \left(\frac{1}{q}\right)\left(\frac{s}{q}\right)^2 g_0$$

$$\vdots$$

Typically,

$$g_i = \left(\frac{1}{q}\right)\left(\frac{s}{q}\right)^{i-1} g_0 \quad \text{for all } i \geq 1 \tag{7.5.3}$$

[1]This method is somewhat similar to the normalization method in Problem 6.5.

Step 2: We substitute these expressions into the left-hand side of (7.5.2):

$$\left[1+\left(\frac{1}{q}\right)\sum_{i=1}^{\infty}\left(\frac{s}{q}\right)^{i-1}\right]g_0 = 1 \tag{7.5.4}$$

If $q > s$, the geometric series in (7.5.4) converges; that is,

$$1+\left(\frac{1}{q}\right)\sum_{i=1}^{\infty}\left(\frac{s}{q}\right)^{i-1} = 1+\frac{1}{q-s} = \frac{2q}{q-s} < \infty$$

(Bear in mind that $q + s = 1$.) Equation (7.5.4) now gives

$$g_0 = \frac{q-s}{2q}$$

Hence, from Equation (7.5.3),

$$g_i = \left(\frac{1}{2q}\right)\left(1-\frac{s}{q}\right)\left(\frac{s}{q}\right)^{i-1} \qquad \text{for all } i > 0$$

Here, all proportions of visits g_is are nonzero, and the expected return times $\mathcal{E}[n_i] = 1/g_i$ are finite.

For a general chain, if (7.5.1) and (7.5.2) yield a nonzero solution as in this case, the chain (and all its states) is said to be *positive-recurrent*.

Like the finite chains, if an infinite positive-recurrent chain is *aperiodic*, its limiting distribution exists and is equal to its long-term visiting rates, or $\boldsymbol{\pi} = \mathbf{g}$; if the chain is periodic, it does not have a limiting distribution.

7.6 NULL-RECURRENCE

In §7.4, we proved that our random walk example is recurrent when $q \geq s$. In §7.5, we said that the chain was positive-recurrent when $q > s$. But how about when $q = s$? In this case, we have a problem: The left-hand side of Equation (7.5.4) does not converge, while its right-hand side is 1!

Generally, to resolve this difficulty, we add the *infinite state* ∞ to the state space S, forming a new state space $S^* \equiv [S \cup \infty]$, changing (7.5.2) to:

$$\sum_{i \in S^*} g_i = 1 \tag{7.6.1}$$

This allows for the following set of solutions, which satisfies both (7.5.1) and (7.6.1) simultaneously: $g_i = 0$ for all $i \in S$, and $g_\infty = 1$. Chains having this property (and all their states) are said to be *null-recurrent*.

Null-recurrence is a *boundary* between transience and positive-recurrence. Although null-recurrent states are said to be recurrent, they share many properties of the transient states.

For the foregoing random walk:

1. As the value of q/s decreases from a value greater than 1 to that less than 1, the chain changes from positive-recurrence to transience. At the boundary, when $q/s = 1$, the chain is still recurrent in the sense that all return probabilities f_{ij}s are still equal to 1. Starting from any state i, the probability q of the chain's moving down to state $i-1$ is the same as its probability s of moving up to state $i+1$. Hence, the chain has no inclination to drift away in either direction but will surely return to state i, making it recurrent.

2. As the value of q/s increases from a value less than 1 to that greater than 1, the chain changes from transience to positive-recurrence. At the boundary, when $q/s = 1$, the chain is still transient in the sense that the visiting rates g_is are still zeros, and the expected return times $\mathcal{E}[n_i]$ are still infinite for all finite states i.

For the infinite transient chain ($q/s < 1$), we said that state ∞ acts as an absorbing state; for the null-recurrent chain ($q/s = 1$), the chain's tendency to stay at state ∞ is equal to its tendency to return to the finite states; hence, state ∞ is not absorbing. However, *it takes forever* for the chain to return from state ∞ to any finite state.

Yes, in the null-recurrence case, the chain will surely return to any finite state i (that is, $f_{ii} = 1$); however, the expected return time $\mathcal{E}[n_i]$ is infinite because the chain might venture into state ∞ in between.

7.7 SUMMARY

1. An infinite absorbing chain may never be absorbed if Equation (7.3.1) has a bounded positive solution.
2. Figure (7.7.1) summarizes the classification method for an infinite irreducible chain, which can be either transient, null-recurrent, or positive-recurrent.

FIGURE 7.7.1 The classification of infinite chains.

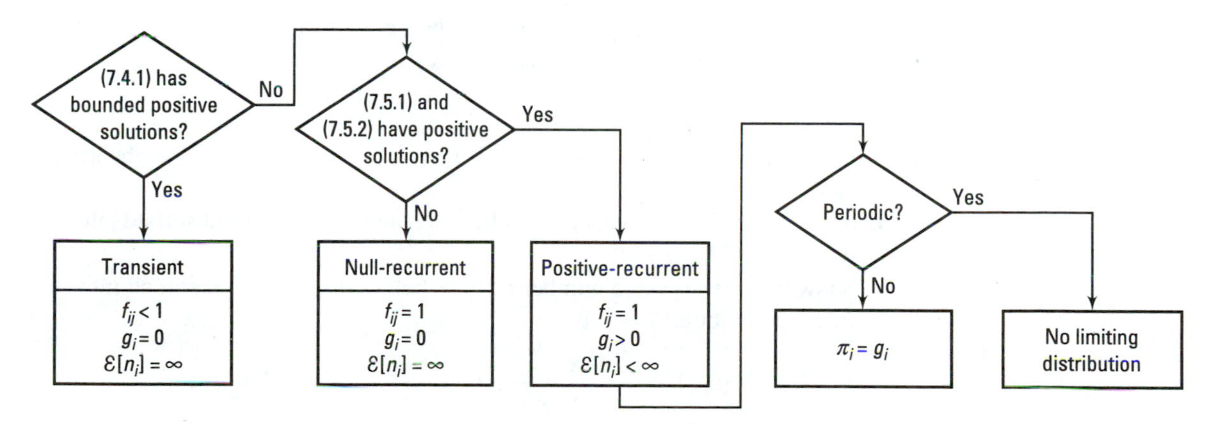

PROBLEMS

7.1 Consider an infinite chain with

$$\mathbf{P} = \begin{array}{c} \\ 1 \\ 2 \\ 3 \\ \vdots \end{array} \begin{array}{c} \begin{array}{ccccc} 1 & 2 & 3 & 4 & \cdots \end{array} \\ \begin{pmatrix} \alpha_1 & \alpha_2 & \alpha_3 & \alpha_4 & \cdots \\ 1 & & & & \cdots \\ & 1 & & & \cdots \\ \vdots & \vdots & \vdots & \vdots & \ddots \end{pmatrix} \end{array}$$

where $\sum_{i \in S} \alpha_i = 1$. This chain can be used to represent the *remaining life* r_n of an instrument at time n. If $r_n > 1$, then $r_{n+1} = r_n - 1$. If $r_n = 1$, then the instrument will fail at time $n + 1$ and is replaced by an equivalent instrument. In this case, $r_{n+1} = L$, where L is the life of the instrument, with $\alpha_i \equiv \Pr\{L = i\}$.

a. Show that the chain is irreducible and recurrent.

b. Show that the chain is positive-recurrent when the expected life is finite; that is, $\mathcal{E}[L] \equiv \sum_{j=1}^{\infty} j\alpha_j < \infty$.

7.2 Refer to Problem 7.1. Assume now that the chain is positive-recurrent.

a. Show that the limiting probability is

$$\lim_{n \to \infty} \Pr\{r_n = i\} = \frac{1}{\mathcal{E}[L]} \sum_{j=i}^{\infty} \alpha_j \quad \text{for all } i \geq 1$$

i. by using Equations (7.5.1) and (7.5.2)

ii. by using a regenerative cycle from state 1

iii. by using a regenerative cycle from state $i > 1$

b. Obtain the limiting expected remaining life $\lim_{n\to\infty} \mathcal{E}[r_n]$. Is the answer what you would expect intuitively? Compare with the finite chain in Problem 6.37.

c. Find the limiting remaining life distribution when the instrument's life has a geometric distribution; that is, $\alpha_i = qp^{i-1}$ for all $i \geq 1$.

7.3 Consider an infinite chain $\{X_n\}_{n=0,1,2,\ldots}$ with

$$\mathbf{P} = \begin{array}{c} \\ 0 \\ 1 \\ 2 \\ 3 \\ \vdots \end{array} \begin{array}{c} \begin{array}{ccccc} 0 & 1 & 2 & 3 & \cdots \end{array} \\ \begin{pmatrix} q_0 & s_0 & & & \cdots \\ q_1 & r_1 & s_1 & & \cdots \\ q_2 & & r_2 & s_2 & \cdots \\ q_3 & & & r_3 & \cdots \\ \vdots & \vdots & \vdots & \vdots & \ddots \end{pmatrix} \end{array}$$

where $q_i + r_i + s_i = 1$ and $r_i < 1$ for all $i = 1, 2, \ldots$. Let state 0 be absorbing, or $q_0 = 1$ and $s_0 = 0$.

a. Show that, starting from state $i > 0$, the chain will surely be absorbed into state 0 if $\lim_{j\to\infty} \prod_{k=1}^{j} s_k = 0$.

b. Show that the expected number of steps before the chain is absorbed into state 0 from state i for all $i > 0$ is

$$\mathcal{E}[n_{i,0}] = \frac{1}{(1 - r_i)\gamma^{(i)}} \left(\gamma^{(i)} + \gamma^{(i+1)} + \gamma^{(i+2)} + \cdots\right)$$

where $\gamma^{(1)} \equiv (1 - r_1)^{-1}$ and

$$\gamma^{(i)} \equiv \frac{1}{(1-r_1)} \frac{s_1}{(1-r_2)} \cdots \frac{s_{i-1}}{(1-r_i)} \qquad \text{for all } i = 2, 3, 4, \ldots$$

Compare with Problem 5.58.

7.4 Refer to Problem 7.3. Now let $q_0 < 1$. Thus, the chain is irreducible.

a. Find the condition for recurrence.

b. Show that the chain is positive-recurrent when $\sum_{i=1}^{\infty} \gamma^{(i)} < \infty$.

c. Show that the positive-recurrence condition implies the recurrence condition.

d. Check that when the chain is positive-recurrent, $\mathcal{E}[n_{i,0}]$ as obtained in Problem 7.3 is finite.

e. Let $q_i = 1/(i+2)$, $r_i = 0$, and $s_i = (i+1)/(i+2)$ for all $i = 0, 1, 2, \ldots$. Show that the chain is null-recurrent.

f. Let $q_i = (i+1)/(i+2)$, $r_i = 0$, and $s_i = 1/(i+2)$ for all $i = 0, 1, 2, \ldots$. Show that the chain is positive-recurrent.

7.5 Refer to Problem 7.4. Assume the chain is positive-recurrent. Show that the limiting probability is

$$\lim_{n\to\infty} \Pr\{X_n = i\} = \delta^{(i)} \left(\sum_{j=0}^{\infty} \delta^{(j)} \right)^{-1} \qquad \text{for all } i = 0, 1, 2, \ldots$$

where $\delta^{(0)} \equiv 1$ and $\delta^{(i)} = s_0 \gamma^{(i)}$ for all $i = 1, 2, \ldots$ (γ_is were defined in Problem 7.3.)

a. by using (7.5.1) and (7.5.2).

b. by using a regenerative cycle from state 0.

c. by using a regenerative cycle from state $i > 0$. *Hint:* Use Problem 5.59.

d. by letting $N \to \infty$ in Problem 6.36.

7.6 Let a_n be the *age* of an instrument at time n, which increases by 1 from time n to $n + 1$. However, if the instrument fails at step n, it will be replaced immediately by an equivalent one, and the age is reset to 0 at time n. Let α_i be the probability that the entire life L of this instrument is i.

a. Find the transition matrix for the chain $\{a_n\}_{n=0,1,2,\ldots}$. *Hint:* Use Problem 2.29.

b. Find the condition for positive-recurrence. *Hint:* Use Problem 7.4.

c. Find the limiting age distribution of $\{a_n\}_{n=0,1,2,\ldots}$. Compare this with the limiting remaining life distribution in Problem 7.2. *Hint:* Use Problem 7.5.

d. Find $\lim_{n\to\infty} \mathcal{E}[a_n]$.

e. Find the limiting age distribution when the instrument's life has a geometric distribution; that is, $\alpha_i = qp^{i-1}$ for all $i \geq 1$.

7.7 Consider an infinite *random walk* chain with transition matrix (7.1.1). Let $q_0 = 1$. Thus, state 0 is absorbing.

a. Use first-step analysis to show that the probability that the chain will eventually be absorbed into state 0 is

$$f_{i0} = \begin{cases} F_i/(1+F_1) & \text{if } F_1 < \infty \\ 1 & \text{if } F_1 = \infty \end{cases}$$

where $\rho^{(i)} \equiv \prod_{k=1}^{i}(q_k/s_k)$ and $F_i \equiv \sum_{k=i}^{\infty} \rho^{(k)}$ for all $i = 1, 2, 3, \ldots$.

b. Use Equation (7.3.1) to verify the result in part a.

c. Let $q_i > s_i$ for all $i = 1, 2, \ldots$. Show that the chain will be absorbed into state 0.

d. Let $q_i = i/(2i+2)$, $r_i = 0$, and $s_i = (i+2)/(2i+2)$ for all $i = 0, 1, 2, \ldots$. Show that the chain might never be absorbed into state 0. *Hint:* $\sum_{m=0}^{\infty} 1/[(m+2)(m+1)] = 1$.

e. Let $n_{i,j}$ be the number of steps for the chain to go from state i to state j for the first time. Use first-step analysis and the fact that $\mathcal{E}[n_{i,0}] = \mathcal{E}[n_{i,i-1}] + \mathcal{E}[n_{i-1,0}]$ to show that $q_i \mathcal{E}[n_{i,i-1}] = 1 + s_i \mathcal{E}[n_{i+1,i}]$.

f. Hence, show that

$$q_i \mathcal{E}[n_{i,i-1}] = 1 + \frac{s_i}{q_{i+1}} + \frac{s_i}{q_{i+1}}\frac{s_{i+1}}{q_{i+2}} + \frac{s_i}{q_{i+1}}\frac{s_{i+1}}{q_{i+2}}\frac{s_{i+2}}{q_{i+3}} + \cdots$$

7.8 Refer to Problem 7.7. Let $0 < r_0 < 1$. Thus, the chain is irreducible.

a. For all chains, find $\mathcal{E}[n_{i,i+1}]$. *Hint:* Use Problem 5.54.

b. Find the condition for recurrence.

c. Find the condition for recurrence when $s_i = s$ and $q_i = q$ for all $i = 1, 2, 3, \ldots$.

d. Show that the chain is positive-recurrent when $\sum_{j=0}^{\infty} \delta^{(j)} \leq \infty$, where $\delta^{(0)} \equiv 1$ and

$$\delta^{(i)} \equiv \frac{s_0}{q_1}\frac{s_1}{q_2}\cdots\frac{s_{i-1}}{q_i} \quad \text{for all } i = 1, 2, 3, \ldots$$

e. Verify that positive-recurrence implies recurrence.

f. Use Problem 7.7e to verify that positive-recurrence implies $\mathcal{E}[n_{1,0}] < \infty$.

g. Find the condition for positive-recurrence when $s_i = s$ and $q_i = q$ for all $i = 1, 2, 3, \ldots$.

7.9 Refer to Problem 7.8. Assume the chain is positive-recurrent. Show that the limiting probability is

$$\lim_{n\to\infty} \Pr\{X_n = i\} = \delta^{(i)} \left(\sum_{j=0}^{\infty} \delta^{(j)} \right)^{-1} \quad \text{for all } i = 0, 1, 2, \ldots$$

a. by using Equations (7.5.1) and (7.5.2).

b. by using a regenerative cycle from state 0.

c. by using a regenerative cycle from state $i > 0$. *Hint:* Use Problems 7.7 and 5.54.

d. by letting $N \to \infty$ in Problem 6.38.

e. Hence, obtain Equation (7.5.4).

7.10 Consider the following *unrestricted* random walk $\{X_n\}_{n=0,1,2,\ldots}$:

$$\mathbf{P} = \begin{array}{c} \\ \vdots \\ -2 \\ -1 \\ 0 \\ 1 \\ 2 \\ \vdots \end{array} \begin{array}{c} \cdots \; -3 \;\; -2 \;\; -1 \;\; 0 \;\; 1 \;\; 2 \;\; 3 \; \cdots \\ \left(\begin{array}{cccccccccc} \ddots & \vdots & \vdots & \vdots & \vdots & \vdots & \vdots & \vdots & \iddots \\ \cdots & q & r & s & & & & & \cdots \\ \cdots & & q & r & s & & & & \cdots \\ \cdots & & & q & r & s & & & \cdots \\ \cdots & & & & q & r & s & & \cdots \\ \cdots & & & & & q & r & s & \cdots \\ \iddots & \vdots & \vdots & \vdots & \vdots & \vdots & \vdots & \vdots & \ddots \end{array}\right) \end{array}$$

where $q + r + s = 1$.

a. Show that the chain is irreducible.

b. Show that the chain is null-recurrent if $q = s$ and transient otherwise. *Hint:* Use Problem 7.8.

c. From now on, assume $r = 0$. (This chain represents the difference between the number of successes and failures in successive Bernoulli trials.) Show that, for all integers n, i, j such that $(n+j-i)/2$ is also an integer and $0 \leq (n+j-i)/2 \leq n$,

$$p_{ij}^n = \left(\begin{array}{c} n \\ \dfrac{n+j-i}{2} \end{array} \right) q^{(n+j-i)/2} s^{(n-j+i)/2}$$

d. Use *Stirling's approximation* $n! \cong n^n e^{-n} \sqrt{2\pi n}$ to show that

$$p_{00}^{2m+1} = 0 \qquad p_{00}^{2m} \cong \frac{(4qs)^m}{\sqrt{\pi m}} \qquad \text{for all } m = 0, 1, 2, \ldots$$

e. Note that $qs \leq 1/4$. Show that

$$p_{ij}^{2m} \leq \binom{2m}{m} \frac{1}{2^{2m}} \left(\frac{q}{s}\right)^{(j-i)/2}$$

7.11 In an infinite absorbing Markov chain, let a_i $(i \in T)$ be the probability of absorption from a transient state i; that is, $a_i = 1 - \bar{a}_i = \sum_{a \in A} f_{ia}$. Show that $[a_i]_{i \in T}$ is the *minimal positive* solution of the following set of equations:

$$t_i = \sum_{m \in T} p_{im} t_m + \sum_{m \in A} p_{im} \quad \text{for all } i \in T$$

7.12 Let Y be an integer-valued random variable with $\Pr\{Y = n\} = p_n$ for all $n \geq 0$. Consider the following Markov chain $\{X_n\}_{n=0,1,2,\ldots}$ with

$$X_{n+1} = X_n - U_n + Y$$

where

$$U_n = \begin{cases} 1 & \text{if } X_n > 0 \\ 0 & \text{if } X_n = 0 \end{cases}$$

This chain is said to be *skip-free to the left* because it cannot skip any state in between when moving from state i to a lower state $j < i$.

a. Explain why the transition matrix of this chain is

$$P = \begin{array}{c} \\ 0 \\ 1 \\ 2 \\ 3 \\ \vdots \end{array} \begin{array}{c} \begin{array}{ccccc} 0 & 1 & 2 & 3 & \cdots \end{array} \\ \begin{pmatrix} p_0 & p_1 & p_2 & p_3 & \cdots \\ p_0 & p_1 & p_2 & p_3 & \cdots \\ & p_0 & p_1 & p_2 & \cdots \\ & & p_0 & p_1 & \cdots \\ \vdots & \vdots & \vdots & \vdots & \ddots \end{pmatrix} \end{array}$$

b. For the rest of the problem, make state 0 absorbing. Let $a_i \equiv f_{i,0}$ $(i \geq 1)$ be the probability of absorption to state 0 from state i. Show that $a_i = (a_1)^i$.

c. Hence, show that the chain will surely be absorbed into state 0 if $\mathcal{E}[Y] \leq 1$
   i. by using Problem 7.11
   ii. by trying $y_i = 1 - s^i$ in Equation (7.3.1)

d. Assume absorption is certain; that is, $\mathcal{E}[Y] \leq 1$.
   i. Let $n_{i,0}$ be the number of steps until absorption from state i. Show that $\mathcal{E}[n_{i,0}] = i\mathcal{E}[n_{1,0}]$.
   ii. Hence, use first-step analysis to show that

$$\mathcal{E}[n_{1,0}] = 1/(1 - \mathcal{E}[Y])$$

7.13 A gambler plays against an infinitely rich adversary. In each step he can win $\$k$ with probability p or lose \$1 with probability $1 - p$. Use the results obtained in Problem 7.12 to find his probability of ruin and the expected duration of the game.

7.14 Refer to Problem 7.12. Assume now that state 0 is not absorbing; that is, $p_0 < 1$.

a. Find the condition for recurrence.

b. Show that the chain is positive-recurrent when $\mathcal{E}[Y] < 1$.

c. Verify that positive-recurrence implies recurrence.

7.15 Refer to Problem 7.14. Assume that the chain is positive-recurrent.

a. Show that

$$\lim_{n\to\infty} \Pr\{X_n = 0\} = 1 - \mathcal{E}[Y]$$

   i. by using Equations (7.5.1) and (7.5.2)
   ii. by using a regenerative cycle from state 0. *Hint:* Use Problem 7.12.

b. When $p_i = (1 - t)t^i$ with $0 \leq t < .5$ for all $i = 0, 1, 2, \ldots$, show that

$$\lim_{n\to\infty} \Pr\{X_n = i\} = \frac{1 - 2t}{1 - t}\left(\frac{t}{1 - t}\right)^i \quad \text{for all } i = 0, 1, 2, \ldots$$

c. Take the expectations on both sides of $X_{n+1} = X_n - U_n + Y$ and then on both sides of the square of that equation to show that

$$\lim_{n\to\infty} \mathcal{E}[X_n] = \frac{\mathcal{E}[Y] - 2\mathcal{E}^2[Y] + \mathcal{E}[Y^2]}{2(1 - \mathcal{E}[Y])}$$

d. Find $\lim_{n\to\infty} \mathcal{E}[X_n]$ when $p_i = (1-t)t^i$ with $0 \le t < .5$ for all $i = 0, 1, 2, \ldots$
 i. by using the results in part b
 ii. by using the results in part c

7.16 (Liu, 1994) Consider an infinite chain with

$$\mathbf{P} = \begin{array}{c|cccccc} & 0 & 1 & 2 & 3 & 4 & \cdots \\ \hline 0 & q_0 & p_0 & & & & \cdots \\ 1 & q_1 & q_1 & p_1 & & & \cdots \\ 2 & q_2 & q_2 & q_2 & p_2 & & \cdots \\ 3 & q_3 & q_3 & q_3 & q_3 & p_3 & \cdots \\ 4 & q_4 & q_4 & q_4 & q_4 & q_4 & \cdots \\ \vdots & \vdots & \vdots & \vdots & \vdots & \vdots & \ddots \end{array}$$

where

$$p_i \equiv 1 - (i+1)q_i \quad \text{for all } 0 \le i$$

Make state 0 absorbing; that is, $q_0 = 1$. Show that the chain will be absorbed into state 0 when $\sum_{j=1}^{\infty}(q_j/p_1 \cdots p_j) < \infty$.

7.17 Refer to Problem 7.16. Let $0 < q_0 < 1$. Thus, the chain is irreducible.

a. Find the condition for recurrence.

b. Show that the chain is transient when $p_0 = q_0 = 1/2$ and $p_i = 1 - c^i$ for all $0 \le i$.

c. Show that the chain is positive-recurrent when $\sum_{i=0}^{\infty} \delta^{(i)} < \infty$, where $\delta^{(0)} \equiv 1$, and

$$\delta^{(i)} \equiv (i+1)p_0 p_1 \cdots p_{i-1} \quad \text{for all } i = 1, 2, 3, \ldots$$

d. Show that the chain is null-recurrent when $p_i = (i+1)/(i+2)$ for all $i = 0, 1, 2, \ldots$.

e. Show that the chain is positive-recurrent when $p_i = 1/(i+2)$ for all $i = 0, 1, 2, \ldots$.

7.18 Refer to Problem 7.17. Assume the chain is positive-recurrent.

a. Show that the limiting distribution of the chain is

$$\lim_{n\to\infty} \Pr\{X_n = i\} = \frac{\delta^{(i)}}{\sum_{j=0}^{\infty} \delta^{(j)}} \quad \text{for all } i = 0, 1, 2, \ldots$$

b. Find the limiting distribution when $p_i = 1/(i+2)$ for all $i = 0, 1, 2, \ldots$.

7.19 (Foster, 1952) Let $a_1, a_2, a_3, \ldots$ be a sequence of positive numbers and let $A_i \equiv 1 + a_1 + a_2 + \cdots + a_i$. Consider the following chain:

$$\mathbf{P} = \begin{array}{c} \\ 0 \\ 1 \\ 2 \\ 3 \\ 4 \\ \vdots \end{array} \begin{array}{c} \begin{array}{cccccc} 0 & 1 & 2 & 3 & 4 & \cdots \end{array} \\ \begin{pmatrix} \frac{1}{A_1} & \frac{a_1}{A_1} & & & & \cdots \\ \frac{1}{A_2} & \frac{a_1}{A_2} & \frac{a_2}{A_2} & & & \cdots \\ \frac{1}{A_3} & \frac{a_1}{A_3} & \frac{a_2}{A_3} & \frac{a_3}{A_3} & & \cdots \\ \frac{1}{A_4} & \frac{a_1}{A_4} & \frac{a_2}{A_4} & \frac{a_3}{A_4} & \frac{a_4}{A_4} & \cdots \\ \frac{1}{A_5} & \frac{a_1}{A_5} & \frac{a_2}{A_5} & \frac{a_3}{A_5} & \frac{a_4}{A_5} & \cdots \\ \vdots & \vdots & \vdots & \vdots & \vdots & \ddots \end{pmatrix} \end{array}$$

Show that the chain is positive-recurrent when $\sum_{j=0}^{\infty} \delta^{(j)} < \infty$, where $\delta^{(0)} \equiv 1$, $\delta^{(1)} \equiv a_1$, and

$$\delta^{(i)} \equiv \frac{a_1}{1} \frac{a_2}{A_1} \cdots \frac{a_{i-1}}{A_{i-2}} \frac{a_i}{A_{i-1}} \quad \text{for all } 2 \le i \le N$$

7.20 Refer to Problem 7.19. Assume that the chain is positive-recurrent.

a. Show that the limiting distribution of this chain is

$$\lim_{n \to \infty} \Pr\{X_n = i\} = \delta^{(i)} \left(\sum_{j=0}^{\infty} \delta^{(j)} \right)^{-1} \quad \text{for all } i = 0, 1, 2, \ldots$$

i. by using Problem 6.42

ii. by using Equation (7.3.1)

b. Find the limiting distribution when $A_i = i + 1$ for all $i = 0, 1, 2, \ldots$.

c. Find the limiting distribution when $A_i = (1 + p)^i$ for all $i = 0, 1, 2, \ldots$.

CHAPTER 8

Poisson Streams of Events

8.1 CONTINUOUS-TIME CHAINS

So far in this book, we have concentrated on the discrete-time Markov chains. For a continuous-time Markov chain, we noted in §2.18 that there are many ways it can be *partially* studied by using a discrete-time Markov chain embedded in it. The embedded chains can be observed only at the transitions (as in the rat-maze example introduced in §2.10), at equally spaced epochs (as in the taxicab example introduced in §3.9), or at randomly spaced epochs.

Although this method of studying continuous-time chains does provide some information about the system, we cannot be satisfied with it. We want to know something about the chains' behavior all along the continuous-time axis. For the taxicab example, we need the answers to questions such as: If we look for Tex at an arbitrary time, not necessarily at the beginning of each hour, where are we most likely to find him? For the rat-maze example, we want to know the expected total duration the rat stays in room 2 before it finds food or shock as well as the expected number of its visits to that room.

Understanding continuous-time Markov chains is our next focus, from Chapter 8 to Chapter 12.

8.2 MARKOVIAN STREAMS

Let us call an *event* something that occurs at a particular time epoch.[1] Before studying continuous-time chains, let us first analyze in this chapter and the next the *sequences of similar events*, which we call the *streams (of events)*. The arrivals of buses at a bus stop form one stream; lightning in the sky forms another. We normally denote by E_0 the event at time $t = 0$ (if there is one) and by E_i the ith one after time $t = 0$ (Figure 8.2.1).

[1]The term *event* here is not "a subset of the sample space" as defined in Chapter 1.

FIGURE 8.2.1 A stream of events.

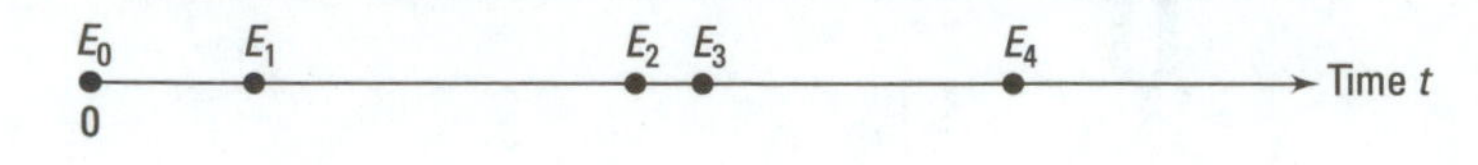

Similar to the Markov chains, a stream is *Markovian* (or *memoryless*) if *its development to the future is independent of its past.* In other words, the information about the times of its past occurrences before any time t does not help predict those in the future after t. If the arrivals of buses at a bus stop are memoryless, the knowledge that a bus just arrived 5 minutes ago does not help us predict the arrival time of the next bus.

An important consequence of the assumption that a stream is memoryless at all times is that *its developments within any two nonoverlapping intervals are independent* of each other. If accidents are memoryless, then the number of accidents occurring today is independent of those during any other day.

Is the memoryless assumption for streams too restrictive to be useful? On the one hand, there are many cases in which the probability of an event's occurrence in the next time unit is dependent on when the last similar event occurred. For example:

1. Consider a machine having a certain component. Whenever this component fails, it is immediately replaced by a similar one. The stream of component failures is memoryless only if the probability that the current component will fail within the next time unit is independent of how long it has been in use. This assumption implies that any old and memoryless component is *always as good as new*. For some components, such as computer chips, such an assumption is reasonable; for others, such as human bodies or cars, it is quite unrealistic. Human fatality varies with age, being higher at birth, during adolescence, and in old age than at other times. (For sale: a memoryless old car. Year? Mileage? Who cares?)
2. The longer a person has been in a particular occupation, the smaller the probability that he or she leaves it next year. Thus the stream of a person's changing occupations is not memoryless.
3. The longer a person has been talking to someone at a party, the higher the probability that he or she talks to someone else in the next minute. Thus the stream of a person's changing conversational partners is not Markovian.

On the other hand, similar to the discrete-time Markov chains, there are many important practical situations that can fit into this model. When a person goes to a bank or supermarket, he or she normally does not check to see when the last person arrived. Similarly, although we remember very well when the last accident occurred in our lives, this information does not help us predict when the next one will strike.

We should also realize that there are not many practical situations that are completely memoryless. With radioactive decay, the disintegration of one atom may lead to a chain reaction, making the next decay dependent on the previous ones. To be fair to public transport authorities, bus arrivals are not that memoryless. In these cases, the realness of the model has to be sacrificed for its *mathematical tractability*: The memoryless assumption, although it may not be quite appropriate, lends itself to some very elegant methods of solution and makes the model much easier to analyze. If we have to choose between a somewhat unrealistic model or a highly realistic one with which we can do nothing, usually we have to make do with the former.

Markovian stream is the simplest model, and much of its use in practice is due to its analytic tractability rather than its close fit with observable data.

8.3 REGULAR STREAMS

A stream of similar events is said to have an *explosion* at a particular time $T < \infty$ if, approaching T, events occur more and more frequently; by time T, an infinite number of occurrences have already taken place. An example is a stream in which event E_1 occurs at time $t = 1$, and the duration between events E_n and E_{n+1} is $(1/3)^n$ for all $n \geq 1$. Here, the events occur faster and faster. Furthermore, since $1 + \sum_{n=1}^{\infty}(1/3)^n = 1.5$, an explosion must take place at time $T = 1.5$.

To have an explosion at time T, it is necessary that the durations between two consecutive occurrences converge to zero, allowing an infinite number of events to occur during the duration $(T - h, T]$ regardless of how small the value of h is.

In this book, we only study *regular*[2] streams; that is, streams having *no explosion*. This requires that, at any time t,

$$\Pr\{\text{two or more occurrences within } (t, t+h]\} = o(h)$$

where $o(h)$ is a function of h that approaches zero *faster than* h or

$$\lim_{h \to 0} \frac{o(h)}{h} = 0$$

8.4 POISSON STREAMS

Let us call a stream that is both Markovian and regular a *Poisson* stream.

Because a Markovian stream develops independently in each time interval, the probability of its having exactly $i \geq 0$ occurrences during $(t, t+h]$ is a function of t and h alone, independent of its past before t and its future after $t + h$. Let us denote it by $m_i(t, h)$.

Because the regularity assumption prevents two or more occurrences from occurring within $(t, t+h]$ as $h \to 0$, let us now consider the probability $m_1(t, h)$ of having exactly one occurrence within this interval.

Assume now that $m_1(t, h)$ does not approach 0 as $h \to 0$. Let us then divide the interval $(t, t+h]$ into two equal subintervals, $(t, t+h/2]$ and $(t+h/2, t+h]$, each having length $h/2$. As $h \to 0$, this assumption also means that the probabilities α of having one occurrence in each subinterval do not approach 0. The probability of two occurrences in the interval $(t, t+h]$, which is α^2, therefore also does not approach 0,

---

[2]We should not confuse this definition with that in §6.3 describing a finite, irreducible, and aperiodic discrete-time chain.

resulting in a contradiction with the regularity property. Regularity thus also implies that $m_1(t, h) \to 0$ as $h \to 0$.

The difference between $m_1(t, h)$ and $m_i(t, h)$ for $i > 1$ is that the latter approaches 0 faster than h, whereas the former approaches 0 with the same order of h. Otherwise, if both approach 0 faster than h, there is no chance for any occurrence during any infinitesimal interval, and hence there is no stream. In other words, in a Poisson stream,

$$\begin{cases} m_1(t, h) = \lambda(t)h + o(h) & \\ m_i(t, h) = o(h) & \text{for all } i > 1 \end{cases} \tag{8.4.1}$$

where $\lambda(t)$ is the partial derivative of $m_1(t, h)$ with respect to h, and the term $o(h)$ indicates that these equalities have some errors that become insignificant as $h \to 0$. This implies

$$m_0(t, h) = 1 - \lambda(t)h + o(h) \tag{8.4.2}$$

8.5 INSTANTANEOUS AND AVERAGE RATES

During any infinitesimal interval h, the number of occurrences in a Poisson stream can only be 0 or 1, with the probability of 1 as $\lambda(t)h$. Since this is a Bernoulli random variable, according to Equation (1.11.1), the probability $\lambda(t)h$ is also the expected number of occurrences within $(t, t + h]$, or

$$\lambda(t) \equiv \lim_{h \to 0} \frac{\mathcal{E}\,[\text{number of occurrences within } (t, t + h]]}{h} \tag{8.5.1}$$

For this reason, $\lambda(t)$ is known as the *instantaneous rate (of occurrence)* of the stream. This term is consistent with the common understanding of rate as the expected number of occurrences *per unit time*.

In a way, the instantaneous rate is the instantaneous speed of the stream. Just as the distance traveled by a car during $(t, t + h]$ is the product of its instantaneous speed and h when $h \to 0$, the expected number of occurrences of a stream during $(t, t + h]$ is the product of its instantaneous rate $\lambda(t)$ and h when $h \to 0$.

The elegance of the memoryless assumption is here: The whole process can be completely specified by $\lambda(t)$, which describes the stream's instantaneous behavior. Since the stream develops independently within each infinitesimal duration, relatively simple calculus can now be used to study the stream's behavior during any period, as it has been used to obtain the distance covered during any time interval from the information about the instantaneous speeds.

The concept of the instantaneous rate leads to the following definition: For all $t > 0$ and for any duration $\tau > 0$, let the *average rate (of occurrence)* of a stream during $(t, t + \tau]$ be

$$\lambda(t, \tau) \equiv \frac{\mathcal{E}\,[\text{number of occurrences within } (t, t + \tau]]}{\tau} \tag{8.5.2}$$

The average rate is analogous to that of the average speed. Generally, $\lambda(t, \tau)$ is a function of both t and τ. For example, if the average number of buses arriving at a bus stop from 7:00 A.M. to 7:00 P.M. ($t = $ 7:00 A.M., $\tau = 12$ hours) is 24, its average rate is $24/12 = 2$ arrivals per hour during this period. The average rate of bus arrivals during the peak hours from 7:00 A.M. to 9:00 A.M. can be much higher than that during the off-peak from 1:00 A.M. to 3:00 A.M.

8.6 TIME-HOMOGENEOUS POISSON STREAMS AND THEIR RATES

We only study the *time-homogeneous* Poisson streams in this book—that is, streams in which $\lambda(t) = \lambda$ for all $t > 0$. Normally, we drop the adjective "time-homogeneous." Again, although the time-homogeneity assumption rules out many important practical applications, it simplifies the analysis a great deal.

If the instantaneous speed of a car is constant with time, it is also equal to the average speed over any duration. Similarly, in a time-homogeneous Poisson stream, the instantaneous rate $\lambda(t)$ as defined in (8.5.1) is equal to the average rate $\lambda(t, \tau)$ within any duration τ as defined in (8.5.2). Both can be referred to simply as *rate*:

$$\lambda = \lambda(t) = \lambda(t, \tau) \quad \text{for all } t > 0 \text{ and } \tau > 0 \tag{8.6.1}$$

To estimate the rate λ, we can conveniently pick any duration τ and use standard statistical techniques to find the expected number of occurrences within that duration. Rate λ completely specifies a time-homogeneous Poisson stream.

8.7 REMAINING LIVES

Let us now study the *remaining life*[3] $r(t)$ of a Poisson stream at time t. This is the duration from time t to the next event occurrence (Figure 8.7.1). This terminology arises from *reliability theory*, which is mainly concerned with the functioning of a certain component. It is normally assumed that when this component fails, it is immediately replaced by a similar one. In this case, $r(t)$ measures the duration from t to the next component failure, and hence the name.

We are interested in the distribution function $\Pr\{r(t) < \tau\}$. First, we divide the interval $(t, t + \tau]$ into k equal nonoverlapping subintervals, each having length τ/k. When $k \to \infty$, the length τ/k of each subinterval approaches zero, and the regularity assumption allows only zero or one occurrence within each subinterval. Thus, each subinterval can be treated as a Bernoulli trial in which an event occurrence yields a success. Because of the time-homogeneity assumption, we have k identically distributed trials, and the probability of a success in each is $\lambda\tau/k$ (Equation 8.4.1). Because of the memoryless property, these k trials are independent of each other.

[3]Other authors call it the *residual life*.

FIGURE 8.7.1 The lives and remaining lives of a stream.

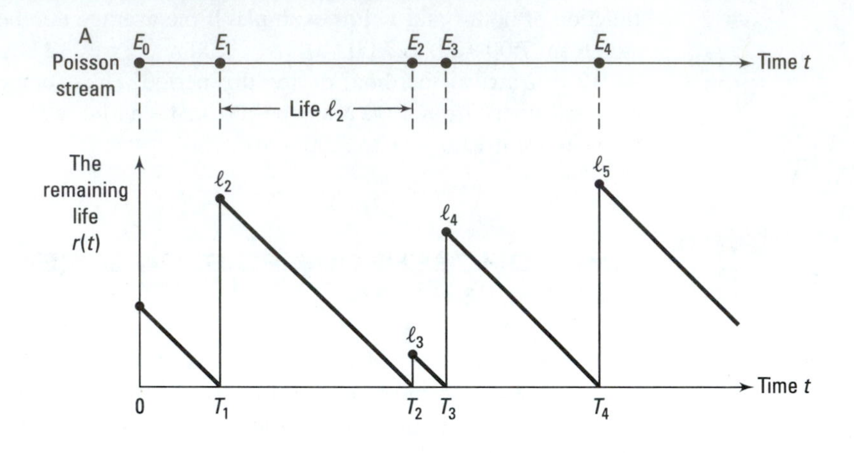

For a Poisson stream, $\Pr\{r(t) > \tau\}$ is the probability of having no occurrence within $(t, t+\tau]$. This is the probability of having no success during all of the foregoing k trials as $k \to \infty$, which is *the continuous version of the geometric distribution* $\lim_{k\to\infty}(1 - \lambda\tau/k)^k = \exp(-\lambda\tau)$ (Appendix A.1).

Thus, for a time-homogeneous, memoryless, and regular stream, the time $r(t)$ until the next occurrence has the *exponential distribution* with rate λ, or

$$\Pr\{r(t) < \tau\} = 1 - \exp\{-\lambda\tau\} \quad \text{for all } t, \tau > 0 \tag{8.7.1}$$

Note that, because of the time-homogeneous assumption, the right-hand side of this equation is independent of t.

Do you remember that any old and memoryless component is always as long-lasting as a new one? Now we know further that, old or new, its remaining life at any time t is always exponentially distributed with rate λ.

8.8 EXPONENTIAL AND HYPEREXPONENTIAL DISTRIBUTIONS

Consider a continuous random variable X having an exponential distribution with *parameter* $\mu > 0$ (Figure 8.8.1). Similar to Equation (8.7.1), the distribution function of X is

$$\Pr\{X < t\} = 1 - e^{-\mu t} \quad \text{for all } t > 0 \tag{8.8.1}$$

Its probability density function is

$$\Pr\{t \le X \le t + dt\} = \mu e^{-\mu t}\, dt \quad \text{for all } t \ge 0$$

FIGURE 8.8.1 The exponential distribution.

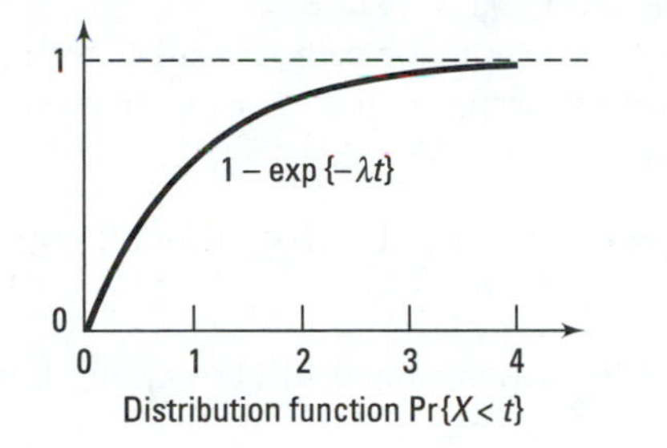

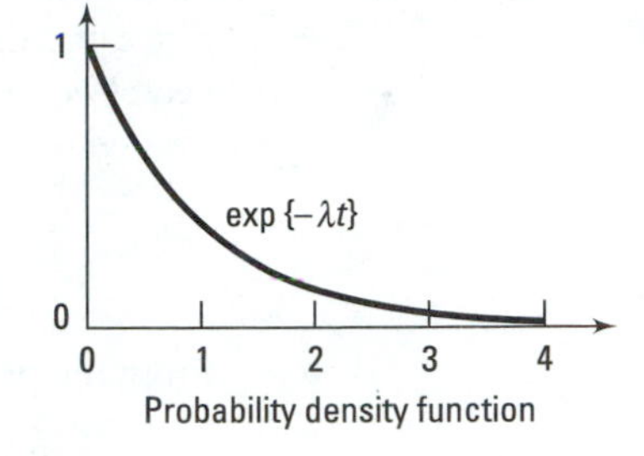

With the help of Appendix (A.17), we can calculate

$$\mathcal{E}[X] = \int_0^\infty t\mu e^{-\mu t}\, dt = \frac{1}{\mu} \tag{8.8.2}$$

$$\mathcal{E}\left[X^i\right] = \int_0^\infty t^i \mu e^{-\mu t}\, dt = \frac{i!}{\mu^i} \tag{8.8.3}$$

$$\text{Var}[X] = \mathcal{E}\left[X^2\right] - \mathcal{E}^2[X] = \frac{1}{\mu^2} \tag{8.8.4}$$

A random variable Y is said to have a *hyperexponential distribution* if it is a *mixture* of k exponential distributions; that is, its probability density function is

$$f(t) = \sum_{i=1}^{k} p_i \mu_i e^{-\mu_i t} \quad \text{for all } t \geq 0 \tag{8.8.5}$$

where $\sum_{i=1}^{k} p_i = 1$ and μ_i is the parameter of the ith exponential distribution.

When $k = 2$, it can be shown (Problem 8.6) that

$$\mathcal{E}[Y] = \frac{p_1}{\mu_1} + \frac{p_2}{\mu_2} \qquad \text{Var}[Y] = \frac{p_1(2 - p_1)}{\mu_1^2} + \frac{p_2(2 - p_2)}{\mu_2^2} - \frac{2p_1p_2}{\mu_1\mu_2} \tag{8.8.6}$$

8.9 LIVES

Let T_n be the occurrence time of event E_n for all $n \geq 0$. Consider now the duration ℓ_n between the $(n - 1)$th and nth events in a Poisson stream; that is, $\ell_n \equiv T_n - T_{n-1}$ (Figure 8.7.1). In *reliability theory*, it is the duration between the $(n - 1)$th and the nth component failures, which is equal to the life of the nth component. This is why it is known as the nth *life*, or *interoccurrence time* of the stream.

Recall that the remaining life $r(t)$ is the duration from time t to the next event occurrence, while the life ℓ_n is the duration from T_{n-1} to the next event occurrence.

Thus, $\ell_n = r(T_{n-1})$, and the discrete-time process $\{\ell_n\}_{n=0,1,2,\ldots}$ is embedded in the continuous-time process $\{r(t)\}_{t\in[0,\infty)}$ (§2.18).

Since the distribution of $r(t)$ is the same for all t, we conclude that all durations between two consecutive occurrences in a Poisson stream must also be *exponentially distributed with rate* λ, or

$$\Pr\{\ell_n < \tau\} = 1 - e^{-\lambda\tau} \quad \text{for all } n \geq 1$$

From the property of the exponential distribution, for all $n \geq 1$,

$$\mathcal{E}[\ell_n] = \frac{1}{\lambda} \qquad \mathcal{E}\left[(\ell_n)^2\right] = \frac{2}{\lambda^2} \qquad \text{Var}[\ell_n] = \frac{1}{\lambda^2}$$

The first result should be intuitively clear: If there is an average of three occurrences per hour, the average interoccurrence time must be one-third hour.

We have shown that, if the remaining lives $r(t)$ of a stream are exponentially distributed for all $t > 0$, so are the lives ℓ_n for all $n = 1, 2, 3, \ldots$. We now show that, if all lives ℓ_n in a stream are exponentially distributed, so are the remaining lives $r(t)$.

Consider an arbitrary time t, assuming that the most recent event E_n occurred before time $t - \theta$. Hence, $\ell_n > \theta$. The distribution function of the remaining life at t is thus

$$\begin{aligned}
\Pr\{r(t) < \tau \mid \ell_n > \theta\} &= \frac{\Pr\{r(t) < \tau \text{ and } \ell_n > \theta\}}{\Pr\{\ell_n > \theta\}} \\
&= \frac{\Pr\{\theta < \ell_n < \theta + \tau\}}{\Pr\{\ell_n > \theta\}} \\
&= \frac{\exp\{-\lambda\theta\} - \exp\{-\lambda(\theta + \tau)\}}{\exp\{-\lambda\theta\}} \\
&= 1 - \exp\{-\lambda\tau\}
\end{aligned}$$

Thus, if all lives ℓ_n are exponentially distributed, the stream is memoryless, because the remaining life at time t is independent of when the most recent event before t is; also, all remaining lives $r(t)$ are exponentially distributed, or $\Pr\{r(t) < \tau\} = 1 - \exp\{-\lambda\tau\}$ for all $t > 0$.

Normally, in analyzing past data, it is easier to make inference about the distribution of the lives than of the remaining lives.

Poisson Streams in Action: Durations between Earthquakes. Analyzing the intervals between earthquakes with magnitude greater than 6 east of the Rockies, Nishenko and Bollinger found that these intervals fit the exponential distribution. As reported by Bowler (1990),

> To confirm that their method was valid, Nishenko suggested they find evidence of earthquakes far back in the geological record. Traces of sediment liquefaction, sandblows (sand volcanoes) and isotopic dating of fault movements gave them intervals between movements of a particular fault over thousands of years. This data showed that they were on the right track.

On this basis, they found the probabilities of further quakes in the next 10, 30, 50 and 100 years, and reckon that the probability of a damaging quake in this region in the next 30 years is between 0.4 and 0.6—surprisingly high. ■

Poisson Streams in Action: Guarding Strategy against Predators. Ostriches (*Struthio camelus*) do not bury their heads in the sand but in the grass and other low vegetation on which they are feeding. Occasionally, they raise their heads for several seconds to detect visually any approaching predator. Bertram (1980) observed the ostriches during their breeding season in the Lake Jipe region of Tsavo West National Park, Kenya, from July to October 1977 and concluded that the intervals between two consecutive head raisings are exponentially distributed.

The exponential distribution of the durations between head raisings suggests a memoryless scanning pattern, which makes it difficult for a predator to predict when the bird will scan next.

> Improved detection of a nearby approaching predator is probably best achieved by frequent and unpredictable raising of the head . . . , rather than by long scans (during which a predator would remain motionless). ■

8.10 THE SHORTER REMAINING LIFE

Consider two independent Poisson streams S_X and S_Y having rates λ_X and λ_Y, respectively. Let $r_X(t)$ and $r_Y(t)$ be the remaining lives of S_X and S_Y at time t, respectively. We are now interested in the probability that the next event in S_X occurs before the next event in S_Y, or $\Pr\{r_X(t) < r_Y(t)\}$.

This quantity can be obtained by using the continuous-time version of the first-step analysis as follows. Consider an infinitesimal duration $(t, t+h]$.

1. Let O_1 be the event that there is one occurrence in S_X and none in S_Y. We have $\Pr\{r_X(t) < r_Y(t) \mid O_1\} = 1$ and

$$\Pr\{O_1\} = \lambda_X h(1 - \lambda_Y h) + o(h) = \lambda_X h + o(h)$$

2. Let O_2 be the event that there is one occurrence in S_Y and none in S_X. We have $\Pr\{r_X(t) < r_Y(t) \mid O_2\} = 0$ and

$$\Pr\{O_2\} = (1 - \lambda_X h)\lambda_Y h + o(h) = \lambda_Y h + o(h)$$

3. Let O_3 be the event that there is no occurrence in both S_Y and S_X. We have $\Pr\{r_X(t) < r_Y(t) \mid O_3\} = \Pr\{r_X(t) < r_Y(t)\}$ and

$$\Pr\{O_3\} = (1 - \lambda_X h)(1 - \lambda_Y h) + o(h) = 1 - \lambda_X h - \lambda_Y h + o(h)$$

4. Let O_4 be the event that there is more than one occurrence in either S_Y or S_X. We have $\Pr\{O_4\} = o(h)$.

Hence,

$$\Pr\{r_X(t) < r_Y(t)\} = \lambda_X h + (1 - \lambda_X h - \lambda_Y h)\Pr\{r_X(t) < r_Y(t)\} + o(h)$$

Letting $h \to 0$, we obtain:

$$\Pr\{r_X(t) < r_Y(t)\} = \frac{\lambda_X}{\lambda_X + \lambda_Y} \tag{8.10.1}$$

which should be intuitive enough: The higher the rate λ_X, the higher the probability that an event in S_X occurs before an event in S_Y. The reader might want to compare this equation with Equation (1.7.2).

Suppose that the times Dr. Long and Dr. Short spend examining each patient are exponentially distributed with rates two and five patients/hour, respectively. Given that both are attending patients now, the probability that Dr. Long finishes before Dr. Short is $2/(2+5)$.

8.11 POISSON (COUNTING) PROCESS

Let us now turn our attention to the continuous-time, discrete state space *counting process* $\{m(t)\}_{t\in[0,\infty)}$, in which $m(t)$ is the random variable counting the number of occurrences in a stream during $(0, t]$, starting with $m(0) = 0$ (Figure 8.11.1).

Note that:

1. Its sample space is the set of integers $\{0, 1, 2, 3, \ldots\}$.
2. It is nondecreasing with time; that is, $m(t) \le m(s)$ if $t < s$.
3. It increases by 1 whenever an event occurs.

A counting process $\{m(t)\}_{t\in[0,\infty)}$ is said to be *Poisson* if all $m(\tau)$ has a *Poisson distribution*; that is,

$$\Pr\{m(\tau) = j\} = \frac{(\lambda\tau)^j}{j!}e^{-\lambda\tau} \quad \text{for all } \tau > 0 \text{ and } j \ge 0 \tag{8.11.1}$$

FIGURE 8.11.1 A Poisson process of a stream.

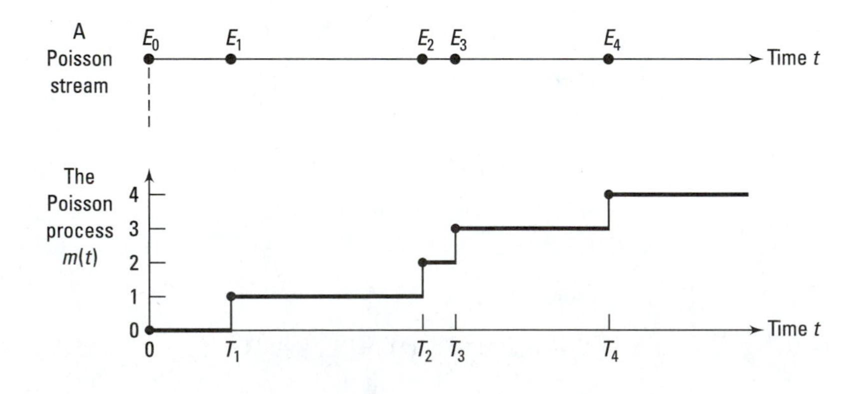

For a Poisson stream, we now show that its counting process is Poisson.[4] But before doing so, let us note that:

1. Because the stream is memoryless, its counting process increases independently in nonoverlapping intervals and is said to possess *independent increments*.
2. Because the stream is time-homogeneous, its counting process is said to have *stationary increments*.
3. Because the stream is time-homogeneous, $m(\tau)$ is not only the number of occurrences during $(0, \tau]$ but also the number of occurrences during any interval $(t, t+\tau]$ for all $t \geq 0$.

We divide the interval $(t, t+\tau]$ into k equal subintervals, each having length τ/k. When $k \to \infty$, the length of each subinterval approaches zero, and the number of occurrences in each subinterval becomes a Bernoulli random variable with the probability of success $\lambda\tau/k$. The total number of occurrences in $(t, t+\tau]$ is thus the number of successes in these k independent and identically distributed trials and has a *continuous version of the binomial distribution*:

$$\begin{aligned}\Pr\{m(\tau) = j\} &= \lim_{k\to\infty} \frac{k!}{(k-j)!j!}\left(\frac{\lambda\tau}{k}\right)^j \left(1 - \frac{\lambda\tau}{k}\right)^{k-j} \\ &= \lim_{k\to\infty} \frac{k!}{(k-j)!k^j}\,\frac{(\lambda\tau)^j}{j!}\left(1 - \frac{\lambda\tau}{k}\right)^{k-j}\end{aligned}$$

Equation (8.11.1) follows because of Appendix (A.1) and

$$\lim_{k\to\infty} \frac{k!}{(k-j)!k^j} = \frac{k(k-1)\cdots(k-j+1)}{k^j} = 1$$

$$\lim_{k\to\infty} \left(1 - \frac{\lambda\tau}{k}\right)^j = 1$$

In Problem 8.42, we show that Equation (8.11.1) gives

$$\mathcal{E}[m(t)] = \lambda t \quad \text{for all } t \geq 0 \tag{8.11.2}$$

which is consistent with (8.5.2), and

$$\operatorname{Var}[m(t)] = \lambda t \quad \text{for all } t \geq 0 \tag{8.11.3}$$

Poisson Streams in Action: Is a Patient's Stream of Seizures Poisson? The analysis of many double-blind trials of anticonvulsion assumes that the stream of seizures of a patient is Poisson. However, according to Hopkins (1985), who analyzed the data of one patient between February 1980 and January 1982,

> The number of seizures each month varied between 13 and 59. If these seizures had occurred . . . [as a Poisson stream], the distribution of the number of seizures per day would correspond to that of the Poisson probability distribution (with mean .993) . . . . When the theoretical Poisson distribution is compared with the actual observed

[4] In the literature, the term *Poisson process* is sometimes used to refer to the Poisson *stream* itself. (For example, "Customers arrive as a Poisson process.") We try to avoid this practice.

> distribution of seizures . . . , it can be seen that this is not so. There are significantly more days with a high number of seizures, or with no seizures at all than would be observed in a . . . [Poisson stream]. ■

We have shown that the number of occurrences in a Poisson stream within any duration has a Poisson distribution. Thus, as in the foregoing stream of seizures, if we can find a length of time in which the number of occurrences is not Poisson distributed, then the stream is not Poisson. We now show that the converse is also true: If the number of occurrences of a stream in all interval lengths is Poisson distributed, we have a Poisson stream.

First, streams having Poisson processes must be regular. This is because, with the help of Appendix (A.2), we write

$$\begin{aligned}
\Pr\{0 \text{ event in}(t, t+h]\} &= \exp\{-\lambda h\} \\
&= \left(1 - \lambda h + (1/2)(\lambda h)^2 + \cdots\right) \\
&= 1 - \lambda h + o(h) \\
\Pr\{1 \text{ event in } (t, t+h]\} &= \lambda h \exp\{-\lambda h\} \\
&= \lambda h \left(1 - \lambda h + (1/2)(\lambda h)^2 + \cdots\right) \\
&= \lambda h + o(h) \\
\Pr\{2 \text{ events in } (t, t+h]\} &= (1/2)(\lambda h)^2 \exp\{-\lambda h\} \\
&= (1/2)(\lambda h)^2 \left(1 - \lambda h + (1/2)(\lambda h)^2 + \cdots\right) \\
&= o(h)
\end{aligned}$$

Second, streams having Poisson processes must have exponentially distributed remaining lives. This is because the event that the remaining life $r(t)$ is longer than τ is equivalent to the event of no occurrence during $(t, t+\tau]$, which is $\exp\{-\lambda\tau\}$. (We can consider Equation (8.7.1) as a special case of (8.11.1) in which $j = 0$.) The exponential distribution of the remaining lives in turn implies the memoryless property (§8.9).

We must be very careful in using the Poisson distribution to prove that a certain stream is Poisson. Normally, we pick a convenient time interval (such as day, month, or year) and then test the hypothesis that the number of occurrences in it fits the Poisson distribution. However, even if it does, it is not sufficient to conclude that the stream is Poisson. To be sufficient, the distribution must fit for all interval lengths, not only a specific one.

Poisson Streams in Action: Is the Stream of Wars Poisson? Richardson (1944) analyzed data on war and peace from 1500 to 1931. He found that there were 223 years having no war outbreak, 142 years having one, 48 years having two, 15 years having three, and 4 years having four. Since this fits quite closely to the Poisson distribution, he suggested that the stream of war outbreaks is Poisson.

However, according to Houweling and Kuné (1984),

> Richardson and his followers inferred from the analysis mentioned above that outbreaks of war . . . [form a Poisson stream]. This inference, however, is questionable. The first weakness is the unwarranted step from the found Poisson distribution . . . to the Poisson [stream] . . . as an underlying mechanism. . . . The finding that the number of years with

k war outbreaks ($k = 0, 1, 2, \ldots$) follows the Poisson distribution does not necessarily mean that outbreaks occur independently of each other . . . and that in any subinterval of length t of the observational period, the number of outbreaks is Poisson distributed. ■

8.12 ERLANG DISTRIBUTIONS

If the duration between events E_n and E_{n+1} (spanning one life) in a Poisson stream is exponentially distributed, what is the distribution of the duration between events E_n and E_{n+k} (spanning across $k > 1$ lives)? If each memoryless component now has $k - 1$ identical *standbys* so that replacement is needed only after all k have failed, what is the distribution of the duration between two replacements? If the durations of both examination and x-ray are exponentially distributed with the same rate, what is the distribution of the total duration in a doctor's office, which includes both activities? In other words, if $\ell_1, \ell_2, \ldots, \ell_k$ are independent and exponentially distributed random variables with a common rate λ, what is the distribution of $T_k \equiv \ell_1 + \ell_2 + \cdots + \ell_k$?

Note that $T_k \leq t$ if and only if there are k or more occurrences during the interval $(0, t]$. Thus, from Equation (8.11.1),

$$\Pr\{T_k \leq t\} = 1 - \sum_{n=0}^{k-1} \frac{(\lambda t)^n}{n!} e^{-\lambda t} \quad \text{for all } k \geq 1 \text{ and } t \geq 0 \tag{8.12.1}$$

It can be shown (Problem 8.56) that the probability density function of T_k is

$$\Pr\{t < T_k < t + dt\} = \lambda e^{-\lambda t} \frac{(\lambda t)^{k-1}}{(k-1)!} dt \quad \text{for all } t \geq 0 \tag{8.12.2}$$

We say that T_k has the *Erlang distribution* with parameters $(k, \mathcal{E}[T_k])$.

Because this is the sum of k independent and identically exponentially distributed random variables, for all $k \geq 1$, we have

$$\mathcal{E}[T_k] = \frac{k}{\lambda} \qquad \mathcal{E}\left[(T_k)^2\right] = \frac{k(k+1)}{\lambda^2} \qquad \text{Var}[T_k] = \frac{k}{\lambda^2}$$

8.13 RANDOM STREAMS

Consider a time interval $(t, t + T]$. Let us divide this interval into three subintervals: $A \equiv (t, t + \tau_1]$ having length τ_1, $B \equiv (t + \tau_1, t + \tau_1 + \tau_2]$ having length τ_2, and $C \equiv (t + \tau_1 + \tau_2, t + T]$ having length $\tau_3 \equiv T - \tau_1 - \tau_2$.

Let I be the event that a Poisson stream with rate λ has exactly one occurrence during $(t, t + T]$. From Equation (8.11.1), $\Pr\{I\} = \lambda T \exp\{-T\lambda\}$.

Let J be the event that this single occurrence falls in interval B. Because the occurrences are independent in disjoint intervals, we have

$$\begin{aligned}
\Pr\{J\} &= \Pr\{0 \text{ event in } A\} \times \Pr\{1 \text{ event in } B\} \times \Pr\{0 \text{ event in } C\} \\
&= \exp\{-\lambda\tau_1\} \times \lambda\tau_2 \exp\{-\lambda\tau_2\} \times \exp\{-\lambda\tau_3\} \\
&= \lambda\tau_2 \exp\{-\lambda T\}
\end{aligned}$$

Thus, given exactly one occurrence in $(t, t+T]$, the probability that this single occurrence falls into subinterval B is

$$\Pr\{J \mid I\} = \frac{\Pr\{J \text{ and } I\}}{\Pr\{I\}} = \frac{\Pr\{J\}}{\Pr\{I\}} = \frac{\lambda\tau_2 \exp\{-\lambda T\}}{\lambda T \exp\{-\lambda T\}} = \frac{\tau_2}{T}$$

which is *proportional* to the length τ_2 of subinterval B. This means that, given exactly one occurrence during $(t, t+T]$, that single occurrence is *randomly*, or *uniformly*, distributed within $(t, t+T]$.

This result can be generalized to the event of having $k < \infty$ occurrences during $(t, t+T]$. Given this happens, each of these k occurrences is independently and uniformly distributed within $(t, t+T]$. This is why Poisson streams are also referred to as *random streams*.

8.14 SUPERPOSITION OF TWO POISSON STREAMS

Consider two independent Poisson streams S_X and S_Y having rates λ_X and λ_Y, respectively. A stream S_Z is called the *superposition* (or *combination*) of S_X and S_Y if an event in S_Z occurs whenever an event in *either* S_X *or* S_Y occurs. Figure 8.14.1 illustrates this definition.

A unique feature of the Poisson streams is that *the superposition of two independent Poisson streams is also Poisson.* While it is easy to see why the combined stream is time-homogeneous and regular, it is also intuitive that combining two forgetful streams cannot make one that remembers. Formally, let $r_X(t)$, $r_Y(t)$, and $r_Z(t)$ be the remaining lives of S_X, S_Y, and S_Z at time t, respectively. Then, for all $t, \tau \geq 0$,

$$\begin{aligned}
\Pr\{r_Z(t) > \tau\} &= \Pr\{\min[r_X(t), r_Y(t)] > \tau\} \\
&= \Pr\{r_X(t) > \tau \text{ and } r_Y(t) > \tau\} \\
&= \exp\{-\lambda_X \tau\} \exp\{-\lambda_Y \tau\} \\
&= \exp\{-(\lambda_X + \lambda_Y)\tau\}
\end{aligned}$$

which shows that the remaining lives of the combined stream are exponentially distributed with a rate equal to the *sum of the two rates*; hence, the combined stream is memoryless.

Note that:

1. From Equation (8.10.1), we know that the probability of the next occurrence in the combined stream coming from stream S_X rather than from stream S_Y is $\lambda_X/(\lambda_X + \lambda_Y)$.

FIGURE 8.14.1 Superposition of two Poisson streams.

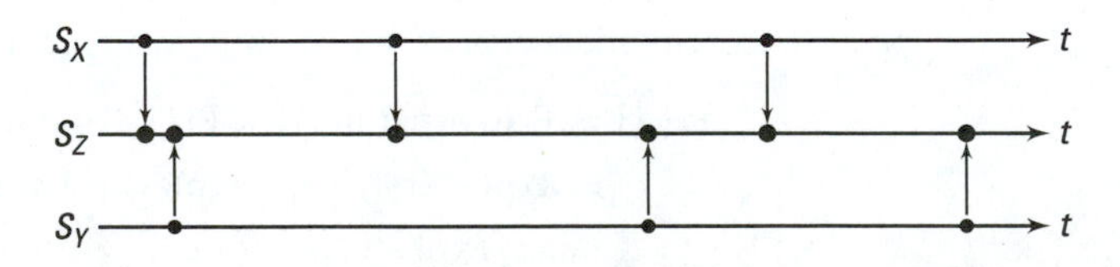

2. In Problem 8.21, we also show that the time until the next event occurrence in the combined stream is independent of whether it comes from stream S_X or stream S_Y.

Suppose buses arrive at an intersection as a Poisson stream with rate 10 arrivals/hour, and cars arrive at the same intersection as another Poisson stream with rate 100 arrivals/hour. Then the combined stream of car or bus arrivals must also be Poisson, with rate 110 arrivals/hour. At any time, the probability that the next arrival is a car rather than a bus is $100/(100+10)$.

This result can be generalized into a superposition of more than two Poisson streams.

8.15 DECOMPOSITION OF A POISSON STREAM

Consider a Poisson stream S having rate λ. Whenever an occurrence takes place in S, an independent Bernoulli trial is performed. If the outcome is a success (with probability p), an occurrence is triggered in a stream S_X; if a failure (with probability $q \equiv 1 - p$), an occurrence is triggered in another stream S_Y. We call the two streams S_X and S_Y the *decompositions* of S (Figure 8.15.1).

For example, suppose the births of babies at a particular hospital form a Poisson stream S. Let each baby be a Bernoulli trial in which a success is a girl. Then stream S can be decomposed into a stream of girl arrivals and a stream of boy arrivals.

We can expect that S_X and S_Y, being randomly selected from a memoryless stream, are both memoryless; hence, they have Poisson counting processes. What we might not be so ready to accept is that S_X and S_Y are *independent*.

To prove this, we note that the Bernoulli trials and the stream S are independent. Thus, in any duration τ:

$$\begin{aligned}
&\Pr\{m \text{ occurrences in } S_X \text{ and } k \text{ occurrences in } S_Y\} \\
&\quad = \Pr\{m+k \text{ occurrences in } S\} \Pr\{m \text{ successes in } m+k \text{ trials}\} \\
&\quad = \frac{(\lambda\tau)^{m+k}}{(m+k)!} e^{-\lambda\tau} \frac{(m+k)!}{m!k!} p^m q^k \\
&\quad = \left(\frac{(p\lambda\tau)^m}{m!} e^{-p\lambda\tau}\right) \left(\frac{(q\lambda\tau)^k}{k!} e^{-q\lambda\tau}\right) \\
&\quad = \Pr\{m \text{ occurrences in } S_X\} \Pr\{k \text{ occurrences in } S_Y\}
\end{aligned}$$

Thus, S_X and S_Y, being Poisson with rates $p\lambda$ and $q\lambda$, respectively, are independent of each other.

FIGURE 8.15.1 Decomposition of a Poisson stream.

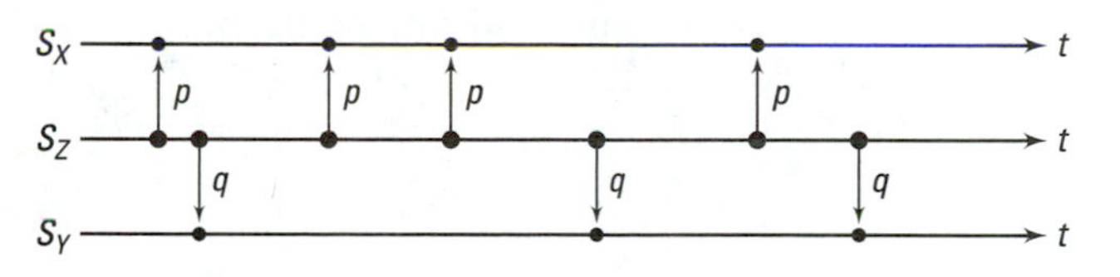

Assume that babies arrive at the hospital at a rate of ten per hour, and the probability of having a girl is .6. The girl arrivals would be Poisson with rate $(.6)(10) = 6$ arrivals/hour; the boy arrivals would be Poisson with rate $(.4)(10) = 4$ arrivals/hour. Furthermore, if 20 girls were born in the last hour, this information would tell us nothing about the number of boys born during the same period: The expected number is still four boys.

The ostriches introduced in §8.9 were observed to raise their heads in accordance with a Poisson stream. Let us now assume that, when they raise their heads, whether they turn to the right first or to the left first is independent of what they did previously. In that case, the stream of ostriches looking to the right first when they raise their heads is also Poisson.

8.16 COMPOUND POISSON PROCESS

Consider a Poisson stream with rate λ. Suppose that whenever event E_i of this stream occurs, there is a *reward* Y_i. These rewards are independent and identically distributed random variables having common mean $\mathcal{E}[Y]$ and variance $\text{Var}[Y]$. They can be discrete or continuous, positive or negative (thus, they do not have to be pleasant). When a customer buys a product, the reward is the dollar amount spent. When a disaster strikes, it is the dollar amount of damage. When a bus arrives, it is the number of passengers getting off that bus.

With $m(t)$ as the number of occurrences with $(0, t]$, consider the random sum (§1.21)

$$n(t) \equiv \sum_{i=0}^{m(t)} Y_i$$

We call the process $\{n(t)\}_{t\in[0,\infty)}$ the *compound Poisson process*. This is the total dollar amount spent by the customers during $(0, t]$ or the total dollar amount of damage caused by all disasters during $(0, t]$ or the total number of passengers getting off the buses during $(0, t]$.

Recall from Equations (8.11.2) and (8.11.3) that $\mathcal{E}[m(t)] = \text{Var}[m(t)] = \lambda t$ for all $t \geq 0$; Equations (1.21.1) and (1.21.2) now yield

$$\mathcal{E}[n(t)] = \mathcal{E}[m(t)]\mathcal{E}[Y] = \mathcal{E}[Y]\lambda t \tag{8.16.1}$$

and

$$\begin{aligned}\text{Var}[n(t)] &= \text{Var}[Y]\mathcal{E}[m(t)] + \mathcal{E}^2[Y]\text{Var}[m(t)] \\ &= \text{Var}[Y]\lambda t + \mathcal{E}^2[Y]\lambda t \\ &= \mathcal{E}\left[Y^2\right]\lambda t\end{aligned}$$

Suppose that claims arrive at an insurance company as a Poisson stream with rate $\lambda = 3$ claims/day. Also assume that the mean and variance of the dollar amount of each claim are \$1000 and 100, respectively. Then the total amount claimed in 5 days is $\$(1000)(3)(5) = 15{,}000$, and its variance is $(100 + 1000^2)(3)(5) = 15{,}001{,}500$.

8.17 SUMMARY

1. A memoryless (or Markovian) and regular stream of events is called a Poisson (or random) stream.
2. A time-homogeneous Poisson stream is completely specified by its rate λ.
3. All lives and remaining lives of a Poisson stream are exponentially distributed with rate λ.
4. A Poisson stream has a Poisson counting process, and a Poisson counting process implies a Poisson stream.
5. The duration between the nth and the $(n+k)$th events has the Erlang distribution with parameters $(k, k/\lambda)$.
6. If a Poisson stream has k occurrences in $(t, t+T]$, each occurrence is uniformly distributed within that interval.
7. The probability that an event in a Poisson stream S_X (having rate λ_X) occurs before an event in a Poisson stream S_Y (having rate λ_Y) is $\lambda_X/(\lambda_X + \lambda_Y)$.
8. The superposition of two Poisson streams having rate λ_X and λ_Y is also Poisson with rate $\lambda_X + \lambda_Y$.
9. A Poisson stream having rate λ can be decomposed into two independent Poisson streams, one having rate $p\lambda$ and the other $(1-p)\lambda$.
10. A compound Poisson process is the random sum totaling the rewards given at each occurrence of a Poisson stream.

PROBLEMS

8.1 John arrives at a bank having one teller and is told that this teller has been serving another customer for 10 minutes. He also knows that this teller serves an average of ten customers per hour and his service times are exponentially distributed. What is the probability that John has to wait for more than 15 minutes?

8.2 Show that, if X is an exponentially distributed random variable with parameter λ, then ξX is exponentially distributed with parameter λ/ξ for all $\xi > 0$.

8.3 Suppose that X is an exponentially distributed random variable with parameter λ. Let Y be the integer portion of X. Show that Y has a geometric distribution. Explain.

8.4 Let X be exponentially distributed with parameter λ. For a constant d, show that

$$\Pr\left\{\left(X + d^2\right)^{1/2} \le x\right\} = 1 - \exp\left\{-\lambda\left(x^2 - d^2\right)\right\}$$

8.5 At time 0, there are N persons in a room. Assume the duration each person stays in this room is exponentially distributed with rate μ. Show that the number of people still in this room at time T has a binomial distribution.

8.6 Obtain the mean and variance of a random variable X having the hyperexponential distribution with $k = 2$, as in Equation (8.8.6).

8.7 Let X be an exponentially distributed random variable having parameter λ. Let T be a constant.

a. Show that

$$\Pr\{X < t \mid X < T\} = \frac{1 - e^{-\lambda t}}{1 - e^{-\lambda T}} \quad \text{for all } 0 < t < T$$

b. Use Problem 1.51 to show that

$$\mathcal{E}[\min(X, T)] = \frac{1 - e^{-\lambda T}}{\lambda}$$

c. Show that

$$\mathcal{E}[X \mid X < T] = \frac{1}{\lambda} - \frac{Te^{-\lambda T}}{1 - e^{-\lambda T}}$$

i. by conditioning on whether $X < T$ or not
ii. by using Problem 1.51a and the result in part a
iii. by using Problem 1.51d

d. Hence show that, and explain why,

$$\mathcal{E}[X \mid X > T] = T + \frac{1}{\lambda}$$

e. Also show that

$$\mathcal{E}[\max(X, T)] = T + \frac{1}{\lambda}e^{-\lambda T}$$

8.8 Assume it takes a fixed duration T to repair a machine. However, if an electrical shock occurs during the repair, the whole repairing process has to be restarted from the beginning immediately. Assume shocks arrive as a Poisson stream with rate λ. Find the expected duration to finish a job by using conditional arguments and Problem 8.7.

8.9 Refer to Problem 8.8. Assume now that, if an electrical shock occurs during repair time, the whole repairing process can only be restarted after a random duration r. Find the expected duration to finish a job.

8.10 Cars pass a point in a road as a Poisson stream with rate λ. Assume that you only cross the road immediately after the nth car if the next car won't pass by at least T seconds later.

a. Let N be the number of cars passing by before you cross the road. What is the distribution of N and its expected value?

b. What is the expected duration you have to wait before crossing the road? *Hint:* Use Problem 8.7.

8.11 Let $x_1, x_2, x_3, \ldots$ be a sequence of independent and identically distributed bids on a certain asset that is offered for sale. Suppose the bids are exponentially distributed with parameter λ. The successful bid is the one that equals or exceeds a predetermined amount M. Find the expected value of the successful bid. Compare with Problem 1.43. *Hint:* Use Problem 8.7.

8.12 Let X be an exponentially distributed random variable with parameter λ. Let T be a constant and $\rho \equiv \exp\{-\lambda T\}$.

a. Find $\text{Var}[X \mid X > T]$.

b. Hence, show that

$$\mathcal{E}[X^2 \mid X > T] = T^2 + \frac{2T}{\lambda} + \frac{2}{\lambda^2}$$

c. Use conditional arguments to show that

$$\mathcal{E}[X^2 \mid X < T] = \frac{2 - \rho\left(\lambda^2 T^2 + 2\lambda T + 2\right)}{\lambda^2 (1 - \rho)}$$

d. Hence, show that

$$\text{Var}[X \mid X < T] = \frac{1}{\lambda^2} - \frac{\rho T^2}{(1 - \rho)^2}$$

8.13

a. The *age* $a(t)$ of a stream at time t is the duration elapsed since the last event occurrence. Find the distribution and the expected value of $a(t)$ for a Poisson stream.

b. The *current life* $c(t)$ of a stream at time t is the sum of its age and its remaining life. Find the expected value of $c(t)$ of a Poisson stream. Show that it is longer than the stream's expected life.

8.14 Arriving at a barbershop, Tom finds two barbers serving two other customers. He will be served as soon as one of these customers leaves. The duration of each haircut is exponentially distributed with rate λ. What is the probability that, of the three customers, Tom is the last to leave?

8.15 Sue and Liz arrive at a beauty salon together and plan to leave together. Sue needs a perm, and Liz a manicure. The duration of a perm is exponentially distributed with rate λ/hour; that of a manicure is exponentially distributed with rate μ/hour. If both are served immediately, what is the expected duration one has to wait for the other?

8.16 Arriving at a barbershop, Tom finds the barber busy and Jim waiting for his turn. Tom and Jim are impatient persons. Jim is only prepared to wait for an exponential duration with rate λ_1; he will leave if he is not served by that time. Similarly, Tom is only prepared to wait for an exponential duration with rate λ_2. Suppose the duration of a haircut is exponentially distributed with rate μ.

a. What is the probability that Jim will receive a haircut?

b. What is the probability that Tom will receive a haircut and Jim will not?

c. What is the probability that both Tom and Jim will receive haircuts?

8.17 Arriving at a beauty salon, Sue finds Liz and Grace being served by two beauticians. Sue will be served as soon as Liz or Grace leaves. Liz is having a perm, and Grace a manicure. The duration of a perm is exponentially distributed with rate λ; that of a maincure is exponentially distributed with rate μ. In the following cases, what is the probability that Sue will be the last one to leave? That Liz will be the last one to leave?

a. Sue needs a perm.

b. Sue needs a manicure.

c. The beautician serving Liz only does perms, and the one serving Grace only does manicures. Sue will go to whomever is available first.

8.18 Customers arrive at an office as a Poisson stream with rate λ. The service time of each customer is exponentially distributed with rate μ. Assume the first customer arrives at time 0 and is served immediately.

a. What is the probability that the second customer has to wait?

b. What is the expected waiting time of the second customer?

8.19 Cars at a washing facility have to go through a washing station and then a drying station. These stations can only serve one car at a time. If an arriving car finds the washing station busy, it has to wait. After being washed, it can move to the drying station immediately if this station is free; if not, it has to remain and wait at the washing station. The washing and drying times are exponentially distributed with rates μ_1 and μ_2, respectively. Suppose car A arrives at the facility and finds another car being washed. What is the expected duration that car A spends at the facility?

8.20 Let X and Y be two exponentially distributed random variables with parameters λ and μ, respectively. Let $Z \equiv |X - Y|$. Find the distribution of Z, its expected value, and its variance. *Hint:* Use Problem 8.6.

8.21 Let X and Y be two exponentially distributed random variables with parameters λ and μ, respectively.

a. Show that the events $\min(X, Y) < t$ and $X < Y$ are independent; that is

$$\Pr\{\min(X, Y) < t \text{ and } X < Y\} = \frac{\lambda}{\lambda + \mu}\left(1 - e^{-(\lambda+\mu)t}\right)$$

b. Hence, show that

$$\mathcal{E}[X \mid X < Y] = \frac{1}{\lambda + \mu}$$

c. Also show that

$$\mathcal{E}[X \mid X > Y] = \frac{1}{\lambda + \mu} + \frac{1}{\lambda}$$

8.22 Assume it takes an exponentially distributed duration T with rate μ to repair a machine. However, if an electrical shock occurs during repair time, the whole procedure has to be restarted immediately from the beginning. Assume shocks arrive as a Poisson stream with rate λ. Use conditional arguments and Problem 8.21 to find the expected duration to finish a job. Explain.

8.23 Refer to Problem 8.22. Assume now that after each shock, the whole procedure cannot be restarted immediately but only after a random time r. Use conditional arguments and Problem 8.21 to find the expected duration to finish a job.

8.24 *Geiger-Müller* counters are used to count the number of impulses emitted by radioactive materials. After an impulse is counted (or registered), some counters have a *dead period*, in which arriving impulses are not counted. In a *nonparalyzable* counter, these dead periods are not affected by the uncounted impulses arriving during them. Suppose the dead periods are exponentially distributed with rate μ and the impulses arrive as a Poisson stream with rate λ.

a. Find the expected duration from a counted impulse to the next counted impulse.

b. Find the expected duration from a counted impulse to the next uncounted impulse. *Hint:* Use Problem 8.21.

c. Hence show that the expected duration from an uncounted impulse to the next uncounted impulse is $(\lambda + \mu)\,\lambda^{-2}$.

d. Show that the expected total dead period from an uncounted impulse to the next uncounted impulse is $1/\lambda$.

8.25 Let X and Y be two exponentially distributed random variables with parameters λ and μ, respectively. Show that

$$\mathcal{E}[\min(X, Y)] = \frac{1}{\lambda + \mu}$$

a. by using Problem 1.52

b. by using Problem 8.7

c. by using Problem 8.21

d. by superpositioning two Poisson streams

8.26 Let X and Y be two exponentially distributed random variables with rates λ and μ, respectively. Show that

$$\mathcal{E}[\max(X, Y)] = \frac{1}{\lambda} + \frac{1}{\mu} - \frac{1}{\lambda + \mu}$$

a. by using Problem 1.52

b. by using Problem 8.7

c. by using Problem 8.21

d. by using Problem 8.25

e. by using the fact that

$$\mathcal{E}[\max(X, Y)] = \mathcal{E}[X] + \mathcal{E}[Y - X \mid Y > X]\,\Pr\{Y > X\}$$

8.27 Let $X_1, X_2, \ldots, X_N$ be N independent and identically distributed exponential random variables with common rate λ. Let $Y_N \equiv \max(X_1, X_2, \ldots, X_N)$.

a. Find the distribution of Y_N. *Hint:* Use Problem 1.52.

b. Find $\mathcal{E}[Y_N] - \mathcal{E}[Y_{N-1}]$. *Hint:* Use Problem 1.51a.

c. Hence, show and explain why

$$\mathcal{E}[Y_N] = \frac{1}{\lambda}\left(1 + \frac{1}{2} + \cdots + \frac{1}{N}\right)$$

d. Let $Z_1, Z_2, \ldots, Z_N$ be N independent and identically distributed exponential random variables with rates $\lambda, 2\lambda, \ldots, N\lambda$, respectively. Let $T_N \equiv Z_1 + Z_2 + \cdots + Z_N$. Explain why T_N and Y_N are identically distributed.

8.28 Let X and Y be two independent exponential random variables with rates λ and μ, respectively.

a. Find Var[max(X, Y)]. *Hint:* Use Problem 8.20.

b. Find Var[max(X, Y)] when $\lambda = \mu$.

8.29 There are N different types of coupons. Coupons of type i $(1 \le i \le N)$ arrive in a person's mailbox as a Poisson stream with rate λ_i. Assume $\lambda_i = \lambda$ for all $1 \le i \le N$. What is the expected time until this person collects r different types of coupons?

8.30 Machine X has two parallel components A and B, the lives of which are exponentially distributed with rates λ_A and λ_B, respectively. It stops functioning when both components fail. Machine Y has one component C, the life of which is exponentially distributed with rate μ. It stops functioning when this component fails. What is the probability that X stops before Y?

8.31 Consider a system having N components. It is called a *k-out-of-N* system because it works only when *at least k* components work. Assume that all components function at time $t = 0$, each component fails independently, the lives of all components are identically exponentially distributed with rate λ, and the components are *nonrepairable*.

a. Find the system's *reliability* $R(t)$, which is the probability that it survives up to time t, or the probability that it functions *continually* in the interval $(0, t]$.

b. The system's mean-time-to-system-failure *MTSF* is the expected duration from time $t = 0$ to the first time the system fails. When $N = 3$ and $k = 2$, use Problem 1.55 to find the *MTSF*.

8.32 Consider a two-out-of-three system which has three components and only functions when at least two of its components function. Assume that all components function at time $t = 0$, each component fails independently, the lives of all components are identically exponentially distributed with rate λ, and the components are *repairable*—that is, when a component fails, it can be repaired immediately by its own repairer with an exponential rate μ.

a. The system's *up-period* t_{12} is the duration from the time the system is brought back into operation (when one of the two down machines is repaired) to its next failure (when one of the two up machines fails). Obtain $\mathcal{E}[t_{12}]$ using conditional arguments. *Hint:* Use Problem 8.21.

b. The system's *down-period* t_{21} is the duration between a system failure and the moment it is brought back into operation. Obtain $\mathcal{E}[t_{21}]$ using conditional arguments.

8.48 Consider a Poisson stream having rate λ. Let N be the number of events occurring during τ, where τ is a random duration having probability density function $h(t)$, with $\mathcal{E}[\tau] < \infty$ and $\mathcal{E}[\tau^2] < \infty$.

a. Find the distribution of N.

b. Show that $\mathcal{E}[N] = \lambda\mathcal{E}[\tau]$.

c. Show that $\text{Var}[N] = \lambda\mathcal{E}[\tau] + \lambda^2\text{Var}[\tau]$.

d. When τ is exponentially distributed with rate μ, show that N has a geometric distribution with parameter $p = \mu/(\lambda + \mu)$. Hence, find $\mathcal{E}[N]$ and $\text{Var}[N]$. *Hint:* Use Appendix (A.17).

e. Find the distribution of N, $\mathcal{E}[N]$, and $\text{Var}[N]$ when τ has an Erlang distribution with parameters (k, μ).

8.49 Passengers arrive at a bus stop as a Poisson stream with rate ten per hour. Buses arrive at the same stop as a Poisson stream with rate two per hour. Each bus takes all waiting customers away. What is the distribution of the number of passengers getting on each bus? What is its expected value?

8.50 Let X be a discrete random variable having a Poisson distribution with parameter θ. The parameter θ in turn is a random variable having an exponential distribution with parameter λ. Show that X has a geometric distribution. *Hint:* Use Appendix (A.17).

8.51 Suppose two random variables X and Y are exponentially distributed with the same parameter λ. Use the convolution equation in Problem 1.56 to show that $X + Y$ has the Erlang distribution with parameters $(2, \lambda/2)$.

8.52 Suppose two random variables X and Y are exponentially distributed with parameters λ and μ, respectively ($\lambda \neq \mu$). Use the convolution equation in Problem 1.56 to show that the distribution function of $X + Y$ is

$$\Pr\{X_1 + X_2 < t\} = 1 - \frac{\lambda}{\lambda - \mu}e^{-\mu t} - \frac{\mu}{\mu - \lambda}e^{-t\lambda} \quad \text{for all } t > 0$$

8.53 Consider a sequence of independent and identically exponentially distributed random variables $X_1, X_2, \ldots$, all with the same parameter λ. We say X_n holds a *record* if $X_n > \max(X_1, X_2, \ldots, X_{n-1})$. (This means that X_1 holds the first record.) Find the distribution of the ith record.

8.54 Consider a random sum $T_N \equiv \ell_1 + \ell_2 + \cdots + \ell_N$, in which all ℓ_i are exponentially distributed with rate λ, and N has a geometric distribution with $\Pr\{N = n\} = \beta(1 - \beta)^{n-1}$.

a. Show that T_N is exponentially distributed with rate $\lambda\beta$. Explain. *Hint:* Use Equation (8.12.2) and Appendix (A.2).

b. Find $\mathcal{E}[T_N]$ and $\text{Var}[T_N]$

i. from the result obtained in part a

ii. from Equations (1.21.1) and (1.21.2)

c. A customer requires an exponential amount of service with rate μ. After each service, if she is not satisfied, she will demand another service of the same rate. Let the probability of dissatisfaction be α. What is the distribution of the total service time of the customer?

8.33 There are two machines and one repairperson. Each machine operates for an exponential duration with rate λ before breaking down. Let the duration to repair a machine be R. If a machine breaks down while the repairperson is repairing the other machine, it has to wait. Assume a machine is down at time 0, let t_{10} be the duration until the first time both machines work again. Use conditional arguments to show that

a. $\mathcal{E}[t_{10}] = R\exp\{\lambda R\}$ when R is a fixed duration

b. $\mathcal{E}[t_{10}] = (\lambda + \mu)/\mu^2$ when R is exponentially distributed with rate μ

8.34 There are two parallel machines A and B, and two repairpersons. Machine i $(i = A, B)$ operates for an exponential duration with rate λ_i before breaking down. If machine i is down, it takes a repairperson an exponential duration with rate μ_i to fix it. Let state 0 indicate that no machine is down at t, state A that machine A is down, state B that machine B is down, and state 2 that both are down.

a. Given that only machine A is down, let P_{A0} be the probability that both machines are up when machine A is brought back into operation. Use conditional arguments to show that

$$P_{A0} = \frac{\mu_A + \mu_B}{\lambda_B + \mu_A + \mu_B}$$

b. Let t_{i0} $(i = 0, A, B, 2)$ be the time until both machines are working for the first time, given that the system is in state i at time 0. Obtain a set of two equations to show that

$$\mathcal{E}[t_{A0}] = \frac{(\lambda_B + \mu_B)(\mu_A + \mu_B + \lambda_A)}{(\lambda_A + \lambda_B + \mu_A + \mu_B)\mu_A\mu_B}$$

and

$$\mathcal{E}[t_{B0}] = \frac{(\lambda_A + \mu_A)(\mu_A + \mu_B + \lambda_B)}{(\lambda_A + \lambda_B + \mu_A + \mu_B)\mu_A\mu_B}$$

c. Hence, show that

$$\mathcal{E}[t_{00}] = \frac{\lambda_A\lambda_B + \lambda_A\mu_B + \lambda_B\mu_A + \mu_A\mu_B}{(\lambda_A + \lambda_B)\mu_A\mu_B}$$

8.35 Refer to Problem 8.34. Assume now that there is only one repairperson. If a machine breaks down while the repairperson is fixing the other machine, it has to wait. Use conditional arguments to show that

$$\mathcal{E}[t_{A0}] = \frac{(\lambda_B\mu_B + \lambda_B\mu_A + \mu_B\mu_A)(\lambda_A + \mu_B)}{\mu_A\mu_B(\lambda_B\mu_B + \lambda_A\mu_A + \mu_A\mu_B)}$$

and

$$\mathcal{E}[t_{B0}] = \frac{(\lambda_A\mu_A + \lambda_A\mu_B + \mu_A\mu_B)(\lambda_B + \mu_A)}{\mu_A\mu_B(\lambda_A\mu_A + \lambda_B\mu_B + \mu_A\mu_B)}$$

8.36 Refer to Problem 8.35. Assume now that machine A has a *higher priority* than machine B. If machine A breaks down when the repairperson is fixing machine B, she will leave machine B to work on machine A and resume working on B only when she finishes with A. If machine B breaks down when she is fixing machine A, machine B will have to wait until she finishes with A.

a. By conditioning on whether machine A breaks down during machine B's repair time, show that

$$\mathcal{E}[t_{B0}] = \frac{\lambda_A + \mu_A}{\mu_A \mu_B}$$

b. By conditioning on whether machine B breaks down during machine A's repair time, show that

$$\mathcal{E}[t_{A0}] = \frac{\lambda_A \lambda_B + \lambda_B \mu_A + \lambda_B \mu_B + \mu_A \mu_B}{(\lambda_B + \mu_A)\, \mu_A \mu_B}$$

c. By conditioning on whether A breaks down before B, show that

$$\mathcal{E}[t_{00}] = \frac{(\lambda_A + \mu_A)\left(\lambda_A \lambda_B + \lambda_B^2 + \lambda_B \mu_A + \mu_B \lambda_B + \mu_B \mu_A\right)}{(\lambda_B + \mu_A)(\lambda_A + \lambda_B)\, \mu_A \mu_B}$$

8.37 Refer to Problem 8.34. Let us now assume that only one machine is needed at a time. If machine A is working, machine B is used as a *standby redundant* and thus cannot fail. Only when machine A is under repair is machine B put in use and thus can fail.

a. Use Problem 8.34 to show that

$$\mathcal{E}[t_{00}] = \frac{(\lambda_A + \mu_A)\left(\mu_B \mu_A + \lambda_A \mu_B + \lambda_B \mu_B + \lambda_A \lambda_B + \mu_B^2\right)}{\lambda_A \mu_A \mu_B (\lambda_A + \lambda_B + \mu_A + \mu_B)}$$

b. Given that machine A is down and machine B is up, what is the probability that both machines are down when machine B fails?

c. Given that machine B is down and machine A is up, what is the probability that both machines are down when machine A fails?

d. Let t_{i2} ($i = 0, A, B, 2$) be the time until both machines are down for the first time, given that the system is in state i at time 0. Show that

$$\mathcal{E}[t_{A2}] = \frac{\lambda_A + \mu_A}{\lambda_A \lambda_B}$$

$$\mathcal{E}[t_{B2}] = \frac{\lambda_A \lambda_B + \lambda_B \mu_B + \lambda_A \mu_B + \mu_A \mu_B}{\lambda_A \lambda_B (\lambda_A + \mu_B)}$$

and

$$\mathcal{E}[t_{22}] = \frac{(\lambda_A + \mu_A)\left(\lambda_A \lambda_B + \lambda_A \mu_B + \lambda_B \mu_B + \mu_A \mu_B + \mu_B^2\right)}{\lambda_A \lambda_B (\mu_A + \mu_B)(\lambda_A + \mu_B)}$$

8.38 Consider a Poisson stream with rate $\lambda = 4$ occurrences/hour. Find the joint probability of having five occurrences in the first 2.5 hours, seven in the first 4 hours, and ten in the first 5 hours.

8.39 Consider a Poisson stream with rate 3 occurrences/hour.

a. Use a computer to draw a graph of $\Pr\{r(t) \geq \tau\}$ for $0 \leq \tau \leq 2$ hours. Compare this graph with a geometric distribution

i. having step increment of .1 hour and $p = .03$

ii. having step increment .05 hour and $p = .015$

b. Use a computer to draw a graph of $\Pr\{m(t) = n\}$ for $t = 1$ hour and $1 \leq n \leq 10$. Compare this graph with a binomial distribution

i. having step increment .1 hour and $p = .03$

ii. having step increment .05 hour and $p = .015$

8.40 Ulmer (1982) studied 191 years of appointments to the U.S. Supreme Court (from 1790 to 1980) and found 114 years with no appointment, 58 years with one appointment, 17 years with two appointments, and 2 years with three appointments. Test the hypothesis that the appointments follow a Poisson distribution. Verify that

> When Jimmy Carter stepped down as president on January 20, 1981, he carried with him at least one unique distinction: he is the only president serving four or more years who failed to make a single appointment to the U.S. Supreme Court. . . . On the . . . assumption that Supreme Court vacancies are independent events, the probability that President Carter would serve four years without an appointment to the Court is .13. Thus, the Carter experience was not unlikely to occur by chance at conventional probability levels. Viewed in that manner, surprise at Carter's barren four years would be considerably diminished.

8.41 Test the hypothesis that the war data obtained by Richardson (1944) in §8.11 follow a Poisson distribution.

8.42 Prove Equations (8.11.2) and (8.11.3). *Hint:* Find $\mathcal{E}[X(X-1)]$ and use Appendix (A.2).

8.43 Let X be a discrete random variable having a Poisson distribution with parameter λ. What is the *mode* of X—that is, the value of X that has the highest probability?

8.44 Let X be a discrete random variable having a Poisson distribution. Show that it is more likely to be even than odd. (Consider 0 even.)

8.45 Let Z be the superposition of two independent Poisson streams X and Y having rates λ and μ, respectively. Let $m_Z(\tau)$ be the number of occurrences in stream Z during $(0, \tau]$. Find the distribution of $m_Z(\tau)$. Explain.

8.46 For a Poisson stream, show and explain why for all $s < t$, $n > 0$, and $k = 0, 1, 2, \ldots, n$,

$$\Pr\{m(s) = k \mid m(t) = n\} = \binom{n}{k}\left(\frac{s}{t}\right)^k \left(1 - \frac{s}{t}\right)^{n-k}$$

8.47 Assume there are n occurrences in a Poisson stream during $(0, \tau]$; that is, $m(\tau) = n$. We divide the interval $(0, \tau]$ into k disjoint subintervals $I_1, I_2, \ldots, I_k$ having lengths $\tau_1, \tau_2, \ldots, \tau_k$, respectively ($\tau_1 + \tau_2 + \cdots + \tau_k \cdots = \tau$). Show that the probability of having $n_1, n_2, \ldots$ and n_k occurrences ($n_1 + n_2 + \cdots + n_k = n$) in the intervals $I_1, I_2, \ldots$ and I_k, respectively, is

$$\frac{n!}{n_1! n_2! \cdots n_k!}\left(\frac{\tau_1}{\tau}\right)^{n_1}\left(\frac{\tau_2}{\tau}\right)^{n_2}\cdots\left(\frac{\tau_k}{\tau}\right)^{n_k}$$

8.55 Let X be a random variable having a Poisson distribution with rate θ. The parameter θ in turn is a random variable having an Erlang distribution with parameters $(k, \lambda/k)$. Show that X has the negative binomial distribution with probability mass function (1.14.2) in which $p \equiv \lambda/(1-\lambda)$. *Hint:* Use Appendix (A.17).

8.56 Derive Equation (8.12.2) from Equation (8.12.1).

8.57 Assume there are k occurrences in a Poisson stream during $(0, T]$; that is, $m(T) = k$. Let T_1 be the time of the first occurrence. Use Problem 1.53 to find $\Pr\{T_1 < s \mid m(T) = k\}$ and $\mathcal{E}[T_1 \mid m(T) = k]$.

8.58 Let $Y(T)$ be the time of last occurrence before T.

a. Assume there are k occurrences in a Poisson stream during $(0, T]$; that is, $m(T) = k$. Use Problem 1.54 to find $\Pr\{Y(T) < s \mid m(T) = k\}$ and $\mathcal{E}[Y(T) \mid m(T) = k]$.

b. Hence, find $\mathcal{E}[Y(T)]$.

8.59 A random variable X, taking values on the interval $[0, T]$, is said to have a *beta distribution* with parameters $(v > 0, w > 0)$ if its probability density function is

$$f(x) = \begin{cases} \dfrac{1}{T}\dfrac{\Gamma(v+w)}{\Gamma(v)\Gamma(w)}\left(\dfrac{x}{T}\right)^{v-1}\left(1-\dfrac{x}{T}\right)^{w-1} & \text{for all } 0 \le x \le T \\ 0 & \text{otherwise} \end{cases}$$

where, for all $0 \le x < \infty$, $\Gamma(x) \equiv \int_0^\infty t^{x-1}e^{-t}\,dt$ is known as the *gamma function*. Note that when x is an integer, $\Gamma(x) = (x-1)!$. It can be shown that $\mathcal{E}[X] = (vT)/(v+w)$.

a. Consider a Poisson stream having rate λ. Assume there are k occurrences during $(0, T]$; that is, $m(T) = k$. Show that, for all $i \le k$, the occurrence time T_i has the beta distribution with parameters $(v = i, w = k - i + 1)$. Hence, find $\mathcal{E}[T_i \mid m(T) = k]$. Verify the results in Problems 8.57 and 8.58.

b. Assume the kth occurrence in a Poisson stream takes place at time T; that is, $T_k = T$. Show that, for all $i \le k$, the occurrence time T_i has the beta distribution with parameters $(v = i, w = k - i)$. Hence, find $\mathcal{E}[T_i \mid T_k = T]$.

8.60 Consider a Poisson stream having rate λ.

a. Assume there is one occurrence during $(0, T]$ at time $T_1 < T$; that is, $m(T) = 1$. Let $f(t)$ be any nonnegative function. Show that

$$\mathcal{E}\left[f(T_1) \mid m(T) = 1\right] = \frac{1}{T}\int_0^T f(x)\,dx$$

b. Hence, show that, for any fixed duration T,

$$\mathcal{E}\left[\sum_{i=1}^{m(T)} f(T_i)\right] = \lambda \int_0^T f(x)\,dx$$

where T_i is the time of the ith occurrence.

c. Hence, prove Equation (8.11.2).

8.61 Messages arrive at a clearing system as a Poisson stream with rate λ arrivals per hour. The system is inspected every T hours, at which time all messages are cleared. Let

ω be the total waiting time (before being cleared) of all messages arriving between two consecutive clearings.

a. Use Problem 8.60 to show that $\mathcal{E}[\omega] = \lambda T^2/2$ when T is a fixed value.

b. Hence, find $\mathcal{E}[\omega]$ when T is a random variable.

c. Find $\mathcal{E}[\omega]$ when T is exponentially distributed with rate μ.

8.62 It costs \$$c$ to replace a Markovian instrument having exponentially distributed life with rate λ. Suppose \$1 spent at time t has a present value of \$$e^{-\alpha t}$, where α is the *continuous-time interest rate*. Let $C(T)$ be the present value of the total replacement cost during $(0, T]$.

a. Given k replacements within duration $(0, T]$, use Problem 8.60 to show that

$$\mathcal{E}[C(T) \mid m(T) = k] = ck\frac{1 - e^{-\alpha T}}{T\alpha}$$

b. What is $\mathcal{E}[C(T)]$? Compare with Problem 1.19.

c. If T is exponentially distributed with rate μ, what is $\mathcal{E}[C(T)]$?

d. If T is exponentially distributed with rate μ, what is $\lim_{T\to\infty} \mathcal{E}[C(T)]$?

e. Find $\lim_{T\to\infty} \mathcal{E}[C(T)]$ when $c = \$100$, $\lambda = 1$/month, and $\alpha = \$15$ annually.

8.63 Electrical pulses arrive at a counter as a Poisson stream with rate λ. At their arrivals, their amplitudes A are independent and identically distributed. After their arrivals, their amplitudes decrease exponentially; that is, if a pulse arrives at time t with the initial amplitude a, its amplitude will be $a\exp\{-\alpha\tau\}$ at time $(t+\tau)$. Assume no pulse at time 0.

a. Given one impulse having amplitude a arriving during $(0, T]$, find the expected amplitude at time T. *Hint:* Use Problem 8.60.

b. Given one impulse having a random amplitude A arriving during $(0, T]$, find the expected amplitude at time T.

c. Hence, find the expected total amplitude at time T of all impulses arriving during $(0, T]$.

8.64 Cars arrive at a parking lot as a Poisson stream with rate λ. They stay there for an exponential duration with rate μ. Let $N(T)$ be the number of cars in the lot at time T.

a. Use Problem 8.60 to show that $\mathcal{E}[N(T) \mid N(0) = 0] = \lambda(1 - e^{-\mu T})/\mu$.

b. Find $\mathcal{E}[N(T) \mid N(0) = k]$.

8.65 A man collected some kind of instruments. They came as a Poisson stream with rate λ. He took good care of them and thus all functioned and none deteriorated. Then he died at time T and the instruments started to deteriorate independently, with each remaining life exponentially distributed with rate δ. There were no more additions to the collection. Let $X(t)$ be the number of working instruments at time t.

a. Show that $X(t)$ has the Poisson distribution for all $t > 0$. *Hint:* Use Problem 8.5.

b. Hence, find $\mathcal{E}[X(t)]$.

8.66 Let Z be the superposition of two independent Poisson streams X and Y having rates λ and μ, respectively. Let $m_X(T)$ and $m_Z(T)$ be the number of occurrences in streams X and Z, respectively, during $(0, T]$.

a. Given $m_Z(T) = k$, show that $m_X(T)$ has the binomial distribution.

b. Hence, find $\mathcal{E}[m_X(T) \mid m_Z(T) = k]$.

FIGURE 8.18.1 **Problem 8.69.**

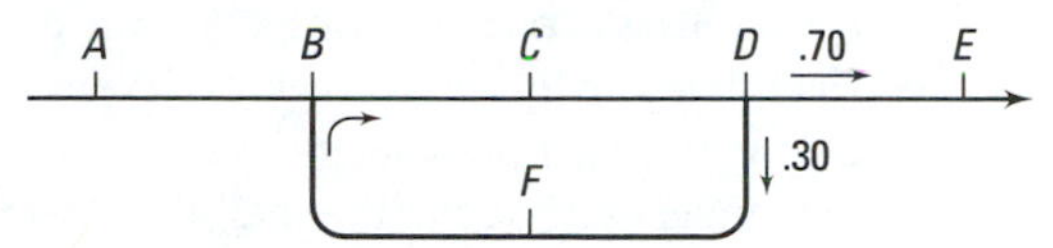

8.67 Let stream S be the superposition of k Poisson streams $S_1, S_2, \ldots, S_k$ having rates $\lambda_1, \lambda_2, \ldots, \lambda_k$, respectively.

a. Find the probability that the first event in stream S comes from stream S_i.

b. Find the probability that the kth event in stream S comes from stream S_i.

8.68 Consider a Poisson stream S having rate λ. Whenever an occurrence takes place in S, an independent multinomial trial having k possible outcomes is performed. If the outcome is i (with probability p_i), an occurrence is triggered in a stream S_i. Show that stream S_i is Poisson with rates $p_i\lambda$ and is independent with other streams.

8.69 People pass point A on a one-way street as a Poisson stream with rate $\lambda = 8$/hour. For each person, there is a probability $\alpha = .30$ that he or she turns right at D to travel along segment BD again (Figure 8.18.1). Are the streams of arrivals at points C, E, and F Markovian? If so, at what rates?

8.70 Cars can come directly from outside a city to its three points A, B, and C as Poisson streams with rates three, two, and one arrival/hour, respectively (Figure 8.18.2). Passing a point, the probabilities that a car goes to another point or leaves the city are shown in the following matrix:

$$\begin{array}{c} \\ A \\ B \\ C \end{array}\begin{array}{c} \begin{array}{cccc} A & B & C & \text{leave} \end{array} \\ \begin{pmatrix} 0 & 1/2 & 1/2 & 0 \\ 1/3 & 0 & 1/3 & 1/3 \\ 1/2 & 0 & 0 & 1/2 \end{pmatrix} \end{array}$$

Is the stream of car arrivals at point A from all sources Poisson? If so, at what rate?

8.71 Buses and cars arrive at a mall as a Poisson stream with rate 20/hour. The probability that an arrival is a bus is .20.

a. Given 15 cars arriving during the last hour, what is the probability of five buses arriving during the same period?

FIGURE 8.18.2 **Problem 8.70.**

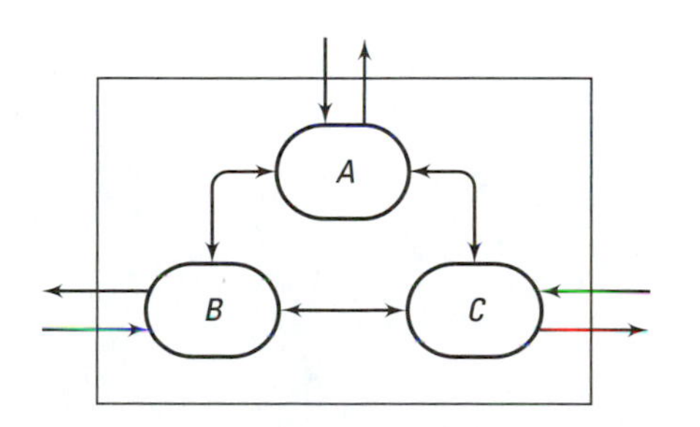

b. Given 25 cars and buses arriving during the last hour, what is the probability of seven buses arriving during the same period?

c. Each bus carries an average of 20 passengers, with variance 4; each car has an average of two passengers, with variance .09. What is the expected number of passengers arriving at the mall in 2 hours? What is its variance?

8.72 Consider a compound Poisson process $\{n(t)\}_{t\in[0,\infty)}$ as defined in §8.16. Let Y_is have the geometric distribution with parameter p, or $\Pr\{Y_i = k\} = p(1-p)^{k-1}$ for all $k \geq 1$. Find $\mathcal{E}[n(t)]$

a. by obtaining the distribution of $n(t)$. *Hint:* Use §1.14 and Appendix (A.13).

b. by using Equation (8.16.1).

8.73 Refer to Problem 8.9. Assume now that, if an electrical shock occurs during repair time, the repairing process can continue after a random duration r (that is, no service is lost). Find the expected value and the variance of the duration to finish a job.

8.74 An instrument is working at time 0. Impacts are applied to it as a Poisson stream with rate λ. Each impact causes an amount of damage, which is exponentially distributed with rate δ. Assume the instrument will fail when the total cumulative damage $d(t)$ at time t exceeds a fixed amount D. Show that the mean duration from time 0 to the time the instrument fails is $(\delta D + 1)/\lambda$. *Hint:* Use Problem 1.55 and Appendix (A.17).

CHAPTER 9

From Renewal Streams to Regenerative Processes

9.1 RENEWAL STREAMS AND RENEWAL (COUNTING) PROCESSES

In the previous chapter, we studied the Poisson streams in which all lives are exponentially distributed. In this chapter, we study more general streams of events in which *all lives are independent and identically distributed*, but not necessarily exponentially (Figure 9.1.1).

For such a stream, we use the following terms:

1. The stream itself is a *renewal stream*.

FIGURE 9.1.1 **A renewal stream and its renewal process.**

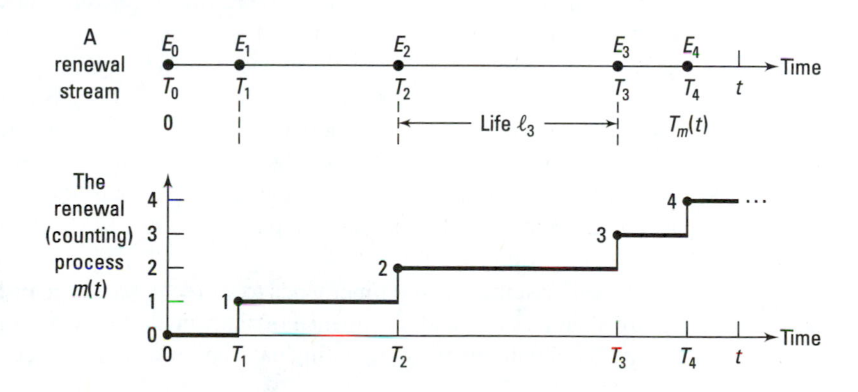

2. The ith event after $t = 0$ is the ith *renewal*. We denote it by E_i. We assume a renewal E_0 occurs at time $t = 0$.
3. The duration ℓ_n between E_{n-1} and E_n is the nth *life* or *interoccurrence time* of the stream.
4. The distribution $L(t) \equiv \Pr\{\ell_1 < t\}$ is the *life distribution* of the stream. Since we assume a renewal at $t = 0$, the first life distribution $\Pr\{\ell_1 < t\}$ is identical to the rest. Hence, we can use this first life as a typical life. For example, $\mathcal{E}[\ell_1]$ can be used to represent any expected life $\mathcal{E}[\ell_i]$. Avoiding triviality, we assume all lives are not identically zero; that is, $L(0) < 1$. This implies that $\mathcal{E}[\ell_1] > 0$. Furthermore, we assume that $L(t)$ is a *proper distribution*; that is, $\Pr\{\ell_1 < \infty\} = 1$ so that the stream will not be terminated by an infinitely long life. Note that the lives can take discrete values.
5. The time $T_i \equiv \ell_1 + \cdots + \ell_i$ of renewal E_i is the ith *renewal epoch*.
6. The counting process $\{m(t)\}_{t\in[0,\infty)}$, where $m(t)$ is the number of renewals during $(0, t]$ (excluding E_0), is the *renewal (counting) process*. Thus, $T_{m(t)}$ is the time of the last renewal before t.
7. The function $\mathcal{E}[m(t)]$ is the *renewal function*.

In this chapter, we pay particular attention to the renewal function $\mathcal{E}[m(t)]$ of a renewal stream. But what can we say about it without assuming any specific form of the life distribution?

To answer this question, let us first note that a Poisson stream with rate λ is a special renewal stream. In this case, its renewal process is a Poisson process, and (8.11.2) shows that its renewal function $\mathcal{E}[m(t)]$ is equal to λt for *any* value of t, even when $t \to 0$ or $t \to \infty$.

The first result we establish in this chapter is that, for a general renewal stream, $\mathcal{E}[m(t)]$ is also a linear function of t, but *only when* $t \to \infty$. This turns out to be an important result. Not only does it allow us to understand how the stream develops during an infinite period of time, but it also provides the foundation for our study of the long-term behavior of a very general class of stochastic processes, namely, the regenerative processes. We have seen how the regenerative argument was used for the discrete-time irreducible Markov chains in Chapter 6. In this chapter, based on the foregoing simple result, we develop this regenerative argument more rigorously and use it extensively for the rest of this book.

Renewal Streams in Action: Pulses from the Hypothalamus. The hypothalamus is a small area at the base of the brain that regulates the body's general level of activity. It produces in pulses the releasing hormones (GnRH), which ultimately influence growth, sexual development, and the rate at which the body changes food into energy and living tissue. The durations between these pulses are not constant, but fluctuate depending on the steroid milieus, the sleep/wake state, and so forth.

Assuming that these pulses are regulated by a timer, there are at least two ways we can model a stream of hypothalamic pulses:

1. We can assume that the timer tends to correct a sequence of long interpulse intervals by a shorter one and a sequence of short intervals by a longer one. This requires a hypothalamic memory persisting past the immediately preceding pulse.

2. We can assume that the pulses form a renewal stream. In this case, the timer only remembers when the immediately preceding pulse occurs.

According to Butler et al. (1986), who analyzed the timing of these pulsative events of hormones secretion in 20 normal adult men every 10 minutes for 24 hours, "[W]e cannot reject, even at a 85% confidence, a null hypothesis that the GnRH pulse pattern is a renewal process." ■

9.2 ELEMENTARY RENEWAL THEOREM

According to the *elementary renewal theorem*, the expected number of renewals in $(0, t]$ is *asymptotically* a linear function of t, or

$$\lim_{t \to \infty} \frac{\mathcal{E}[m(t)]}{t} = \frac{1}{\mathcal{E}[\ell_1]} \tag{9.2.1}$$

While we omit the formal proof of this theorem, the following observations will help us appreciate it:

1. The Poisson streams are the only streams in which this linear property is applicable for all t, finite or infinite. For a general renewal stream, it is only true when $t \to \infty$. Consider a service facility at which customers are scheduled to arrive exactly 30 minutes apart. Excluding the customer who arrives at 9:00 A.M., the expected number of arrivals cannot be linear during the first few hours after 9:00 A.M. because it jumps from 0 to 1 at 9:30 A.M.

2. Note that the *actual shape* of the life distribution $L(t)$ loses its influence on $\mathcal{E}[m(t)]$ when $t \to \infty$. This fact bears an intimate relationship with the *central limit theorem*, one of the most remarkable results in probability theory that forms the foundation of statistical inference. According to this theorem, if $\{\ell_1, \ell_2, \ldots\}$ is a sequence of independent and identically distributed random variables with common mean $\mathcal{E}[\ell_1]$, then regardless of the shape of the common distribution, the sum $T_n \equiv \ell_1 + \ell_2 + \cdots + \ell_n$ will be normally distributed with mean $n\mathcal{E}[\ell_1]$ when $n \to \infty$.

This theorem explains why many random variables encountered in nature have distributions close to the normal distribution. It enables us to infer some information about the mean of any random variable from a sufficiently large sample size.

Suppose we do not know the exact value of the expected life $\mathcal{E}[\ell_1]$ and want to estimate it. From the central limit theorem, we would observe the stream and obtain the duration from time 0 to a renewal epoch T_n. When $n \to \infty$, this duration T_n, divided by n, can be used to estimate $\mathcal{E}[\ell_1]$. For example, if $T_{10{,}000} = 23{,}123$ hours, then $\mathcal{E}[\ell_1] \approx (23{,}123/10{,}000)$ hours.[1]

[1] The relation "$\approx$" means "approximately equal to."

Stating this procedure slightly differently, with $m(t)$ as the number of renewals during $(0, t]$ and $T_{m(t)}$ the last renewal epoch before t,

$$\lim_{n\to\infty} \frac{T_n}{n} = \lim_{t\to\infty} \frac{T_{m(t)}}{m(t)} = \mathcal{E}[\ell_1] \tag{9.2.2}$$

Because the lives are finite, when $t \to \infty$, the difference between t and $T_{m(t)}$ becomes insignificant in comparison with t or $m(t)$. Hence,

$$\lim_{t\to\infty} \frac{t}{m(t)} = \mathcal{E}[\ell_1]$$

Inverting this equation and then taking expectations on both sides would yield (9.2.1).

Equation (9.2.1) now yields another method of estimating $\mathcal{E}[\ell_1]$ for our example: Instead of following the stream continuously for 23,123 hours (nearly 1000 days), we observe the number of occurrences each day for 1000 days and calculate its daily average $\mathcal{E}[m(t)]$ with $t = 24$ hours. Dividing 24 hours by this average gives an estimate of $\mathcal{E}[\ell_1]$ in hours.

9.3 RENEWAL RATES

We have a reason for presenting Equation (9.2.1) upside down, in terms of $1/\mathcal{E}[\ell_1]$ rather than $\mathcal{E}[\ell_1]$: Consistent with Equation (8.5.2), $\mathcal{E}[m(t)]/t$ is the *average rate of occurrence* of the stream during t. Its unit is the number of renewals per unit time. When $t \to \infty$, we call it the *(long-term) (renewal) rate*.

Now we see that Equation (8.11.2) is a special case of (9.2.1). However, for a Poisson stream, λ is the average rate during any time period; for a general renewal stream, $1/\mathcal{E}[\ell_1]$ is only the long-term average rate.

In daily life, we use the term *rate* very often: accident rate, birthrate, death rate, crime rate, and so on. We use it without ever worrying about the underlying life distribution. Now we know that, except for the Poisson streams, what we routinely refer to as rate can only be the long-term rate. (We should also remember that *time* is a very general term here. As discussed in §2.18, it could refer to measurements such as distance. Similarly, *unemployment rate* can refer to the average number of unemployed persons per thousand people in the work force.)

Consider the following machine-failure example: A machine works for an independent and identically distributed random duration U before it fails. If it fails, it takes an independent and identically distributed random duration D to repair it. The stream of machine failures is a renewal stream with each life equal to $U + D$. The elementary renewal theorem now states that the long-term failure rate of this machine is $(\mathcal{E}[U] + \mathcal{E}[D])^{-1}$ failures per unit time.

As another example, let us consider a discrete-time irreducible Markov chain $\{X_n\}_{n=0,1,2,\ldots}$. The stream of its visits to a particular state i is a discrete-time renewal stream. The life of the stream is the number of steps between its two consecutive visits to state i, which is the return times n_i as defined in §6.6. According to the elementary renewal theorem, the long-term rate that this chain visits state i is $1/\mathcal{E}[n_i]$. This is consistent with our definition of the visiting rates g_i in (6.4.2) and with Equation (6.6.1).

9.4 RENEWAL-REWARD PROCESSES

The renewal process $\{m(t)\}_{t\in[0,\infty)}$ is a counting process, the value of which increases by 1 whenever a renewal occurs. A natural generalization of the renewal process is the *renewal-reward* process $\{n(t)\}_{t\in[0,\infty)}$ that starts with $n(0) = 0$ and increases its value by a *finite, independent, and identically distributed random variable* whenever a renewal occurs. In other words, when renewal E_i occurs, $n(t)$ receives a *random reward* of size Y_i. This reward can be either discrete or continuous, negative or positive (Figure 9.4.1).

Because $m(t)$ is a random variable counting the number of renewals during $(0, t]$, $n(t)$ is a *random sum* (§1.21):

$$n(t) = \sum_{i=0}^{m(t)} Y_i$$

If the number of wars during $(0, t]$ is a renewal process, the total number of resulting casualties during $(0, t]$ is a renewal-reward process. If the number of transactions during $(0, t]$ is a renewal process and if each business transaction can bring a loss or a profit, the net profit during $(0, t]$ is a renewal-reward process.

Q: How many probabilists does it take to change a light bulb?
A: 1.25.

Q: How come?
A: One to do the work, and another who helps occasionally to make it a random variable.

Q: Suppose the duration of each bulb replacement is a random variable with mean 5 minutes, and the average life of a bulb is 1.5 months. What is the long-term labor rate for replacing bulbs?

FIGURE 9.4.1 A renewal stream and its renewal-reward process.

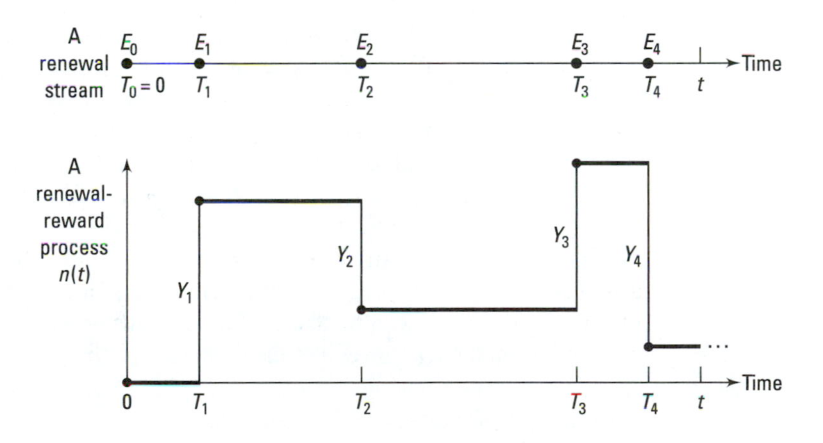

Here, the bulb failures define a renewal stream with $\mathcal{E}[\ell_1] = 1.5$ months.[2] At each renewal, there is a labor reward with an average of $1.25 \times 5/60$ person-hour. The total person-hours needed during $(0, t]$ is thus a renewal-reward process.

The calculation of the labor rate of replacing bulbs lies in the following theorem: The long-term expected reward per unit time of a renewal-reward process $\{n(t)\}_{t\in[0,\infty)}$, or its *(long-term) reward rate*, is

$$\lim_{t\to\infty} \frac{\mathcal{E}[n(t)]}{t} = \frac{\mathcal{E}[Y_1]}{\mathcal{E}[\ell_1]} \tag{9.4.1}$$

This result, being a simple generalization of (9.2.1), is intuitively pleasing: As $t \to \infty$, the expected number of renewals during $(0, t]$ is $t/\mathcal{E}[\ell_1]$. If each renewal in turn brings an expected reward of $\mathcal{E}[Y_1]$, the expected total reward during $(0, t]$ must be $\mathcal{E}[Y_1]t/\mathcal{E}[\ell_1]$ when $t \to \infty$ (Equation 1.21.1). The expected labor rate for changing light bulbs is thus $(1.25 \times 5/60)/1.5$ person-hour per month.

Note that the compound Poisson process is a special form of the renewal-reward process, and (8.16.1) is a special case of (9.4.1).

9.5 GRADUAL-REWARD PROCESSES

We now generalize the renewal-reward processes further by removing the assumption that they can only increase their values at the renewals.

Consider a process $\{g(t)\}_{t\in[0,\infty)}$ built on a renewal stream having T_i as the ith renewal epoch. Assume $g(0) = 0$. Let $g(T_{i+1}) - g(T_i) \equiv Z_i$. The process $\{g(t)\}_{t\in[0,\infty)}$ is said to be a *gradual-reward process* if all Z_is are independent and identically distributed having finite common expected value $\mathcal{E}[Z_1]$. Like the reward Y_is of the renewal-reward processes, the *net rewards* Z_is of the gradual-reward processes can be either discrete or continuous, negative or positive; unlike the reward Y_is, which have to be given as a lump sum at the renewals, the rewards Z_is can be given *gradually between* the renewals (Figure 9.5.1).

If the number of buses arriving during $(0, t]$ is a renewal process, the total number of passengers getting off the buses at their arrivals is a renewal-reward process. The total fare sold to the *waiting* passengers between the bus arrivals is a gradual-renewal process.

Consider now a renewal-reward process $\{s(t)\}_{t\in[0,\infty)}$ defined on the same renewal stream as that of the gradual-reward process $\{g(t)\}_{t\in[0,\infty)}$. Assume that both receive the same net reward Z_i during life ℓ_i. The only difference between the two is that, while $\{s(t)\}_{t\in[0,\infty)}$ receives the reward at the end of life ℓ_i, $\{g(t)\}_{t\in[0,\infty)}$ receives it gradually during ℓ_i. As an example, $\{g(t)\}_{t\in[0,\infty)}$ can be the total income during $(0, t]$ of a business owner who demands that customers pay immediately after each service rendered; $\{s(t)\}_{t\in[0,\infty)}$ can be that of the same owner who bills customers and receives payments from them at the end of each month.

---

[2]To be precise, $\mathcal{E}[\ell_1]$ should be 1.5 months plus 5 minutes.

FIGURE 9.5.1 A renewal stream and its gradual-reward process.

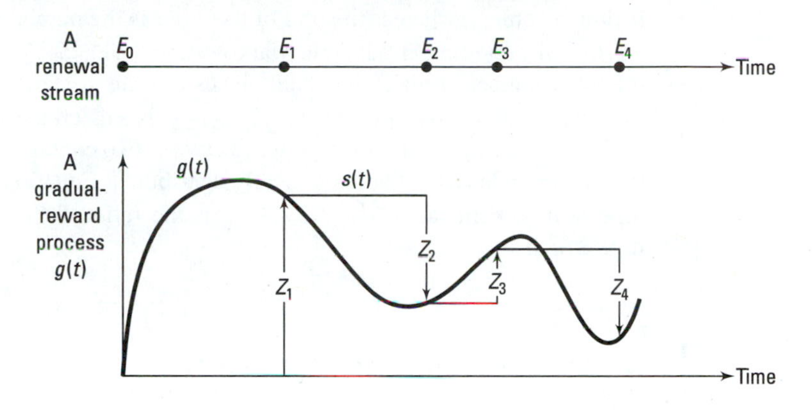

Since we assumed that the rewards are finite, the difference $g(t) - s(t)$ is always finite. As $t \to \infty$, this finite difference, further divided by t, becomes insignificant compared with the value of $g(t)$ or $s(t)$. This shows that the long-term expected reward per unit time of a gradual-reward process, or its *(long-term) reward rate*, is the same as if it were a renewal-reward process, or

$$\lim_{t\to\infty} \frac{\mathcal{E}[g(t)]}{t} = \frac{\mathcal{E}[Z_1]}{\mathcal{E}[\ell_1]} \tag{9.5.1}$$

Let us return to the machine-failure example in which the machine works for a random duration U before failing and then requires a random repair duration D. Now let $d(t)$ be the total downtime during $(0, t]$. As shown in Figure 9.5.2, the process $\{d(t)\}_{t\in[0,\infty)}$ is a gradual-reward process. Its value increases only when the machine is repaired. Between two consecutive failures, its total increment is the random repair time D. Equation (9.5.1) gives the long-term downtime rate of $\{d(t)\}_{t\in[0,\infty)}$ as $\mathcal{E}[D]/(\mathcal{E}[U]+\mathcal{E}[D])$,

FIGURE 9.5.2 The total downtime of a machine as a gradual-reward process.

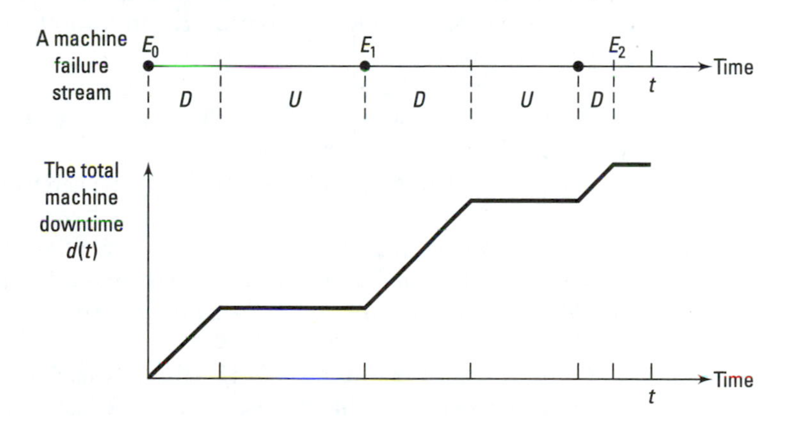

which is different than the long-term failure rate $1/(\mathcal{E}[U]+\mathcal{E}[D])$. (Note that the former is dimensionless, whereas the unit of the latter is the number of failures per unit time.)

In a discrete-time irreducible Markov chain $\{X_n\}_{n=0,1,2,\ldots}$, $v_{ij}^{(n)}$ is defined in §3.10 as the total number of visits the chain makes to state j, starting from state i, from step 0 to step n inclusive. The process $\{v_{ij}^{(n)}\}_{n=0,1,2,\ldots}$ is a discrete-time gradual-reward process because the numbers of visits to state j between two consecutive visits to state i (denoted by ν_{ik}) are independent and identically distributed. Equation (9.5.1) now states that the long-term reward rate of $\{v_{ij}^{(n)}\}_{n=0,1,2,\ldots}$ is the ratio of $\mathcal{E}[\nu_{ij}]$ and the expected return time $\mathcal{E}[n_i]$.

9.6 REGENERATIVE PROCESSES

Consider now a very general stochastic process, the parameter space of which can be either discrete or continuous and the state space of which can be either qualitative or quantitative. This process is said to be *regenerative* if there is a finite and random time c_1 after the starting time 0 that it can *restart* in an independent and identical manner.

This means that the process also restarts at another random time $T_2 \equiv c_1 + c_2$, and hence at an infinite sequence of random times $\{T_i\}_{i=1,2,\ldots}$, where $T_i \equiv c_1 + c_2 + \ldots + c_i$. These restarting events form a renewal stream. We call the renewal epochs T_i the *regenerative points* and the duration c_i between any two consecutive regenerative points the *regenerative cycle*. Note that we assumed all c_is are independent and identically distributed and $\mathcal{E}[c_1] < \infty$.

Since the process restarts at the beginning of each cycle, its developments within the cycles are *independent and probabilistically identical*. These regenerative cycles therefore can be considered the building blocks of the entire history of the process. Even when $t \to \infty$, the process is still in a cycle, and its development within this cycle is probabilistically the same as its development within the first cycle.

Regenerative processes have many practical applications. Actually, we only study this kind of process in this book. For example, as discussed in §6.7, the discrete-time irreducible Markov chain is regenerative: Starting from state i, whenever the chain visits this state again, its development afterward is the same as its development from step 0.

The simplest regenerative process is a sequence of independent and identically distributed random variables: The process restarts at each step, and each step is therefore a regenerative point.

In the machine-failure example, consider the process $\{\sigma(t)\}_{t\in[0,\infty)}$, which describes the machine by two states: up and down. This process is regenerative: It restarts whenever the machine fails. However, the gradual-reward process $\{d(t)\}_{t\in[0,\infty)}$, which records the machine's total downtime during $(0, t]$, is not regenerative because its value can only increase.

We now demonstrate that most *long-term* quantities of a regenerative process (the state space of which can be qualitative) can be expressed as the long-term reward rate of a *suitably defined gradual-reward process* (the state space of which must be quantitative).

Consider a regenerative process $\{\rho(t)\}_{t\in[0,\infty)}$ having state space S, which can be either quantitative or qualitative. Let A be a subset of S. We are interested in the process's *(long-term) visiting rate*, the *(long-term) proportion of visits* to A, which can be defined as

$$\gamma_A \equiv \lim_{t\to\infty} \frac{1}{t}\mathcal{E}[J(t)] \tag{9.6.1}$$

where $J(t)$ is the total duration the process visits A during (0,t].

The right-hand side of this definition suggests that γ_A is the long-term reward rate of the gradual-reward process $\{J(t)\}_{t\in[0,\infty)}$, the value of which increases by an independent and identically distributed random variable within each regenerative cycle. From Equation (9.5.1), we obtain

$$\gamma_A = \frac{\mathcal{E}[\omega_A]}{\mathcal{E}[c_1]} \tag{9.6.2}$$

where ω_A is the total duration that the process $\{\rho(t)\}_{t\in[0,\infty)}$ visits A during a regenerative cycle.

For the machine-failure example, the process $\{\sigma(t)\}_{t\in[0,\infty)}$, with states up and down, is regenerative with regenerative points at the machine failures. Its long-term visiting rate to state down is the long-term reward rate of the graduate-reward process $\{d(t)\}_{t\in[0,\infty)}$, which is $\mathcal{E}[D]/(\mathcal{E}[U]+\mathcal{E}[D])$, as obtained in §9.5.

For a discrete-time irreducible Markov chain $\{X_n\}_{n=0,1,2,\ldots}$, consistent with Equation (9.6.1), Equation (6.4.2) defines the long-term visiting rate to state j as the reward rate of the gradual-reward process $\{v_{ij}^{(n)}\}_{n=0,1,2,\ldots}$, where $v_{ij}^{(n)}$ is the number of visits a chain makes to state j from state i between steps 0 and $n-1$ inclusive. In the previous section, this reward rate was shown to be $\mathcal{E}[\nu_{ij}]/\mathcal{E}[n_i]$, where ν_{ij} is the number of visits to state k within a regenerative cycle from state i and n_i is the return time to state i. This is Equation (6.7.1).

Consider now a regenerative process $\{\rho(t)\}_{t\in[0,\infty)}$, the state space of which is a set of real numbers. Because it has to restart occasionally, its value cannot drift to $\pm\infty$. We are interested in its *long-term time-average*, which is defined as

$$\rho \equiv \lim_{t\to\infty} \frac{1}{t}\mathcal{E}[R(t)] \tag{9.6.3}$$

where $R(t)$ is the area under the curve $\rho(t)$ from 0 to t.

The right-hand side of this definition says that the long-term time-average ρ of the process $\{\rho(t)\}_{t\in[0,\infty)}$ is the long-term reward rate of the process $\{R(t)\}_{t\in[0,\infty)}$. Now $\{R(t)\}_{t\in[0,\infty)}$ is a gradual-reward process because it increases its value by an independent and identically distributed random variable within a regenerative cycle. From Equation (9.5.1), we can write

$$\rho = \frac{\mathcal{E}[r]}{\mathcal{E}[c_1]} \tag{9.6.4}$$

where r is the area under the curve of $\rho(t)$ within a regenerative cycle.

We now present an interesting application of this result.

9.7 REMAINING LIVES

Similar to the definition in §8.7, the *remaining life* of a renewal stream at time t is the duration from t to the next renewal after t, or

$$r(t) \equiv T_{m(t)+1} - t$$

Let us now consider the process $\{r(t)\}_{t\in[0,\infty)}$. Figure 9.7.1 shows a realization of this process, which is regenerative at each renewal. We are interested in its long-term time-average ρ.

If the expected life is $\mathcal{E}[\ell_1]$, we would intuitively suspect that the expected remaining life is $\mathcal{E}[\ell_1]/2$. Suppose buses arrive at a bus stop as a renewal stream on an average of 20 minutes apart. If a person randomly arrives at this bus stop, it is reasonable to expect that he or she has to wait an average of 10 minutes for the next bus.

But this is not always the case. Because $\{r(t)\}_{t\in[0,\infty)}$ is regenerative at each renewal, we know from §9.6 that its long-term time-average ρ is the ratio of the expected area under the curve $r(t)$ within a cycle and the expected cycle length $\mathcal{E}[\ell_1]$. Now, given that the cycle length is τ, the area under the curve $r(t)$ within this cycle is $\tau^2/2$. The unconditional expected area under the curve $r(t)$ within a cycle is therefore $\mathcal{E}[(\ell_1)^2]/2$, and the long-term time-average of the remaining lives is thus[3]

$$\rho = \frac{\mathcal{E}\left[(\ell_1)^2\right]}{2\mathcal{E}[\ell_1]} = \frac{\mathcal{E}[\ell_1]}{2} + \frac{\mathrm{Var}[\ell_1]}{2\mathcal{E}[\ell_1]} \tag{9.7.1}$$

Contrary to our intuition, unless all lives are constant, the average remaining life of a renewal stream is not $\mathcal{E}[\ell_1]/2$, but *longer*. If there is any fluctuation in the bus interarrival times, the average duration a passenger has to wait for the next bus is longer than 10 minutes.

FIGURE 9.7.1 The remaining lives.

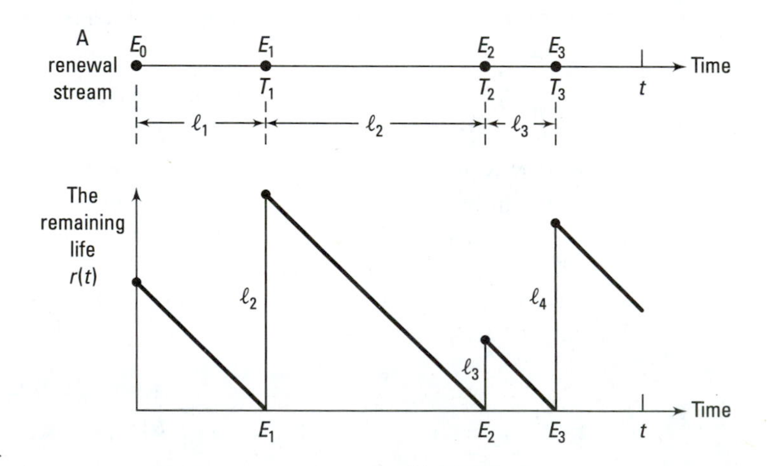

[3]For a direct proof for the discrete case, see Problems 6.37 and 7.2b.

The worst scenario a person would expect is that the bus leaves just before he arrives. In this case, his waiting time is the time until the next bus arrival. We would therefore surmise that a person's average waiting time for a bus must be less than the buses' expected interarrival time. However, we don't see this restriction in Equation (9.7.1) either. In fact, not only may the average remaining time be longer than $\mathcal{E}[\ell_1]/2$, it may also be longer than $\mathcal{E}[\ell_1]$. This seeming inconsistency is known as the *inspector's paradox* in the literature.

To understand the paradox, we have to distinguish between the *time-average* and the *renewal-average*.

1. The renewal-average is the division of a certain quantity by the number of renewals $m(t)$. Equation (9.2.2) shows the expected life $\mathcal{E}[\ell_1]$ as a renewal-average. In the calculation of $\mathcal{E}[\ell_1]$, all lives are given the same weight, whether they are long or short.

2. The time-average is the division of a certain quantity by the time t. Equation (9.6.3) shows the expected remaining life ρ as a time-average. Suppose now that $\text{Var}[\ell_1]$ is large, allowing a very long life to occur now and then. Consider one such long life ℓ_j. The expected remaining life during life ℓ_j is $\ell_j/2$, which is also large. In the calculation of the time-average ρ, this large average $\ell_j/2$ is given even more weight because there is more time during ℓ_j. That is why ρ can be larger than $\mathcal{E}[\ell_1]/2$ or even $\mathcal{E}[\ell_1]$.

In the bus example, normally there are *more customers waiting for a bus that is late*. Thus, the average of the customers' waiting times (time-average) can be larger than the average of the buses' interarrival times (renewal-average).

9.8 LATTICE RANDOM VARIABLES

The question we now seek to address is whether the distribution of a regenerative process converges in any sense as $t \to \infty$. In preparation for the discussion of this question, let us start with the following definition: A random variable X is said to be *lattice*[4] if it takes value only on the set of integral multiples of a number $d > 0$; that is,

$$\sum_{n=-\infty}^{n=\infty} \Pr\{X = nd\} = 1 \quad \text{for all } n = 0, \pm 1, \pm 2, \ldots$$

where d is called its *span* and nd its *lattice epochs*. For example, X is lattice with span 0.6 if it only takes the value in the set $\{1.2, 1.8, 48, 60\}$.

A necessary condition for a random variable to be lattice is that it be discrete. However, the converse is not true. The following random variable is discrete, but it is not lattice: Z, taking value at either $\sqrt{2}$ or $\sqrt{3}$.

Suppose signals are scheduled to be sent to machine A exactly 30 minutes apart to trigger a certain operation having a random duration. However, if an operation takes longer than 30 minutes, the signal will be sent 1 hour after the previous one. Starting from 9:00 A.M., one realization of the stream of signal arrivals is shown in Figure 9.8.1.

---

[4]Some authors call it *arithmetic*.

FIGURE 9.8.1 The arrivals of signals at machine *A*.

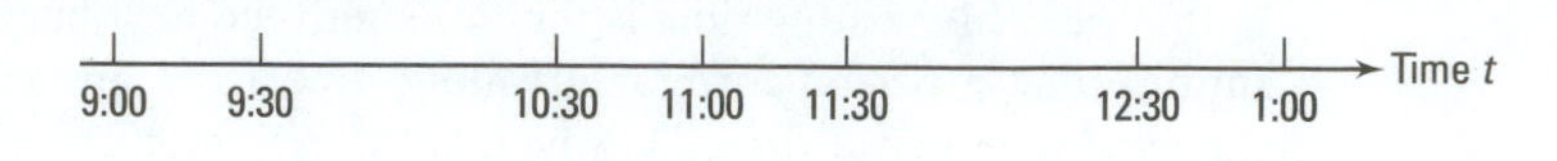

The interarrival time of the signals to machine A are lattice with a 30-minute span.

Note that there is no arrival at 10:00 A.M. or 12:00 P.M. (If there must be one arrival every 30 minutes, the lives of the stream must be constant, and the stream is said to be *deterministic*.)

9.9 SPREAD-OUT RANDOM VARIABLES

We also need the following definition for the discussion of the limiting distribution: A random variable Y and its distribution function $F(x)$ are said to be *spread-out* if $F(x)$ *has a derivative* $f(x)$ within some interval (a, b). We also say that Y has a *spread-out component* (a, b).

Note that a spread-out random variable does not have to be absolutely continuous, as defined in (1.16.1). While the latter requires that $F(x)$ have a derivative throughout $(-\infty, \infty)$, the former says nothing about the distribution function $F(x)$ outside (a, b).

Suppose signals are scheduled to be sent to machine B not exactly 30 minutes apart, but anytime between 29 minutes and 31 minutes after the previous one to trigger a certain operation. However, if an operation takes longer than 30 minutes, the next signal has to arrive exactly 1 hour after the previous one. The interarrival times of the signals to machine B are spread-out because it has a spread-out component $(29, 31)$.

Note also that a spread-out random variable has to be nonlattice, but a nonlattice random variable does not have to be spread-out. For example, the random variable Z in §9.8, taking value only in the set $\{\sqrt{2}, \sqrt{3}\}$, is neither lattice nor spread-out.

9.10 LIMITING DISTRIBUTIONS

In §4.12, we showed that the periodic discrete-time Markov chains do not converge in distribution. Similarly, for any subset A of the state space S, if a regenerative process $\{\rho(t)\}_{t\in[0,\infty)}$ has *lattice regenerative cycles* with span d, then $\Pr\{\rho(t) \subset A\}$ does not converge when $t \to \infty$ *along the continuous-time axis*. Here, even after a very long duration, the probability that a regenerative point occurs at time t is still dependent on whether t is an integral multiple of d or not. For example, if the arrivals of signals at machine A are lattice with a 30-minute span, the probability that there is an arrival when the long hand of a watch points to 12 or 6 is always different from that at other times, even after the machine has been operating for a very long duration.

It would be reasonable to conjecture that $\Pr\{\rho(t) \subset A\}$ will converge if the regenerative cycles are nonlattice. This condition, unfortunately, is not sufficient. Consider, for example, a process in which the cycle length can only be rational and hence nonlattice. The behavior of the process at time t is still dependent on whether t is rational or not, even when $t \to \infty$.

It turns out that the sufficient condition for convergence is only slightly stronger than nonlatticeness: It has been shown that, if its regenerative cycles c_is are spread-out, a regenerative process $\{\rho(t)\}_{t\in[0,\infty)}$ will *converge in distribution* when $t \to \infty$ *along the continuous-time axis*; that is, $\lim_{t\to\infty} \Pr\{\rho(t) \subset A\}$ exists for any subset A of the state space.

If each cycle has a spread-out component, then the duration from time 0 to the nth regenerative point, which is the sum of n cycles, also has one. Furthermore, as n increases, the spread-out component of this duration becomes wider and wider. Return to machine B, where the interarrival times of the signals can take any value within the interval (29, 31), besides the discrete value of 60 minutes. Here, there is a nonzero probability that the second signal arrives at any time point within the interval (58, 62), which is wider than the interval (29, 31).

Thus, regardless of how small the spread-out component of a cycle is, that of the nth regenerative epoch will grow infinitely large as $n \to \infty$ or as $t \to \infty$. The spread-out condition hence ensures that, as the regenerative process develops to ∞, its behavior becomes asymptotically the same at all times.

It has also been shown that for a regenerative process $\{\rho(t)\}_{t\in[0,\infty)}$, which converges in distribution, *its pointwise limits are the same as its time-averages*, or

$$\lim_{t\to\infty} \Pr\{\rho(t) \subset A\} = \gamma_A$$

where the time-average γ_A was defined in (9.6.1) and can be calculated by Equation (9.6.2).

Remember that, beyond the first regenerative point, the initial conditions become irrelevant. Thus, like the long-term averages, the limiting results should be independent of the initial conditions.

9.11 SUMMARY

1. The lives ℓ_is of a renewal stream are independent and identically distributed.
2. The renewal rate of a renewal stream is $1/\mathcal{E}[\ell_1]$.
3. A renewal-reward process increases its value only at the renewals. The increments are independent and identically distributed with common mean $\mathcal{E}[Y_1]$. Its long-term reward rate is $\mathcal{E}[Y_1]/\mathcal{E}[\ell_1]$.

4. A gradual-reward process increases its value gradually as long as the net rewards between any two consecutive renewals are independent and identically distributed with common mean $\mathcal{E}[Z_1]$. Its long-term reward rate is $\mathcal{E}[Z_1]/\mathcal{E}[\ell_1]$.
5. A regenerative process restarts at a sequence of regenerative points. It develops independently and identically within different regenerative cycles. All of its long-term quantities are independent of the initial conditions and can be obtained as the long-term reward rates of some suitably defined gradual-reward processes. If its cycle lengths are spread-out, it converges in distribution, and its limiting probabilities are the same as its time-averages.
6. The long-term time-average of the remaining lives of a renewal stream is $\mathcal{E}[\ell_1]/2$ only when the lives are constant; otherwise, it is longer.

PROBLEMS

9.1 The average duration that John holds a job is 2 years. The average duration that he spends looking for a new job is 2 months. The probability that he has to relocate for each new job is .30. His average moving cost is $10,000.

a. What is his long-term rate of new jobs?
b. What is the long-term proportion of time that he is employed?
c. Do his relocations form a renewal stream? If so, what is his long-term relocation rate?
d. What is his long-term relocation-cost rate?

9.2 Mary writes computer programs. On average, she can produce a bug-free program on the 20th revision. Each revision takes 2 days, 5 minutes of which is computer time.

a. What is her rate of producing bug-free programs?
b. What is her rate of utilizing computer time?

9.3 Assume that the durations in which Harry is available to answer phone calls are independent and identically distributed random variables with common mean A. Assume also that the durations between these availability periods are independent and identically distributed random variables with common mean U. If telephone calls arrive as a Poisson stream with rate λ, what is the proportion of calls that can reach Harry?

9.4 The duration between two consecutive telephone calls has an Erlang distribution with parameters $(2, \lambda/2)$. Let $m(t)$ be the number of calls during $(0, t]$.

a. Find the distribution of $m(t)$.
b. Hence, show that

$$\mathcal{E}[m(t)] = \frac{1}{2}\lambda t - \frac{1}{4}\left(1 - e^{-2\lambda t}\right)$$

Hint: Use Appendix (A.4).

c. Draw a graph of $\mathcal{E}[m(t)]$ versus t.
d. Verify the elementary renewal theorem (9.2.1).

9.5 The life of a machine is exponentially distributed with rate λ/hour. It is inspected every T hours, where T is a constant. Thus, there are durations in which it is down and unnoticed. If an inspection finds that it is down, it is immediately replaced by an identical machine.

a. Find the proportion of time that the machine is functioning

i. by studying the stream of inspections. *Hint:* Use Problem 8.7.

ii. by studying the stream of replacements.

b. Does the probability that the machine is functioning converge as $t \to \infty$?

9.6 Refer to Problem 9.5. Assume now that the machine is inspected randomly with rate μ/hour.

a. Find the proportion of time that the machine is functioning

i. by studying the stream of inspections. *Hint:* Use Problem 8.21.

ii. by studying the stream of replacements.

b. Does the probability that the machine is up converge as $t \to \infty$?

c. Each inspection costs $\$i$, and the cost rate for each down period is $\$d$/hour. Find the total cost rate.

9.7 Refer to Problem 9.6. Assume now that, if an inspection finds that the machine is down, it will be repaired immediately rather than replaced. Let r be the random duration of each repair.

a. What is the proportion of time that the machine is down but not repaired?

b. What is the proportion of time that the machine is being repaired?

9.8 Refer to Problem 9.7. Assume now that there is a probability α that the machine failure is not detected at an inspection and thus remains down until a subsequent inspection detects its failure. What is the proportion of time that the machine is down and not repaired? *Hint:* Use Problem 8.54.

9.9 An instrument has the life distribution $H(t)$. It is replaced whenever it fails (unplanned replacement) or when it reaches the age of T (planned replacement), whichever comes first. Each replacement costs $\$\alpha$. There is also an additional cost of $\$\beta$ if it fails before reaching age T.

a. Show that the long-term replacement rate of this instrument is

$$\left(T - \int_0^T H(x)\,dx\right)^{-1}$$

i. by studying the stream of replacements. *Hint:* Use Problem 1.51.

ii. by studying the stream of unplanned replacements.

iii. by studying the stream of planned replacements.

b. Find the long-term failure rate of this instrument.

c. Show that the long-term cost rate is

$$[\alpha + \beta H(T)]\left(T - \int_0^T H(x)\,dx\right)^{-1}$$

d. When the life of the instrument is uniformly distributed over [0, 1 year] and $T < 1$ year, show that the total cost is minimal when $T = \rho + (\rho^2 + 2\rho)^{1/2}$, where $\rho \equiv \alpha/\beta$. Obtain this value when $\alpha = \$5$ and $\beta = \$10$.

e. When the life of the instrument is exponentially distributed with rate λ, show that the optimal value of T is ∞. Explain why.

9.10 The life distribution of an instrument is exponential with rate λ. If it fails before age T, we repair it (with a random duration r); if it reaches age T, we service it (with a random duration s), and it will become like new.

a. Find the long-term failure rate of this instrument

i. by studying the stream of repair or service completions

ii. by studying the stream of repair completions

iii. by studying the stream of service completions

Hint: Use Problem 8.7.

b. What is the long-term proportion of time that this instrument is under repair?

c. What is the long-term proportion of time that this instrument is under service?

9.11 *Geiger-Müller* counters are used to count the number of impulses emitted by radioactive materials. After an impulse is counted (or registered), some counters have a *dead period* D in which arriving impulses are not counted. The dead periods are independent and identically distributed. In a *nonparalyzable* counter, these dead periods are not affected by the uncounted impulses arriving during them. Suppose the impulse arrives as a Poisson stream with rate λ.

a. Find the rate of counted impulses.

b. Find the proportion of time that the counter is having a dead period.

9.12 Refer to Problem 9.11. Assume now that the dead periods D are exponentially distributed with rate μ.

a. Find the rate of counted impulses

i. by using Problem 9.11

ii. by using Problem 8.24

b. Find the proportion of time that the counter is having a dead period

i. by using Problem 9.11

ii. by using Problem 8.24

9.13 In a *paralyzable* counter, even when the arriving impulse is not counted, it still gives rise to a dead period. Assume the impulses arrive as a Poisson stream with rate λ, and the dead period resulting from each arriving impulse has a fixed length D.

a. Show that the rate of counting impulses is $\lambda \exp\{-\lambda D\}$

i. by studying the stream of impulses

ii. by studying the stream of counted impulses

iii. by studying the stream of uncounted impulses

b. Show that the proportion of time that the counter is having a dead period is $1 - \exp\{-\lambda D\}$.

9.14 A system has two components arranged in *series*. It functions as long as both components work. The lives of both are exponentially distributed with rate λ. There is only one repairperson. If a component fails while the other functions, the system is shut down and the failed component is repaired immediately. Assume that the system is *noninterruptive*; that is, a functioning component still can breakdown while the other component is down. In this case, the former must wait until the latter is repaired before it can be repaired. Find the limiting probability that the system is up

a. when the repair time is a fixed duration R. *Hint:* Use Problem 8.33a.

b. when the repair time is exponentially distributed with rate μ. *Hint:* Use Problem 8.33b.

9.15 Buses arrive at a bus stop every T minutes. Prospective passengers arrive as a Poisson stream with rate λ. Upon arrival, a passenger waits for a duration that is exponentially distributed with rate μ. (This means that the probability of a customer taking a taxi rather than waiting for more than t minutes is $1-\exp\{-\mu t\}$.) Suppose it costs \$$B$ for each bus arrival, and the fare of each passenger is \$$f$. Show that the value of T that maximizes the rate of profit is the unique solution of $(T\mu + 1)e^{-\mu T} = 1 - (\mu B)/(f\lambda)$, provided that $(f\lambda)/\mu > B$. *Hint:* Use Problem 8.64.

9.16 In the *periodic-review inventory policy*, there are no items kept in inventory. The time is divided into fixed periods of equal length $(0, T]$, $(T, 2T]$, $(2T, 3T]$, ... All demands arriving during $((i-1)T, iT]$ are kept in backlog until time iT; then an order is placed to satisfy all of them simultaneously. Suppose the demands arrive as a Poisson stream with rate λ. Assume the cost of holding each demand per unit time is \$$h$ and the cost of purchasing n items is \$ $(K + nc)$. Find the period T that minimizes the total cost rate. *Hint:* Use Problem 8.61.

9.17 Messages arrive at a clearing system as a Poisson stream with rate λ/hour. The system is inspected every T hours, at which time all messages are cleared. Let $\ell(t)$ be the number of messages at time t and $L \equiv \lim_{t\to\infty} \mathcal{E}[\ell(t)]$. Let w_n be the duration message n spends in the system and $W \equiv \lim_{n\to\infty} \mathcal{E}[w_n]$.

a. Let T be a random variable. Find W, L and show that $L = \lambda W$. *Hint:* Use Problem 8.61.

b. Let T be a fixed duration. Let \$$b$ be the cost of each inspection and \$$a$ the cost of storing each message per hour. What is the value of T that minimizes the total cost rate?

9.18 Customers arrive at the entrance of an amusement park as a renewal stream with interarrival times τ hours. They then sit and wait in a bus, which will take them inside the park whenever there are N waiting customers.

a. Find a suitable regenerative process to obtain the limiting expected waiting time of each customer.

b. Each bus trip costs \$$b$. The waiting cost of each customer is \$$a$/hour. Find the value of N that minimizes the total cost rate.

9.19 A system has two components, A and B. The life of A is exponentially distributed with rate μ and that of B is a fixed duration β. Both components are replaced immediately upon failure. Regardless of the number of components being replaced, each replacement requires a fixed duration ρ and costs \$$r$. Component A costs \$$a$ each; component B costs \$$b$ each.

a. One policy is to replace B whenever A is replaced. Find the total cost rate of this policy

i. by studying the stream of simultaneous replacements of both A and B.

ii. by studying the stream of replacements of B.

Hint: Use Problem 8.7.

b. The other policy is to replace the components independently. Assume $\rho = 0$. Show that it is more economical to replace B with A only when

$$\frac{r}{b} > \frac{\lambda\beta - 1 + e^{-\lambda\beta}}{1 - e^{-\lambda\beta} - \lambda\beta e^{-\lambda\beta}}$$

c. Without any calculation, what can we say about the policy of replacing both items whenever one fails?

9.20 A company has a large number of identical instruments. One policy is to replace each instrument independently whenever it fails. The other (which is called the *block replacement policy*) is to replace all of them every T time units in addition to replacing each instrument whenever it fails. Let \$$c_1$ be the cost of replacing all instruments at times nT ($n = 1, 2, 3, \ldots$) and \$$c_2$ be the cost of replacing each instrument individually whenever it fails. Assume the duration between two consecutive failures has an Erlang distribution with parameters $(2, \lambda/2)$. Find the optimal value of T. *Hint:* Use Problem 9.4.

9.21 Consider a renewal stream having life ℓ and remaining life $r(t)$. Show that

$$\lim_{t\to\infty} \mathcal{E}\left[r^k(t)\right] = \frac{1}{k+1}\frac{\mathcal{E}\left[\ell^{k+1}\right]}{\mathcal{E}[\ell]}$$

9.22 Consider a renewal stream having life ℓ and remaining life $r(t)$.

a. Use the regenerative property to show that

$$\lim_{t\to\infty} \Pr\{r(t) \le x\} = \frac{1}{\mathcal{E}[\ell]}\int_0^x \Pr\{\ell > t\}\,dt$$

Hint: Use Problem 1.51c.

b. Hence, obtain Equation (9.7.1). *Hint:* Use Problem 1.51b.

c. When ℓ is a constant, show that $r(t)$ has the uniform distribution.

d. When ℓ is exponentially distributed, show that $r(t)$ has the same distribution as ℓ.

e. Compare with Problem 7.2.

9.23 When a life of an instrument is discrete, use regenerative arguments to find the expected remaining life. Compare with Problem 7.2.

9.24 The *age* $a(t)$ of a renewal stream at time t is the duration since its last arrival.

a. Draw a realization of the age $a(t)$ of an instrument at time t.

b. From this, show that the limiting distribution of $a(t)$ is the same as that of $r(t)$. Compare with Problem 8.13.

c. Thus, show that the average life of the component currently in use is always greater than that of a typical instrument. Explain heuristically why.

9.25 In §9.7, we tacitly implied that the average waiting time for the next bus of a randomly arriving passenger is the same as the bus's average remaining life. Use Problem 8.61 to prove this property, which is known in the literature as *PASTA* (*P*oisson *A*rrivals *S*ee *T*ime *A*verage).

9.26 Customers arrive at a barbershop as a Poisson stream with rate λ. Since there is no waiting room in this shop, if a customer finds the barber busy, he will leave immediately without being served. Let the distribution of each haircut be $H(t)$. Show that the long-term proportion of potential customers who actually receive haircuts and the proportion of time the barber is idle are the same. Can we explain this in terms of *PASTA* (Problem 9.25)?

9.27 An instrument is installed at time $t = 0$. Its life ℓ has distribution $F(t)$ and probability density function $f(t)$. Let $\lambda(t)$ be such that

$$\Pr\{t \le \ell \le t + dt \mid \ell > t\} = \frac{\Pr\{t \le \ell \le t + dt\}}{\Pr\{\ell > t\}} = \frac{f(t)\,dt}{1 - F(t)} \equiv \lambda(t)\,dt$$

We call $\lambda(t)$ the *hazard rate function*, or the *failure rate function*, of the instrument.

a. Show that $\lambda(t)$ is a constant for a Markovian instrument.

b. If ℓ is uniformly distributed within [0, 1], then obtain $\lambda(t)$.

c. Show that the life distribution can be derived from $\lambda(t)$ as

$$F(t) = 1 - \exp\left\{-\int_0^t \lambda(u)\,du\right\}$$

d. Obtain the life distribution when $\lambda(t) = \lambda$.

e. Obtain the life distribution when $\lambda(t) = 1/(t + a)$ for all $t \ge 0$.

f. Suppose $\lambda(t) = (\beta/\alpha)(t/\alpha)^{\beta-1}$. Show that the life has the following *Weibull distribution*

$$F(t) = 1 - \exp\left\{-(t/\alpha)^{\beta}\right\}$$

9.28

a. In a renewal stream with life distribution $F(t)$, show that

$$\Pr\{m(t) = i\} = F^{*i}(t) - F^{*i+1}(t) \quad \text{for all } t > 0$$

where $F^{*1}(t) \equiv F(t)$, $F^{*2}(t) \equiv F(t) * F(t)$ as defined in Problem 1.56, and $F^{*i}(t)$ is the *i-fold convolution* of $F(t)$; that is,

$$F^{*i}(t) \equiv \underbrace{F(t) * \cdots * F(t)}_{i \text{ times}}$$

b. Hence, show that

$$\mathcal{E}[m(t)] = \sum_{i=1}^{\infty} F^{*i}(t) \quad \text{for all } t > 0$$

c. From this, prove that

$$\mathcal{E}[m(t)] = F(t) + \mathcal{E}[m(t)] * F(t) \quad \text{for all } t > 0$$

d. For the Poisson streams, obtain Equation (8.11.2)

i. by finding $(d/dt)\mathcal{E}[m(t)]$ from the result in part b. *Hint:* The i-fold convolution of an exponential distribution is an Erlang distribution with parameters $(i, \lambda/i)$.

ii. by using the result in part c. *Hint:* A solution of $f'(t) = \lambda e^{-\lambda t} - \lambda f(t)$ is $f(t) = \lambda t e^{\lambda t}$.

e. Assume the lives have the Erlang distribution with parameters $(2, \lambda/2)$.

i. Use the result in part b to find $\mathcal{E}[m(t)]$. *Hint:* Use Appendix (A.4).

ii. Compare with the result in Problem 9.4b.

9.29 Let $h(x)$ be a function such that $h(x) \geq 0$ for all $x \geq 0$, $h(x)$ is nonincreasing, and $\int_0^\infty h(t)\,dt < \infty$. The *key renewal theorem* states that, for a renewal stream having a nonlattice life ℓ,

$$\lim_{t\to\infty} h(t) * \mathcal{E}[m(t)] = \frac{1}{\mathcal{E}[\ell]} \int_0^\infty h(t)\,dt$$

a. Use $h(x) = 1$ for all $0 \leq x \leq s$, and $h = 0$ otherwise, to prove the following *Blackwell theorem*:

$$\lim_{t\to\infty} \{[\mathcal{E}[m(t)] - \mathcal{E}[m(t-s)]\} = \frac{s}{\mathcal{E}[\ell]} \quad \text{for all } s > 0$$

That is, as $t \to \infty$, $\mathcal{E}[m(t)]$ increases linearly.

b. Use $h(t) = \int_t^\infty [1 - F(x)]\,dx/\mathcal{E}[\ell]$ to show that

$$\lim_{t\to\infty} \left[m(t) - \frac{t}{\mathcal{E}[\ell]} \right] = \frac{\mathrm{Var}[\ell]}{2\mathcal{E}^2[\ell]} - \frac{1}{2}$$

Hint: Use Problem 9.28c.

c. Verify the result with that obtained in Problem 9.4.

d. Hence, obtain the elementary renewal theorem (9.2.1).

CHAPTER 10

Semi-Markov Chains

10.1 INTRODUCTION

We finished the previous chapter with some important results related to a regenerative process. We now know that most of its long-term results can be obtained if we know how the process behaves within a regenerative cycle. This was as far as we could go because the specification of a regenerative process does not provide any information about its behavior within a cycle.

We need additional assumptions to advance further. In this chapter, to keep the model widely applicable, we try to impose as few assumptions as possible, just enough to provide sufficient additional structure for some numerical and analytical results.

Consider now a continuous-time chain $\{X(t)\}_{t\in[0,\infty)}$, taking value in a *discrete* state space S.

1. We denote by T_n ($n \geq 1$) the time of the nth transition after $t = 0$. (Note that the chain does not have to change states at each transition, as at time T_4 in Figure 10.1.1.)
2. We call the duration between two consecutive transitional epochs T_{n-1} and T_n the chain's nth *holding time*.
3. We assume the chain's sample path is *right-continuous* and we write $X_n \equiv X(T_n)$.

Chain $\{X(t)\}_{t\in[0,\infty)}$ is said to be *semi-Markovian if its development after each transition is independent of its behavior before that time*. This means that the distribution of its holding time $T_n - T_{n-1}$ is independent of its behavior *before* T_{n-1} but may be a function of $X_{n-1} \equiv X(T_{n-1})$ and $X_n \equiv X(T_n)$. If $X_{n-1} = i$ and $X_n = j$, we denote the holding time $T_n - T_{n-1}$ by h_{ij}.

If all holding times in a semi-Markov chain are equal to a constant, the chain can be studied as a discrete-time Markov chain. To describe it completely, we need only have all transition probabilities $p_{ij} \equiv \Pr\{X_n = j \mid X_{n-1} = i\}$.

FIGURE 10.1.1 **A semi-Markov chain $\{X(t)\}_{t\in[0,\infty)}$.**

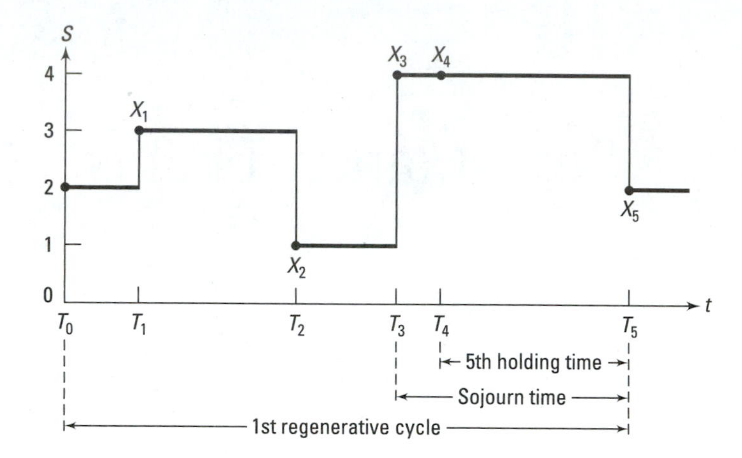

If a semi-Markov chain has only one state, all its holding times can only be a function of this one state; hence, they are independent and identically distributed. The chain therefore can be studied as a renewal stream, with a renewal at each transition. To describe this chain completely, we only need its holding time distribution as the life distribution of the stream.

There are at least three ways a general semi-Markov chain can develop during a holding time $T_n - T_{n-1}$ (and remain independent of its behavior before T_{n-1}):

1. Immediately after entering a state, the chain decides which state it will move into next and how long it will stay at the current state. These decisions are independent of each other.
2. Immediately after entering a state, the chain first decides how long it will stay there. Depending on this decision, it then decides which state it will move into next. The chain's transition probabilities are thus conditioned on the holding times. As one example, the probability of a patient being transferred from an emergency room to outpatient care is normally dependent on whether that person survives beyond a certain critical period. As another example, the longer an individual has been in a given state (place, occupation, status), the more likely he or she will move to the neighboring states rather than to those further away.
3. Immediately after entering a state, the chain first decides which state it will move into next. Depending on this decision, it then decides how long it will stay at the current state. The distribution of the holding time therefore is conditioned on both the current state and the next state.

We should note here that this list is not exhaustive and is not important for the identification of a semi-Markov chain. Understanding the way a chain develops between the transitions helps us to describe it, however. Even so, one kind of description can be translated to the other (Problems 10.1 and 10.2). So without loss of generality, we describe a semi-Markov chain by its transition matrix

$$\mathbf{P} \equiv [p_{ij}]_{i,j\in S} \equiv [\Pr\{X_n = j \mid X_{n-1} = i\}]_{i,j\in S}$$

and its conditional holding time distributions

$$\begin{aligned} H_{ij}(\tau) &\equiv \Pr\{T_n - T_{n-1} < \tau \mid X_{n-1} = i, X_n = j\} \\ &\equiv \Pr\{h_{ij} < \tau\} \qquad \text{for all } i, j \in S \end{aligned}$$

We impose very few restrictions on the conditional holding times. They can be discrete, continuous, or a mixture of both. We assume they are *finite*; that is, $H_{ij}(\infty) = 1$ for all $i, j \in S$. They also have to be such that the stream of the chain's transitions is *regular*, preventing an infinite number of transitions from occurring within a finite duration. This regularity condition can be satisfied if there exists a distribution $H(\tau)$ such that $H(0) < 1$, and $H(\tau) \geq H_{ij}(\tau)$ for all $i, j \in S$ and $\tau \geq 0$.

With such a general model requiring so few assumptions, we might wonder what we can get out of it. It turns out that we can get as many results for the semi-Markov chains as for the discrete-time Markov chains. In fact, we can consider this chapter as a continuous-time version of the materials presented from Chapter 4 to Chapter 7.

As you will see, all results presented in this chapter are dependent only on the *expected values* of the conditional holding times, not on their actual distributions. For our purposes in this chapter, it is sufficient to specify the chain by its two matrices:[1]

1. the transition matrix $\mathbf{P} \equiv [p_{ij}]_{i,j\in S}$
2. the expected conditional holding time matrix $\mathbf{H} \equiv \mathcal{E}[h_{ij}]_{i,j\in S}$

Semi-Markov Chains in Action: Activities of a Predator. Assume the act of predation by a predator is composed of four activities: (1) search, (2) pursuit, (3) handle and eat, and (4) digestion. Holling (1966) used a discrete-time Markov chain to describe how the mantid (*Hierodula crassa*) preys on houseflies (*Musca domestica*). However, Rao and Kshirsagar (1978) suggested that a semi-Markov is a better model for this kind of activity:

> The length of stay in each state depends on the physiological condition of the predator and is a random variable whose distribution function in general may depend on the state being occupied, as well as on the next state to which the process is moved. . . . It will be very unrealistic . . . to assume that the predator will stop searching after a unit of time and go to the pursuit state if the predator has not found any prey. In general, the time interval between any two realistic states will be random. ■

10.2 THE RAT-MAZE EXAMPLE

Let us return to the experiment introduced in §2.10, in which a rat is running inside a maze having six rooms: F (giving food), 2, 3, 4, 5, and S (giving shock). We now

[1] See Problems 10.9, 10.23 and 10.24, in which higher moments of the conditional holding times are needed.

assume further that, between the transitions, the rat can sense the direction toward food and thus, on the average, hesitates more before entering a door leading away from food. Let $R(t)$ be the room that the rat is in at time t. The continuous-time chain $\{R(t)\}_{t\in[0,\infty)}$ is semi-Markovian. To describe this chain completely, besides the canonical transition matrix (5.3.1), we also need to have all the conditional holding time distributions $H_{ij}(\tau)$. However, for illustrative purposes in this chapter, we do not need complete information about the latter, but only the following expected conditional holding time matrix (in minutes):

$$\mathbf{H} = \begin{array}{c} \\ 2 \\ 3 \\ 4 \\ 5 \end{array} \begin{array}{c} \begin{array}{cccccc} F & S & 2 & 3 & 4 & 5 \end{array} \\ \left(\begin{array}{cc|cccc} 1 & * & * & * & 3 & * \\ 1 & * & * & * & 3 & 3 \\ * & 4 & 2 & 2 & * & * \\ * & 4 & * & 2 & * & * \end{array}\right) \end{array} \tag{10.2.1}$$

where * indicates that the corresponding conditional holding time is not applicable because the transition is not possible. (Note that the F-row and the S-row of this matrix do not exist because we stop the experiment after the rat reaches either room F or S.)

10.3 A MARKETING EXAMPLE

In a hypothetical market, a company buys one of three brands X, Y, and Z on a regular basis. Let B_n be the brand it buys at the nth order. We assume that the discrete-time chain $\{B_n\}_{n=0,1,2,\ldots}$ is Markovian with the following transition matrix:

$$\mathbf{P} = \begin{array}{c} \\ X \\ Y \\ Z \end{array} \begin{array}{c} \begin{array}{ccc} X & Y & Z \end{array} \\ \begin{pmatrix} .5 & .2 & .3 \\ .1 & .6 & .3 \\ .2 & .4 & .4 \end{pmatrix} \end{array} \tag{10.3.1}$$

Let $B(t)$ be the brand the company is using at time t. After buying a particular brand, the company uses it continually until the next order, which occurs either when it runs out of it (and may buy the same brand again or try another brand) or decides to try another brand (hence discarding all current inventory). The duration between two consecutive orders is thus dependent not only on how the company feels about the brand it is using (the current state), but also on the strength of the cumulative effect of the advertising of the other brands between the orders (the destination states). The continuous-time chain $\{B(t)\}_{t\in[0,\infty)}$ is semi-Markovian. For our purposes, we need only assume the following expected holding time distribution matrix in weeks:

$$\mathbf{H} = \begin{array}{c} \\ X \\ Y \\ Z \end{array} \begin{array}{c} \begin{array}{ccc} X & Y & Z \end{array} \\ \begin{pmatrix} 1 & 3 & 4 \\ 4 & 2 & 5 \\ 3 & 6 & 1 \end{pmatrix} \end{array} \tag{10.3.2}$$

10.4 HOLDING TIMES

If the expected holding times $\mathcal{E}[h_{ij}]$s are conditional on both the current and the destination states, we can obtain the expected holding times $\mathcal{E}[h_i]$s that are conditional only on the current state as (Problem 10.3):

$$\mathcal{E}[h_i] \equiv \mathcal{E}[T_n - T_{n-1} \mid X_{n-1} = i] = \sum_{j \in S} p_{ij} \mathcal{E}[h_{ij}] \quad \text{for all } i \in S \tag{10.4.1}$$

For the rat-maze example, from the transition matrix (5.3.1) and the expected holding time matrix (10.2.1),

$$\begin{cases} \mathcal{E}[h_2] = \frac{1}{2}(1) + \frac{1}{2}(3) = 2 \text{ minutes} \\ \mathcal{E}[h_3] = \frac{1}{3}(1) + \frac{1}{3}(3) + \frac{1}{3}(3) = \frac{7}{3} \text{ minutes} \\ \mathcal{E}[h_4] = \frac{1}{3}(4) + \frac{1}{3}(2) + \frac{1}{3}(2) = \frac{8}{3} \text{ minutes} \\ \mathcal{E}[h_5] = \frac{1}{2}(4) + \frac{1}{2}(2) = 3 \text{ minutes} \end{cases} \tag{10.4.2}$$

For the marketing example, from the transition matrix (10.3.1) and the expected holding time matrix (10.3.2),

$$\begin{cases} \mathcal{E}[h_X] = (.5)(1) + (.2)(3) + (.3)(4) = 2.3 \text{ weeks} \\ \mathcal{E}[h_Y] = (.1)(4) + (.6)(2) + (.3)(5) = 3.1 \text{ weeks} \\ \mathcal{E}[h_Z] = (.2)(3) + (.4)(6) + (.4)(1) = 3.4 \text{ weeks} \end{cases} \tag{10.4.3}$$

10.5 SOJOURN TIMES

As defined in §2.18, the *sojourn time* s_i is the total duration a chain stays continually at state i before moving out. This is the duration from a transition into state i to the next transition out of it. If $p_{ii} = 0$, as in the rat-maze example, the sojourn time s_i is the same as the holding time h_i. If $p_{ii} > 0$, they may not be the same. In Figure 10.1.1, the sojourn time $T_5 - T_3$ is longer than the holding time $T_4 - T_3$.

We can use first-step analysis to obtain the expected sojourn time as follows. Suppose $X(0) = i$. Then after the first transition,

1. the chain remains at state i with probability p_{ii}, resulting in $\mathcal{E}[s_i \mid X_1 = i] = \mathcal{E}[h_{ii}] + \mathcal{E}[s_i]$.
2. the chain goes to state $j \neq i$ with probability p_{ij}, resulting in $\mathcal{E}[s_i \mid X_1 = j \neq i] = \mathcal{E}[h_{ij}]$.

Thus, from the definition of $\mathcal{E}[h_i]$ in (10.4.1),

$$\mathcal{E}[s_i] = p_{ii}\left(\mathcal{E}[h_{ii}] + \mathcal{E}[s_i]\right) + \sum_{j \neq i} p_{ij} \mathcal{E}[h_{ij}] = p_{ii} \mathcal{E}[s_i] + \mathcal{E}[h_i]$$

or

$$\mathcal{E}[s_i] = \frac{\mathcal{E}[h_i]}{1 - p_{ii}} \quad \text{for all } i \in S \tag{10.5.1}$$

Actually, we can obtain this equation by noting that the sojourn time s_i is the random sum of n holding times, where n is the number of steps the chain stays at state i at each visit. In §2.19, we showed that n has a geometric distribution with mean $\mathcal{E}[n] = 1/(1 - p_{ii})$. Applying Equation (1.21.1) should yield Equation (10.5.1).

For the marketing example,

$$\mathcal{E}[s_X] = \frac{2.3}{1 - .5} = 4.6 \text{ weeks} \qquad \mathcal{E}[s_Y] = \frac{3.1}{1 - .6} = 7.75 \text{ weeks}$$

$$\mathcal{E}[s_Z] = \frac{3.4}{1 - .4} = 5.67 \text{ weeks}$$

10.6 CLASSIFICATION OF STATES

Whenever a semi-Markov chain $\{X(t)\}_{t\in[0,\infty)}$ visits a state, its embedded chain $\{X_n\}_{n=0,1,2,\ldots}$ also visits the same state. Thus, the classification of the former is primarily determined by the latter.

Some definitions used for the discrete-time Markov chains can readily be extended to the semi-Markov chains. We say two states of a semi-Markov chain *communicate*, and hence belong to the same *class*, if they do so in its embedded chain. Similarly, a state of a semi-Markov chain is said to be *recurrent* (*transient* or *absorbing*) if it is so in its embedded chain. Like its embedded chain, a semi-Markov chain is *irreducible* if it has only one class and *absorbing* if all its states are either transient or absorbing.

As before, a recurrent state in a semi-Markov chain is said to be *positive-recurrent* if the expected return time is finite; otherwise, it is *null-recurrent*. In this chapter, since we assume all holding times are finite, a state is positive-recurrent if it is so in its embedded chain; since we assume that the stream of transitions is regular, a state is null-recurrent if it is so in its embedded chain. (Why?)

For the rest of the chapter, we study the absorbing semi-Markov chains first. We then discuss the positive-recurrent irreducible semi-Markov chains.

10.7 ABSORBING SEMI-MARKOV CHAINS

Consider an absorbing semi-Markov chain, which can be either finite or infinite. Let A be the set of all its absorbing states and T be the set of all its transient states. Its state space is organized so that the transition matrix $\mathbf{P}$ is in the canonical form (5.1.1). All the conditional holding time distributions $H_{ij}(\tau)$ corresponding to the absorbing states i are not defined. The rat-maze chain $\{R(t)\}_{t\in[0,\infty)}$ in §10.2 is an example of absorbing semi-Markov chains.

We are interested in calculating $\mathcal{E}[\psi_{ij}]$, where ψ_{ij} is the *total duration* the chain spends in a transient state j throughout its life, starting from a transient state i. This is the semi-Markov counterpart of the total number of visits v_{ij} a discrete-time absorbing chain makes to state j before absorption, starting from state i, as discussed in §5.3.

Regardless of whether an absorbing semi-Markov chain is finite or infinite, we can use first-step analysis. Assume the chain starts from a particular state $i \in T$.

1. With probability p_{ia}, the chain goes from state i to an absorbing state a in the first transition, resulting in

$$\mathcal{E}[\psi_{ii} \mid X_1 = a \in A] = \mathcal{E}[h_{ia}]$$

and

$$\mathcal{E}[\psi_{ij} \mid X_1 = a \in A] = 0 \quad \text{for all } j \in T \text{ and } j \neq i$$

2. With probability p_{it}, the chain goes from state i to a transient state t in the first transition, resulting in

$$\mathcal{E}[\psi_{ii} \mid X_1 = t \in T] = \mathcal{E}[h_{it}] + \mathcal{E}[\psi_{ti}]$$

and

$$\mathcal{E}[\psi_{ij} \mid X_1 = t \in T] = \mathcal{E}[\psi_{tj}] \quad \text{for all } j \in T \text{ and } j \neq i$$

Thus, for a particular transient state j, we have the following system of simultaneous linear equations:

$$\begin{cases} \mathcal{E}[\psi_{jj}] = \sum_{a \in A} p_{ja}\mathcal{E}[h_{ja}] + \sum_{t \in T} p_{jt}\left(\mathcal{E}[h_{jt}] + \mathcal{E}[\psi_{tj}]\right) = \mathcal{E}[h_j] + \sum_{t \in T} p_{jt}\mathcal{E}[\psi_{tj}] \\ \mathcal{E}[\psi_{ij}] = \sum_{t \in T} p_{it}\mathcal{E}[\psi_{tj}] \quad \text{for all } i \in T \text{ and } i \neq j \end{cases} \tag{10.7.1}$$

In the rat-maze example, recall that Equation (10.4.2) gives $\mathcal{E}[h_3] = 7/3$ minutes. When $j = 3$, we have the following semi-Markov version of Equation (5.3.2):

$$\begin{cases} \mathcal{E}[\psi_{23}] = & & + (1/2)\mathcal{E}[\psi_{43}] \\ \mathcal{E}[\psi_{33}] = 7/3 & & + (1/3)\mathcal{E}[\psi_{43}] + (1/3)\mathcal{E}[\psi_{53}] \\ \mathcal{E}[\psi_{43}] = & (1/3)\mathcal{E}[\psi_{23}] + (1/3)\mathcal{E}[\psi_{33}] & \\ \mathcal{E}[\psi_{53}] = & (1/2)\mathcal{E}[\psi_{33}] & \end{cases}$$

Hence,

$$\mathcal{E}[\psi_{23}] = 2/3 \qquad \mathcal{E}[\psi_{33}] = 10/3 \qquad \mathcal{E}[\psi_{43}] = 4/3 \qquad \mathcal{E}[\psi_{53}] = 5/3$$

For finite chains, there is a better way: It can be verified that the following equation, which is similar to Equation (5.3.6), combines all different values of j in Equation (10.7.1) in a matrix form:

$$\mathbf{\Psi} = \mathbf{Y}_d + \mathbf{T}\mathbf{\Psi}$$

where

1. $\mathbf{\Psi} \equiv [\mathcal{E}[\psi_{ij}]]_{i,j \in T}$
2. $\mathbf{Y}_d$ is a diagonal matrix, the (j, j)-element of which equals $\mathcal{E}[h_j]$, and other elements equal zero
3. $\mathbf{T}$ is the transient-to-transient portion of the transition matrix $\mathbf{P}$ as defined in (5.1.1)

Hence,

$$\mathbf{\Psi} = (\mathbf{I} - \mathbf{T})^{-1}\mathbf{Y}_d = \mathbf{U}\mathbf{Y}_d \tag{10.7.2}$$

where $\mathbf{U} \equiv [\mathcal{E}[v_{ij}]]_{i,j \in T} = (\mathbf{I} - \mathbf{T})^{-1}$ is the fundamental matrix.

In other words,

$$\mathcal{E}[\psi_{ij}] = \mathcal{E}[v_{ij}]\mathcal{E}[h_j] \quad \text{for all } i, j \in T \tag{10.7.3}$$

This should be intuitive: Starting from state i, the expected total duration an absorbing chain visits state j before absorption, $\mathcal{E}[\psi_{ij}]$, is a random sum of v_{ij} visits to state j, and the expected duration of each visit is $\mathcal{E}[h_j]$.

For the rat-maze example, from Equations (5.3.8) and (10.4.2),

$$\mathbf{\Psi} = \begin{array}{c} \\ 2 \\ 3 \\ 4 \\ 5 \end{array} \begin{array}{c} \begin{array}{cccc} 2 & 3 & 4 & 5 \end{array} \\ \begin{pmatrix} \frac{26}{21} & \frac{2}{7} & \frac{5}{7} & \frac{2}{21} \\ \frac{4}{21} & \frac{10}{7} & \frac{4}{7} & \frac{10}{21} \\ \frac{10}{21} & \frac{4}{7} & \frac{10}{7} & \frac{4}{21} \\ \frac{2}{21} & \frac{5}{7} & \frac{2}{7} & \frac{26}{21} \end{pmatrix} \end{array} \begin{array}{c} \begin{array}{cccc} 2 & 3 & 4 & 5 \end{array} \\ \begin{pmatrix} 2 & 0 & 0 & 0 \\ 0 & \frac{7}{3} & 0 & 0 \\ 0 & 0 & \frac{8}{3} & 0 \\ 0 & 0 & 0 & 3 \end{pmatrix} \end{array} = \begin{array}{c} \begin{array}{cccc} 2 & 3 & 4 & 5 \end{array} \\ \begin{pmatrix} \frac{52}{21} & \frac{2}{3} & \frac{40}{21} & \frac{2}{7} \\ \frac{8}{21} & \frac{10}{3} & \frac{32}{21} & \frac{10}{7} \\ \frac{20}{21} & \frac{4}{3} & \frac{80}{21} & \frac{4}{7} \\ \frac{4}{21} & \frac{5}{3} & \frac{16}{21} & \frac{26}{7} \end{pmatrix} \end{array} \tag{10.7.4}$$

The 3-column of this matrix agrees with the results obtained for state 3 previously.

Let ψ_i be the *total duration before the chain is absorbed*, starting from state i. This is the semi-Markovian counterpart of the absorbing time u_i, which is the total number of steps before a discrete-time absorbing chain is absorbed, as discussed in §5.4.

While we can obtain $\mathcal{E}[\psi_i]$ directly from first-step analysis (Problem 10.8), we can see that it is simply the sum of the i-row of $\mathbf{\Psi}$; that is,

$$\boldsymbol{\psi} \equiv [\mathcal{E}[\psi_i]]_{i \in T} = \mathbf{\Psi}\mathbf{e}$$

where all elements of column $\mathbf{e}$ are equal to 1.

For example, starting from room 2, the expected duration before the rat finds either food or shock is

$$\mathcal{E}[\psi_2] = \frac{52}{21} + \frac{2}{3} + \frac{40}{21} + \frac{2}{7} = \frac{16}{3} \text{ minutes}$$

Semi-Markov Chains in Action: Coronary Patients' Lengths of Stay in a Hospital. Coronary patients can be transferred among the following units within a hospital: (1) coronary care, (2) postcoronary care, (3) intensive care, (4) the medical unit, (5) the surgical unit, and (6) ambulatory care. They can also be moved from the hospital to one of the following absorbing states: (7) extended care facility, (8) home, and (9) morgue.

Kao (1974a, b) followed 471 patients and obtained the following transition matrix:

$$
\mathbf{P} = \begin{array}{c} \\ 7 \\ 8 \\ 9 \\ 1 \\ 2 \\ 3 \\ 4 \\ 5 \\ 6 \end{array}
\begin{array}{c}
\begin{array}{ccc|cccccc} 7 & 8 & 9 & 1 & 2 & 3 & 4 & 5 & 6 \end{array} \\
\left(\begin{array}{ccc|cccccc}
1 & 0 & 0 & 0 & 0 & 0 & 0 & 0 & 0 \\
0 & 1 & 0 & 0 & 0 & 0 & 0 & 0 & 0 \\
0 & 0 & 1 & 0 & 0 & 0 & 0 & 0 & 0 \\
\hline
0 & .0063 & .0962 & 0 & .7447 & .0084 & .1339 & .0042 & .0063 \\
.0577 & .8298 & .0495 & .0192 & 0 & .0137 & .0247 & .0027 & .0027 \\
0 & 0 & .1667 & 0 & .5833 & 0 & .1667 & .0833 & 0 \\
.0811 & .7028 & .1216 & 0 & .0135 & .0405 & 0 & .0135 & .0270 \\
0 & 1 & 0 & 0 & 0 & 0 & 0 & 0 & 0 \\
0 & 1 & 0 & 0 & 0 & 0 & 0 & 0 & 0
\end{array}\right)
\end{array}
$$

and the following expected conditional holding time matrix in days:

$$
\mathbf{H} = \begin{array}{c} \\ 1 \\ 2 \\ 3 \\ 4 \\ 5 \\ 6 \end{array}
\begin{array}{c}
\begin{array}{ccc|cccccc} 7 & 8 & 9 & 1 & 2 & 3 & 4 & 5 & 6 \end{array} \\
\left(\begin{array}{ccc|cccccc}
* & 2.84 & 1.92 & * & 4.17 & 4.17 & 4.17 & 4.17 & 2.84 \\
14.98 & 14.98 & 6.82 & 14.98 & * & 4.17 & 14.98 & 4.17 & 4.17 \\
* & * & 1.92 & * & 14.98 & * & 4.17 & 4.17 & * \\
14.98 & 14.98 & 6.82 & * & 14.98 & 14.98 & * & 14.98 & 14.98 \\
* & 14.98 & * & * & * & * & * & * & * \\
* & 14.98 & * & * & * & * & * & * & *
\end{array}\right)
\end{array}
$$

Depending on which unit a patient is admitted to, the expected lengths of stay in various units and the expected total length of stay in the hospital can be calculated as

$$
\boldsymbol{\Psi} = \begin{array}{c} \\ 1 \\ 2 \\ 3 \\ 4 \\ 5 \\ 6 \end{array}
\begin{array}{c}
\begin{array}{cccccc} 1 & 2 & 3 & 4 & 5 & 6 \end{array} \\
\left(\begin{array}{cccccc}
3.99 & 11.10 & 0.26 & 2.23 & 0.16 & 0.19 \\
0.08 & 14.71 & 0.16 & 0.43 & 0.07 & 0.06 \\
0.05 & 8.67 & 10.26 & 2.60 & 1.33 & 0.10 \\
0 & 0.55 & 0.42 & 14.10 & 0.26 & 0.41 \\
0 & 0 & 0 & 0 & 14.98 & 0 \\
0 & 0 & 0 & 0 & 0 & 14.98
\end{array}\right)
\end{array}
\rightarrow \boldsymbol{\psi} = \begin{array}{c} 1 \\ 2 \\ 3 \\ 4 \\ 5 \\ 6 \end{array}
\left(\begin{array}{c} 17.93 \\ 15.50 \\ 23.01 \\ 15.73 \\ 14.98 \\ 14.98 \end{array}\right)
$$

Knowing the amount of time a patient is expected to spend in each care unit during his stay in the hospital enables us to project the demands for various resources. They in turn form the data base under which planning and scheduling activities are carried out. ■

10.8 IRREDUCIBLE SEMI-MARKOV CHAINS

We now turn our attention to the irreducible semi-Markov chains, both finite and infinite. The reader should be able to extend the following analysis to the reducible chains in a manner similar to that in §6.9.

While all states in a finite irreducible chain must be positive-recurrent, Chapter 7 shows that the states in an infinite irreducible chain may also be all transient or all null-recurrent. In this chapter, we restrict our discussion to the *positive-recurrent* irreducible semi-Markov chains. This means that, finite or infinite, the chain will always return to each state, and the duration between any two consecutive returns is finite.

Assume now that a semi-Markov chain starts at time $t = 0$ from state $i \in S$. Whenever it returns to this state, its development in the future is a function of i but is independent of its past, and thus is probabilistically independent and identical to its development from time $t = 0$. The chain therefore is *regenerative* with regenerative points at each return to state i. (Figure 10.1.1 shows a regenerative cycle from state 2.) It does not matter how long the chain has been going, it is still in one of the regenerative cycles, and its behavior within this cycle is identical to that within the first cycle.

Note that, starting from state i, this chain does not regenerate at all transitions, but only at the transitions into state i.

We use this regenerative property to derive the limiting probabilities. But before that, we need a few intermediate and very useful results.

10.9 EXPECTED FIRST REACHING TIMES

The *first reaching time*[2] t_{ij} is the duration between a transition to state i and the next transition to state j. In the marketing example, this is the duration between the company's ordering brand i and the first time it orders brand j. This term unfortunately is the same as that for n_{ij} introduced in §6.5 for the discrete-time chains. We just have to be careful not to confuse them.

We are especially interested in the expected value $\mathcal{E}[t_{ij}]$. As in §6.5, one way to obtain $\mathcal{E}[t_{ij}]$ is to make state j absorbing and then to find the expected total duration $\mathcal{E}[\psi_i]$ before this modified chain is absorbed, starting from state i. A slightly different method is to use first-step analysis. Suppose the chain starts from state i.

1. It goes to state j in the first transition with probability p_{ij}, resulting in $\mathcal{E}[t_{ij} \mid X_1 = j] = \mathcal{E}[h_{ij}]$.
2. It goes to state $k \neq j$ in the first transition with probability p_{ik}, resulting in $\mathcal{E}[t_{ij} \mid X_1 = k \neq j] = \mathcal{E}[h_{ik}] + \mathcal{E}[t_{kj}]$.

Thus, for a particular state j, we now have an semi-Markovian version of Equation (6.5.1):

$$\begin{aligned}\mathcal{E}[t_{ij}] &= p_{ij}\mathcal{E}[h_{ij}] + \sum_{k \neq j} p_{ik}\left(\mathcal{E}[h_{ik}] + \mathcal{E}[t_{kj}]\right) \\ &= \mathcal{E}[h_i] + \sum_{k \neq j} p_{ik}\mathcal{E}[t_{kj}] \quad \text{for all } i \in S \end{aligned} \tag{10.9.1}$$

[2]Some authors call it the *first passage time*.

For the marketing example, with $j = X$ and with the values of $\mathcal{E}[h_i]$s obtained in (10.4.3), this equation yields

$$\mathcal{E}[t_{XX}] = 2.3 + (.2)\mathcal{E}[t_{YX}] + (.3)\mathcal{E}[t_{ZX}]$$

$$\begin{cases} \mathcal{E}[t_{YX}] = 3.1 + (.6)\mathcal{E}[t_{YX}] + (.3)\mathcal{E}[t_{ZX}] \\ \mathcal{E}[t_{ZX}] = 3.4 + (.4)\mathcal{E}[t_{YX}] + (.4)\mathcal{E}[t_{ZX}] \end{cases}$$

giving

$$\mathcal{E}[t_{XX}] = \frac{68}{5} \text{ weeks} \qquad \mathcal{E}[t_{YX}] = 24 \text{ weeks} \qquad \mathcal{E}[t_{ZX}] = \frac{65}{3} \text{ weeks.}$$

Like Equation (6.5.3), Equation (10.9.1) can be written in a matrix form as

$$\mathbf{\Theta} = \mathbf{Y} + \mathbf{P}[\mathbf{\Theta} - \mathbf{\Theta}_d] \tag{10.9.2}$$

where

1. $\mathbf{\Theta} \equiv [\mathcal{E}[t_{ij}]]_{i,j \in S}$
2. $\mathbf{\Theta}_d$ is a diagonal matrix in which the (i, i)-element equals $\mathcal{E}[t_{ii}]$, and all other elements equal zero
3. $\mathbf{Y}$ is a matrix in which all elements of the i-row are equal to $\mathcal{E}[h_i]$

10.10 EXPECTED RETURN TIMES

Of the first reaching times t_{ij}s, we are interested in the *return time* $t_i \equiv t_{ii}$, especially its expected value $\mathcal{E}[t_i]$. This is the duration between two consecutive transitions into state i. In the marketing example, it is the duration from the time the company orders brand i to next time it orders brand i again. Once more, we should be careful not to confuse the return time t_i of a semi-Markov chain with that of a discrete-time Markov chain, denoted by n_i in §6.6.

We are talking about a semi-Markov chain having regenerative points at each return to state i. The expected return time $\mathcal{E}[t_i]$ is therefore the expected duration of its regenerative cycles. That is why it is important to find this quantity first.

Although we can calculate $\mathcal{E}[t_i]$ from Equation (10.9.1), there is a better way: Recall from §6.3 and §7.4 that, if our embedded chain is positive-recurrent, the visiting rate vector $\mathbf{g} \equiv [g_i]_{i \in S}$ satisfies $\mathbf{gP} = \mathbf{g}$. Premultiplying Equation (10.9.2) by $\mathbf{g}$, we have

$$\mathbf{g\Theta} = \mathbf{gY} + \mathbf{gP}[\mathbf{\Theta} - \mathbf{\Theta}_d] = \mathbf{gY} + \mathbf{g\Theta} - \mathbf{g\Theta}_d$$

Hence,

$$\mathbf{g\Theta}_d = \mathbf{gY} \tag{10.10.1}$$

The right-hand side of (10.10.1) is a row in which all elements are equal to

$$h \equiv \sum_{k \in S} g_k \mathcal{E}[h_k] \tag{10.10.2}$$

Since g_k is the long-term proportion of visits to state k and $\mathcal{E}[h_k]$ is the expected holding time after each transition to state k, the quantity h can be interpreted as the long-term transition-average of the holding times, or the *(long-term) average spacing* of the transitions.

Since the left-hand side of (10.10.1) is a row, the i-element of which is $g_i\mathcal{E}[t_i]$, Equation (10.10.1) now yields the following semi-Markov version of Equation (6.6.1):

$$\mathcal{E}[t_i] = \frac{h}{g_i} = h\mathcal{E}[n_i] \quad \text{for all } i \in S \tag{10.10.3}$$

Remember that the chain is regenerative at each return to state i. Thus, not only is h the long-term average spacing of the transitions, it is also the average spacing of all transitions within a regenerative cycle. Equation (10.10.3) simply says that the expected duration of a regenerative cycle (that is, $\mathcal{E}[t_i]$) is the product of the expected number of transitions in it (that is, $\mathcal{E}[n_i] = 1/g_i$) and the average spacing between the transitions (that is, h).

For the marketing example, we can obtain the visiting rates as $g_X = 2/9$, $g_Y = 4/9$, and $g_Z = 1/3$ (Problem 6.6). The values of $\mathcal{E}[h_i]$s in (10.4.3) now yield the average spacing as

$$h = \frac{2}{9}(2.3) + \frac{4}{9}(3.1) + \frac{1}{3}(3.4) = \frac{136}{45} \text{ weeks}$$

Hence,

$$\mathcal{E}[t_X] = \frac{136/45}{2/9} = \frac{68}{5} \text{ weeks} \qquad \mathcal{E}[t_Y] = \frac{136/45}{4/9} = \frac{34}{5} \text{ weeks} \tag{10.10.4}$$

$$\mathcal{E}[t_Z] = \frac{136/45}{1/3} = \frac{136}{15} \text{ weeks}$$

The value of $\mathcal{E}[t_X]$ agrees with the result obtained in the previous section.

Semi-Markov Chains in Action: Shock Rates for Experimental Rats. Because the chain's transitions into state i form a renewal stream, $1/\mathcal{E}[t_i]$ is its *long-term rate of entrance* into this state. This rate is of particular interest in the following experiment on unsignaled avoidance by Cotton and Wood (1984). Each of four female albino Wistar rats was put in a two-compartment box and given an electrical shock.

1. If it made a premature response by running to the other compartment within α seconds ($\alpha < 20$) after a shock, it would be given another shock α seconds after the response.
2. If it failed to run to the other compartment within 20 seconds after a shock, it would be given another shock 20 seconds after that shock.
3. If it ran to the other compartment within the time window from α seconds to 20 seconds after a shock, it could delay the next shock. After a delay, it could have another delay if it ran to the other compartment within 20 seconds. Thus, it could postpone the shock indefinitely.

We say the rat enters into state S when it receives a shock, into state N when it runs prematurely to the other compartment within α seconds after the shock, and into state R when it runs to the other compartment within the duration of effective response $(\alpha, 20)$.

When $\alpha = 10$ seconds, empirical data gave the following matrices:

$$\mathbf{P} = \begin{array}{c} \\ R \\ S \\ N \end{array} \!\!\begin{array}{c} \begin{array}{ccc} R & S & N \end{array} \\ \begin{pmatrix} .928 & .072 & 0 \\ .746 & .047 & .207 \\ 0 & 1 & 0 \end{pmatrix} \end{array} \qquad \mathbf{H} = \begin{array}{c} \\ R \\ S \\ N \end{array} \!\!\begin{array}{c} \begin{array}{ccc} R & S & N \end{array} \\ \begin{pmatrix} 12.37 & 20 & * \\ 14.29 & 20 & 4.18 \\ * & 10 & * \end{pmatrix} \end{array}$$

What is of interest here is not the probability of the rat being in a certain state, but the mean duration $\mathcal{E}[t_S]$ between two consecutive shocks, or the long-term rate of receiving shocks $1/\mathcal{E}[t_S]$. The experiment gave an observed mean of 139.86 seconds, which is very comparable with the theoretical mean of $\mathcal{E}[t_S] = 148.39$ seconds.

Note that $\mathcal{E}[t_S]$ is a long-term average, assuming that the rat behaves in a consistent manner over a long period of time. This can explain why the discrepancy between $\mathcal{E}[t_S]$ and its data became greater as the value of α increased, or the duration of effective response $(\alpha, 20)$ decreased, making it harder for the rats to discriminate.

> The agreement between theory and experiment for $\alpha = 15$ is poor. This is not unexpected on the basis of what is already well-known about the process of discrimination. Discriminations of any kind (including of course temporal) may be very easy or very difficult. At one end of the continuum, easy discriminations are rapidly acquired while at the other end of the continuum very difficult discriminations can lead to deterioration of performance once a certain point is reached and ultimately to a breakdown in behavior which Pavlov (1927) aptly referred to as "experimental neurosis." ■

10.11 VISITING RATES

Let γ_i be the long-term rate that a semi-Markov chain visits state i; that is, $\gamma_i \equiv \lim_{t\to\infty} \mathcal{E}[\psi_{ji}(t)]/t$, where $\psi_{ji}(t)$ is the total duration it visits state i during $(0, t]$, starting with state j. (Note that this long-term quantity is independent of the initial conditions.) Again, we should distinguish between the visiting rate γ_i of a semi-Markov chain and that of a discrete-time chain, denoted by g_i in §6.4.

Suppose the chain starts from state i; hence, it is regenerative at each return to state i, having $h\mathcal{E}[n_i]$ as the expected cycle time. In §6.7, we denote the number of visits to state j within a regenerative cycle from state i as ν_{ij}. The expected total duration the chain visits state j within a regenerative cycle from state i is therefore $\mathcal{E}[\nu_{ij}]\mathcal{E}[h_j]$. From the continuous-time regenerative property (9.6.2), we have $\gamma_j = \mathcal{E}[\nu_{ij}]\mathcal{E}[h_j]/(h\mathcal{E}[n_i])$ for all $j \in S$. Because $g_j = \mathcal{E}[\nu_{ij}]/\mathcal{E}[n_i]$ from the discrete-time regenerative property (6.7.1), we now have from Equation (10.10.3)

$$\gamma_j = \frac{g_j \mathcal{E}[h_j]}{h} = \frac{\mathcal{E}[h_j]}{\mathcal{E}[t_j]} \quad \text{for all } j \in S \tag{10.11.1}$$

It is intuitive to see that the visiting rate γ_j of a semi-Markov chain $\{X(t)\}_{t\in[0,\infty)}$ is proportional to both the visiting rate g_j of its embedded chain and the expected visiting time $\mathcal{E}[h_j]$ to state j at each visit. In fact, $h = \sum_{k\in S} g_k h_k$ can be considered simply as a normalizing factor to ensure that $\sum_{i\in S} \gamma_i = 1$.

In the marketing example, from Equations (10.10.4),

$$\gamma_X = \frac{2.3}{68/5} = \frac{23}{136} \qquad \gamma_Y = \frac{3.1}{34/5} = \frac{31}{68} \qquad \gamma_Z = \frac{3.4}{136/15} = \frac{3}{8} \tag{10.11.2}$$

10.12 LIMITING RESULTS

As discussed in §9.9 and §9.10, as a regenerative process, an irreducible positive-recurrent semi-Markov chain $\{X(t)\}_{t\in[0,\infty)}$ will converge in distribution if its regenerative cycle or its return time t_i is spread-out.

The return time t_i is spread-out if there are two states $j, k \in S$ such that $p_{jk} > 0$ and the holding time distribution $H_{ij}(\tau)$ has a derivative within a certain interval (a, b). This is because the chain is assumed to be irreducible, and thus, there is a positive probability that, starting from state i, it will visit state j and then immediately go to state k before it returns to state i. If the holding time between the visits to states j and k is spread-out, so is the return time t_i.

In this case, the chain's limiting probability $\lim_{t\to\infty} \Pr\{X(t) = j\}$ exists, is independent of the initial condition, and is equal to the long-term visiting rate γ_j for all $j \in S$.

10.13 SUMMARY

1. A chain is said to be semi-Markovian if its development after each transition is independent on its behavior before that time.
2. In an absorbing semi-Markov chain, Equation (10.7.3) shows that the expected total duration $\mathcal{E}[\psi_{ij}]$ a chain visits state j throughout its life is the product of the expected number of times $\mathcal{E}[v_{ij}]$ it visits that state and the expected duration $\mathcal{E}[h_j]$ it stays there at each visit.
3. In a positive-recurrent irreducible semi-Markov chain, Equation (10.10.3) shows that the expected return time $\mathcal{E}[t_i]$ is the product of the expected number of transitions between any two consecutive returns to state i (which is $1/g_i$) and the average spacing h of the transitions.
4. In a positive-recurrent irreducible semi-Markov chain, Equation (10.11.1) shows that the long-term visiting rate γ_i to state i is proportional to both the expected holding time $\mathcal{E}[h_i]$ and the long-term visiting rate g_i of its embedded chain.
5. A positive-recurrent irreducible semi-Markov chain converges in distribution if one of its conditional holding times is spread-out. Its limiting probabilities are then equal to its long-term proportions of visits.

PROBLEMS

10.1 Suppose a semi-Markov chain with state space $S \equiv \{A, B\}$ is specified by

$$\Pr\{T_n - T_{n-1} < \tau \mid X_{n-1} = A\} = 1 - e^{-3\tau}$$
$$\Pr\{T_n - T_{n-1} < \tau \mid X_{n-1} = B\} = 1 - e^{-5\tau}$$

and

$$[\Pr\{X_n = j \mid X_{n-1} = i, T_n - T_{n-1} < \tau\}]_{i,j \in S} = \begin{matrix} & A & B \\ A & .3(1 - e^{-4\tau}) & .7 + .3e^{-4\tau} \\ B & .5(1 - e^{-2\tau}) & .5 + .5e^{-2\tau} \end{matrix}$$

Find all $H_{ij}(\tau)$ and p_{ij} as defined in §10.1.

10.2 Suppose a semi-Markov chain with state space $\{A, B\}$ is specified by

$$\mathbf{P} = \begin{matrix} & A & B \\ A & .4 & .6 \\ B & .7 & .3 \end{matrix} \qquad [H_{ij}(\tau)]_{i,j \in S} = \begin{matrix} & A & B \\ A & 1 - e^{-2\tau} & 1 - e^{-3\tau} \\ B & 1 - e^{-5\tau} & 1 - e^{-4\tau} \end{matrix}$$

Find all $\Pr\{T_n - T_{n-1} < \tau \mid X_{n-1} = i\}$ and $\Pr\{X_n = j \mid X_{n-1} = i, T_n - T_{n-1} < \tau\}$.

10.3 Consider a semi-Markov chain.

a. Prove Equation (10.4.1).

b. Also prove that

$$\mathcal{E}\left[(h_i)^2\right] = \sum_{j \in S} p_{ij} \mathcal{E}\left[(h_{ij})^2\right]$$

c. Find all $\mathcal{E}[(h_i)^2]$s in the rat-maze example in §10.2, assuming

$$\left[\text{Var}[h_{ij}]\right]_{i \in T,\, j \in S} = \begin{matrix} & F & S & 2 & 3 & 4 & 5 \\ 2 & .1 & * & * & * & .3 & * \\ 3 & .2 & * & * & * & .2 & .1 \\ 4 & * & .3 & .3 & .2 & * & * \\ 5 & * & .2 & * & .1 & * & * \end{matrix}$$

d. Find all $\mathcal{E}[(h_i)^2]$s in the marketing example in §10.3, assuming

$$\left[\text{Var}[h_{ij}]\right]_{i,j \in S} = \begin{matrix} & X & Y & Z \\ X & .2 & .1 & .3 \\ Y & .1 & .2 & .2 \\ Z & .3 & .1 & .2 \end{matrix}$$

10.4 Consider the sojourn times s_i of a semi-Markov chain as defined in §10.5.

a. Use first-step analysis to prove that

$$\mathcal{E}\left[(s_i)^2\right] = \frac{\mathcal{E}\left[(h_i)^2\right] + 2p_{ii}\mathcal{E}[h_{ii}]\mathcal{E}[s_i]}{1 - p_{ii}} \quad \text{for all } i \in S$$

b. Hence, show that

$$\text{Var}[s_i] = \frac{\text{Var}[h_i]}{1 - p_{ii}} + \frac{p_{ii}\mathcal{E}[h_i]\,(2\mathcal{E}[h_{ii}] - \mathcal{E}[h_i])}{(1 - p_{ii})^2}$$

c. For the marketing example in §10.3 and in Problem 10.3, find $\mathcal{E}[(s_i)^2]$ and $\text{Var}[s_i]$ for all $i = X, Y, Z$.

10.5 The surface of a bacterium consists of sites that are susceptible to invasion by foreign molecules. Suppose foreign molecules arrive as a Poisson stream with rate λ. Some molecules have a particular composition that makes them acceptable to the bacterium and thus will be permanently affixed to the site upon arrival. Other molecules only remain at the site for a random duration having mean η before being rejected. Let α be the probability that an arriving molecule is acceptable. Find the expected duration before an unoccupied site becomes permanently occupied

a. by using an absorbing semi-Markov chain

b. by using first-step analysis

10.6 Verify the results obtained by Kao (1974) for coronary patients in §10.7.

10.7 Moore (1990) studied the vegetation dynamics of a forest community in northwestern Montana. Six different combinations of species were labeled 1, 2, 3, 4, 10, 11, and extinction was labeled E. Besides the deterministic successional changes, there are stochastic natural disturbances such as floods or fires, which affect the state transitions.

a. Let $g_{iq}(n)$ be the probability that a disturbance of type q occurs if the vegetation has been in state i for n years.

> For climatically induced natural disturbances such as cyclones or floods, $g_{iq}(n)$ is likely to be constant for all states and times. For fires, it is likely to increase with time as fuel builds up, and communities are differentially flammable.

Explain why this makes an equally spaced discrete-time Markov chain inappropriate.

b. The following transition matrix was obtained:

$$\mathbf{P} = \begin{array}{c} \\ E \\ 1 \\ 2 \\ 3 \\ 4 \\ 10 \\ 11 \end{array} \begin{array}{c|cccccc} E & 1 & 2 & 3 & 4 & 10 & 11 \\ 1 & & & & & & \\ \hline .095 & & .905 & & & & \\ & .669 & & .331 & & & \\ & & & & .668 & .332 & \\ .545 & & & & & .455 & \\ .095 & & & & & & .905 \\ & & & .331 & & .669 & \end{array}$$

The mean holding times in years were also obtained as: $\mathcal{E}[h_1] = 19.5$, $\mathcal{E}[h_2] = 70.0$, $\mathcal{E}[h_3] = 18.2$, $\mathcal{E}[h_4] = 24.8$, $\mathcal{E}[h_{(10)}] = 19.5$,[3] and $\mathcal{E}[h_{(11)}] = 70.0$. Verify Moore's calculations that the mean durations until extinction in years are $\mathcal{E}[\psi_1] = 456$, $\mathcal{E}[\psi_2] = 482$, $\mathcal{E}[\psi_3] = 325$, $\mathcal{E}[\psi_4] = 232$, $\mathcal{E}[\psi_{(10)}] = 456$, and $\mathcal{E}[\psi_{(11)}] = 482$.

10.8 Consider an absorbing semi-Markov chain.

a. Use first-step analysis to show that

$$\mathcal{E}[\psi_i] = \mathcal{E}[h_i] + \sum_{j \in T} p_{ij}\mathcal{E}[\psi_j] \quad \text{for all } i \in T$$

Compare with Problem 5.44.

b. Hence, find $\mathcal{E}[\psi_i]$ for all $i = 2, 3, 4, 5$ in the rat-maze example. Compare the results with those obtained from matrix $\mathbf{\Psi}$ in §10.7.

---

[3]The parentheses for (10) are to say that this is $\mathcal{E}[h_i]$, not $\mathcal{E}[h_{ij}]$.

10.9 Consider an absorbing semi-Markov chain.

a. Use first-step analysis to show that for all $i \in T$

$$\mathcal{E}\left[(\psi_i)^2\right] = \mathcal{E}\left[(h_i)^2\right] + 2\sum_{j\in T} p_{ij}\mathcal{E}\left[h_{ij}\right]\mathcal{E}\left[\psi_j\right] + \sum_{j\in T} p_{ij}\mathcal{E}\left[(\psi_j)^2\right]$$

Compare with Problem 5.46.

b. Hence, calculate $\mathcal{E}[(\psi_i)^2]$ for all $i = 2, 3, 4, 5$ in the rat-maze example using the variances in Problem 10.3.

10.10 Kao (1974b) obtained the second moments of the holding times of coronary patients discussed in §10.7 as

$$\left[\mathcal{E}\left[(h_{ij})^2\right]\right]_{i\in T,\, j\in S} = \begin{array}{c} \\ 1 \\ 2 \\ 3 \\ 4 \\ 5 \\ 6 \end{array}\begin{array}{c} \begin{array}{ccc|cccccc} 7 & 8 & 9 & 1 & 2 & 3 & 4 & 5 & 6 \end{array} \\ \left(\begin{array}{ccc|cccccc} 0 & a & b & * & c & c & c & c & a \\ d & d & e & d & * & c & d & c & c \\ * & * & b & * & d & * & c & c & * \\ d & d & e & * & d & d & * & d & d \\ * & d & * & * & * & * & * & * & * \\ * & d & * & * & * & * & * & * & * \end{array}\right) \end{array}$$

where $a = 9.88$, $b = 10.13$, $c = 18.39$, $d = 271.86$, and $e = 65.14$. Verify that the standard deviations of ψ_1, ψ_2, ψ_3, ψ_4, ψ_5, and ψ_6 are 9.91, 8.36, 14.12, 9.81, 6.90, and 6.90 days, respectively.

10.11 In the marketing example in §10.3, find $\mathcal{E}[t_{ij}]$ for all $i, j = X, Y, Z$

a. by using Equation (10.9.1)

b. by making the chain absorbing

10.12 A health service system has the following four components of care: (0) no use of service, (1) ambulatory care, (2) hospital care, and (3) domiciliary care. Assume $\mathcal{E}[h_0] =$ 200 days, $\mathcal{E}[h_1] = 40$ days, $\mathcal{E}[h_2] = 10$ days, $\mathcal{E}[h_3] = 5$ days, and the following transition matrix:

$$\mathbf{P} = \begin{array}{c} \\ 0 \\ 1 \\ 2 \\ 3 \end{array}\begin{array}{c} \begin{array}{cccc} 0 & 1 & 2 & 3 \end{array} \\ \left(\begin{array}{cccc} 0 & .8 & .1 & .1 \\ .7 & 0 & .2 & .1 \\ .1 & .8 & 0 & .1 \\ .4 & .4 & .2 & 0 \end{array}\right) \end{array}$$

Find the limiting distribution.

10.13 Using the following data (Rao and Kshirsagar, 1978), find the limiting distribution of the activities of the mantids introduced in §10.1:

$$\mathbf{P} = \begin{array}{c} \\ 1 \\ 2 \\ 3 \\ 4 \end{array}\begin{array}{c} \begin{array}{cccc} 1 & 2 & 3 & 4 \end{array} \\ \left(\begin{array}{cccc} .3757 & .6243 & & \\ .4247 & & .5753 & \\ .9755 & & & .0245 \\ 1 & & & \end{array}\right) \end{array}$$

(continued)

$$\mathbf{H} = \begin{array}{c} \\ 1 \\ 2 \\ 3 \\ 4 \end{array} \begin{array}{c} \begin{array}{cccc} 1 & 2 & 3 & 4 \end{array} \\ \begin{pmatrix} 11.779 & 11.779 & * & * \\ 0.735 & * & 0.735 & * \\ 1.098 & * & * & 1.098 \\ 22.407 & * & * & * \end{pmatrix} \end{array}$$

10.14 A machine has two components A and B in series. Component i ($i = A, B$) works for an exponential duration with rate λ_i before breaking down. It then requires a fixed duration D_i to repair it. When one component is down, the machine does not function; hence, the other does not break down. Obtain the proportion of time that the machine is down due to component i

a. by defining a suitable semi-Markov chain

b. by defining a suitable regenerative process

10.15 An instrument works for a random duration ℓ before failure. It is randomly inspected with rate μ to see whether it is still working. If its failure is discovered within H units of time after failure, it needs a random duration τ_1 for repair. Otherwise, it needs an additional random duration τ_2 for repair. Find the proportion of time that this instrument is undergoing the additional repair

a. by defining a suitable semi-Markov chain

b. by defining a suitable regenerative process

10.16 (Sheps and Menken, 1973) The weather at a certain location can be in one of the following three states: (1) clear, (2) cloudy, and (3) rainy. Its development can be described as a semi-Markov chain in which

$$\mathbf{P} = \begin{array}{c} \\ 1 \\ 2 \\ 3 \end{array} \begin{array}{c} \begin{array}{ccc} 1 & 2 & 3 \end{array} \\ \begin{pmatrix} & 1 & \\ p_{21} & & p_{23} \\ p_{31} & p_{32} & \end{pmatrix} \end{array} \qquad \mathbf{H} = \begin{array}{c} \\ 1 \\ 2 \\ 3 \end{array} \begin{array}{c} \begin{array}{ccc} 1 & 2 & 3 \end{array} \\ \begin{pmatrix} * & \mathcal{E}[h_{12}] & * \\ \mathcal{E}[h_{21}] & * & \mathcal{E}[h_{23}] \\ \mathcal{E}[h_{31}] & \mathcal{E}[h_{32}] & * \end{pmatrix} \end{array}$$

a. Verify that the expected first reaching time matrix is

$$[\mathcal{E}[t_{ij}]]_{i,j\in S} = \begin{array}{c} \\ 1 \\ 2 \\ 3 \end{array} \begin{array}{c} \begin{array}{ccc} 1 & 2 & 3 \end{array} \\ \begin{pmatrix} \dfrac{\eta}{\delta} & \mathcal{E}[h_1] & \dfrac{\mathcal{E}[h_1] + \mathcal{E}[h_2]}{p_{23}} \\ \dfrac{\mathcal{E}[h_2] + p_{23}\mathcal{E}[h_3]}{\delta} & \eta & \dfrac{p_{21}\mathcal{E}[h_1] + \mathcal{E}[h_2]}{p_{23}} \\ \dfrac{p_{32}\mathcal{E}[h_2] + \mathcal{E}[h_3]}{\delta} & p_{31}\mathcal{E}[h_1] + \mathcal{E}[h_3] & \dfrac{\eta}{p_{23}} \end{pmatrix} \end{array}$$

where $\delta \equiv 1 - p_{23}p_{32}$ and $\eta \equiv (1 - p_{23}p_{32})\mathcal{E}[h_1] + \mathcal{E}[h_2] + p_{23}\mathcal{E}[h_3]$.

b. Hence, find the limiting distribution.

10.17 Perrin and Sheps (1964) proposed a model of fecundity in which a woman can be in one of the following states: (0) nonpregnant, fecundable; (1) pregnant; (2) postpartum sterile associated with abortion or fetal loss; (3) postpartum sterile associated with stillbirth; (4) postpartum sterile associated with live birth. Assume

$$\mathbf{P} = \begin{array}{c} \\ 0 \\ 1 \\ 2 \\ 3 \\ 4 \end{array} \begin{array}{c} \begin{array}{ccccc} 0 & 1 & 2 & 3 & 4 \end{array} \\ \begin{pmatrix} & 1 & & & \\ & & \theta_2 & \theta_3 & \theta_4 \\ 1 & & & & \\ 1 & & & & \\ 1 & & & & \end{pmatrix} \end{array} \qquad \mathbf{H} = \begin{array}{c} \\ 0 \\ 1 \\ 2 \\ 3 \\ 4 \end{array} \begin{array}{c} \begin{array}{ccccc} 0 & 1 & 2 & 3 & 4 \end{array} \\ \begin{pmatrix} * & \mathcal{E}[h_{01}] & * & * & * \\ * & * & \nu_2 & \nu_3 & \nu_4 \\ \mathcal{E}[h_{20}] & * & * & * & * \\ \mathcal{E}[h_{30}] & * & * & * & * \\ \mathcal{E}[h_{40}] & * & * & * & * \end{pmatrix} \end{array}$$

The holding times are in months.

a. Let the probability of a woman getting pregnant each month be ρ. Find $\mathcal{E}[h_{01}]$ in terms of ρ.

b. Express the mean duration $\mathcal{E}[t_4]$ between two successive live births in terms of $\rho, \theta_i, \nu_i, \mathcal{E}[h_{i0}]$

i. by using Equation (10.10.3).

ii. by making state 4 absorbing and then verifying that the resulting fundamental matrix is

$$\mathbf{U} = \begin{array}{c} \\ 0 \\ 1 \\ 2 \\ 3 \end{array} \begin{array}{c} \begin{array}{cccc} 0 & 1 & 2 & 3 \end{array} \\ \begin{pmatrix} 1 & 1 & \theta_2 & \theta_3 \\ \theta_2 + \theta_3 & 1 & \theta_2 & \theta_3 \\ 1 & 1 & 1 - \theta_3 & \theta_3 \\ 1 & 1 & \theta_2 & 1 - \theta_2 \end{pmatrix} \end{array} \frac{1}{\theta_4}$$

c. Find the mean duration between two miscarriages $\mathcal{E}[t_2]$.

d. Find the mean duration between two stillbirths $\mathcal{E}[t_3]$.

e. Find the long-term proportion of time that a woman is in each state.

10.18 Let us return to the Cotton and Wood (1984) experiment with the Wistar rats in §10.10. Calculate $\mathcal{E}[t_S]$, $\mathcal{E}[t_R]$, and $\mathcal{E}[t_N]$ when $\alpha = 10$ seconds. How do they compare with the observed values of 139.86, 10.19, and 674.16 seconds, respectively? What is the theoretical shock rate?

10.19 Consider a Markov chain $\{X_n\}_{n=0,1,2,\ldots}$ having state space $S \equiv \{0, 1, 2, \ldots, N\}$. At each transition, it always moves to a different state with equal probability. In other words, for all $0 \le i \le N$

$$p_{ij} = \begin{cases} 0 & \text{for } j = i \\ \dfrac{1}{N} & \text{for all } j \neq i \end{cases}$$

Let the holding time at state i be $h_{ij} = h_i$ for all $0 \le i, j \le N$.

a. Find the expected first reaching times $\mathcal{E}[t_{ij}]$. *Hint:* Use Problem 5.35.

b. Hence, show that the expected return time to all states is $\sum_{k=0}^{N} \mathcal{E}[h_k]$.

c. Find the limiting distribution

i. by using the results obtained in part b

ii. by using the results obtained in Problem 6.28

10.20 Consider a Markov chain with state space $S \equiv \{0, 1, 2, \ldots, N\}$. If it is at state i, it may remain at state i or move to another state with equal probability p_i; that is,

$$p_{ij} = \begin{cases} 1 - Np_i & \text{for } j = i \\ p_i & \text{for all } j \neq i \end{cases}$$

where $0 \le Np_i \le 1$ for all $0 \le i \le N$. Let the holding time at state i be $h_{ij} = h_i$ for all $0 \le i, j \le N$.

a. Find the expected first reaching times $\mathcal{E}[t_{ij}]$. *Hint:* Use Problem 5.36.

b. Hence, show that the expected return time to state i is $p_i \sum_{k=0}^{N} (\mathcal{E}[h_k]/p_k)$.

c. Find the limiting distribution

i. by using the results obtained in part b

ii. by using the results obtained in Problem 6.29

iii. by using the results obtained in Problem 10.19

10.21 There are $N + 1$ competing brands in the market labeled $0, 1, 2, \ldots, N$. Let X_n be the brand bought at time n. It was proposed that chain $\{X_n\}_{n=0,1,2,\ldots}$ has the following transition probabilities:

$$p_{ij} = \begin{cases} p_i w_j & \text{for } i \neq j \\ 1 - p_i(1 - w_i) & \text{for } i = j \end{cases}$$

where $0 \le p_i \le 1$ for all $i \in S$ and $\sum_{j=0}^{N} w_i = 1$. Let the holding time at state i be $h_{ij} = h_i$ for all $0 \le i, j \le N$.

a. Find the expected first reaching times $\mathcal{E}[t_{ij}]$. *Hint:* Use Problem 5.38.

b. Hence, show that the expected return time to state i is

$$\mathcal{E}[t_i] = \frac{p_j}{w_j} \sum_{k \in S} \frac{\mathcal{E}[h_k] w_k}{p_k}$$

c. Find the limiting distribution

i. by using the results obtained in part b

ii. by using the results obtained in Problem 6.30

10.22 Three people A, B, and C are playing an infinite set of games as follows. At game n, two players play against each other while one watches. Whoever loses at game n will watch the other two playing at game $n + 1$. Let p_{ij} be the probability that player i beats player j in a game ($i, j = A, B, C$; $p_{ij} + p_{ji} = 1$) and h_i be the duration that players j and k play in one game ($j, k \neq i$).

a. Use Problem 5.21 to show that the rate of A's being defeated is

$$\frac{1 - p_{AB} p_{AC}}{(1 - p_{AB} p_{AC})\, \mathcal{E}[h_A] + (1 - p_{BA} p_{BC})\, \mathcal{E}[h_B] + (1 - p_{CA} p_{CB})\, \mathcal{E}[h_C]}$$

b. Hence, find the proportion of time that A is playing a game

i. by using the results obtained in part a

ii. by using the results obtained in Problem 6.31

10.23 Consider a positive-recurrent irreducible semi-Markov chain.

a. Use first-step analysis to show that for all $i, j \in S$

$$\mathcal{E}\left[(t_{ij})^2\right] = \mathcal{E}\left[(h_i)^2\right] + 2\sum_{k \neq j} p_{ik}\mathcal{E}[h_{ik}]\mathcal{E}[t_{kj}] + \sum_{k \neq j} p_{ik}\mathcal{E}\left[(t_{kj})^2\right]$$

Compare with Problems 6.47 and 10.9.

b. Hence, use the assumption in Problem 10.3 and the results in Problem 10.11 to find $\mathcal{E}[(t_{ij})^2]$ for all $i, j = X, Y, Z$ in the marketing example.

10.24 Consider a positive-recurrent irreducible Markov chain.

a. Use Problem 10.23 to show that for all $j \in S$

$$g_j\mathcal{E}\left[(t_j)^2\right] = \sum_{i \in S} g_i\mathcal{E}\left[(h_i)^2\right] + 2\sum_{i \in S}\sum_{k \neq j} g_i p_{ik}\mathcal{E}\left[h_{ik}\right]\mathcal{E}\left[t_{kj}\right]$$

Compare with Problem 6.48.

b. Hence, use the assumption in Problem 10.3 and the results in Problem 10.11 to find $\mathcal{E}[(t_i)^2]$ for all $i = X, Y, Z$ in the marketing example.

10.25 In a positive-recurrent semi-Markov chain, define an appropriate gradual-reward process to show that

a. the long-term proportion of time that the chain spends in state i and will visit state j next is $p_{ij}\mathcal{E}[h_{ij}]/\mathcal{E}[t_i]$.

b. the long-term proportion of time that the chain spends in state i, will visit state j next, and the duration until this transition is less than τ units of time is

$$\frac{p_{ij}}{\mathcal{E}[t_i]}\int_0^{\tau}(1 - H_{ij}(t))\,dt$$

Hint: Use Problem 1.51d.

10.26 In a positive-recurrent semi-Markov chain, use Equation (10.10.3) to prove that

$$\sum_{i \in S}\frac{p_{ij}}{\mathcal{E}[t_i]} = \frac{1}{\mathcal{E}[t_j]}$$

Explain this heuristically.

10.27 Find the expected return times and the visiting rates of a semi-Markov chain having transition matrix **P** as in Problem 6.55 and expected holding time matrix as

$$\mathbf{H} = \begin{array}{c} \\ 1 \\ 2 \\ 3 \\ 4 \\ 5 \\ 6 \\ 7 \\ 8 \end{array} \begin{array}{c} \begin{array}{cccccccc} 1 & 2 & 3 & 4 & 5 & 6 & 7 & 8 \end{array} \\ \left(\begin{array}{ccc|cc|ccc} 2 & 4 & 1 & * & * & * & * & * \\ * & 3 & 5 & * & * & * & * & * \\ 1 & 4 & 3 & * & * & * & * & * \\ \hline * & * & * & 3 & 1 & * & * & * \\ * & * & * & 2 & 4 & * & * & * \\ \hline 1 & 3 & 2 & 3 & 2 & 4 & 3 & 2 \\ 2 & * & 1 & * & 4 & 2 & * & 3 \\ 1 & 4 & 3 & 4 & * & 1 & 5 & 3 \end{array}\right) \end{array}$$

Does this chain have a limiting distribution?

CHAPTER 11

Continuous-Time Markov Chains

11.1 INTRODUCTION

In the previous chapter, we studied the continuous-time chains that are Markovian only at the transitions but not necessarily in between. Each chain can be completely specified by the transition matrix **P** and the conditional holding time distributions $H_{ij}(\tau)$s, which can be very general. In this chapter, we study the same chain $\{X(t)\}_{t\in[0,\infty)}$, but assume further that *all the holding times are exponentially distributed and are independent of the chain's destination state*; that is, for all $i, j \in S$, and $\tau > 0$,

$$H_{ij}(\tau) \equiv \Pr\{T_n - T_{n-1} < \tau \mid X_{n-1} = i, X_n = j\} = 1 - \exp\{-q_i \tau\}$$

where q_i is called the *transition rate* at state i. From Equation (8.8.2), the expected holding time at state i is therefore

$$\mathcal{E}[h_i] \equiv \mathcal{E}[T_n - T_{n-1} \mid X_{n-1} = i]\frac{p_{ij}}{h} = \frac{1}{q_i}$$

We call this kind of chain the *continuous-time Markov chain*. It is fully described if all transition probabilities p_{ij}s and all transition rates q_is are known.

The chain is *time-homogeneous*. The stream of its transitions is *regular* if there exists a constant q such that $q_i < q$ for all $i \in S$ (which we shall henceforth assume).

Recall that we allowed a semi-Markov chain to remain at state i at each transition. Let us now consider a semi-Markov chain having a state i such that $p_{ii} > 0$ and the holding time at this state is exponentially distributed with rate λ_i. In this case, it can be shown that the sojourn time at state i is exponentially distributed with rate $\lambda_i(1 - p_{ii})$.[1] If we are interested in this chain's behavior all along the continuous-time axis, not only

[1]This can be proved by applying Problem 8.54 and by showing that the sojourn time s_i is the random sum of n_i holding times and that n_i is geometrically distributed with mean $1/(1 - p_{ii})$.

at the transitional epochs, we can conveniently ignore all transitions at which the chain does not change states and consider the holding time at state i as if it were exponentially distributed with rate $q_i \equiv \lambda_i(1 - p_{ii})$.

For this reason, we also assume without loss of generality in this chapter that a continuous-time Markov chain *must move to another state at each transition*; that is, $p_{ii} = 0$ for all $i \in S$.

11.2 MEMORYLESS PROPERTY

The continuous-time Markov chain $\{X(t)\}_{t\in[0,\infty)}$ is memoryless all along the continuous-time axis: At each transition, it is memoryless because it is semi-Markovian; between the transitions, the exponential distribution of the holding times ensures that the time until the next transition is dependent only on the state it currently visits (its present), not how long it has been in that state (its past). In other words, the development of the chain after any time point t may be dependent on its position at time t, but not on its history before that time; or for all $t \geq 0$ and $\tau \geq 0$,

$$\Pr\{X(t+\tau) = j \mid X(u), 0 \leq u \leq t\} = \Pr\{X(t+\tau) = j \mid X(t) = i\}$$

The process is not only semi-Markovian, but completely Markovian.

11.3 TRANSITION RATES

Assume a continuous-time Markov chain is visiting state i at time t. Consider now an infinitesimal duration $\lim_{h\to 0}(t, t+h)$. Because the chain's holding time is exponentially distributed with rate q_i, there can only be zero or one transition out of state i during this infinitesimal duration. Equation (8.4.1) gives the probability of one transition as $q_i h$. Given the chain moves out of state i, the probability that it moves into state j is p_{ij}. Thus, as $h \to 0$, the probability of the chain moving from state i to state $j \neq i$ is $q_i p_{ij} h$.

As the number of transitions from state i to state j during an infinitesimal duration h is a Bernoulli random variable, the probability $q_i p_{ij} h$ is also its expected value. Consistent with the material in §8.5, the *(instantaneous) transition rate*[2] *from state i to state j* ($j \neq i$) is now defined as

$$q_{ij} \equiv \lim_{h\to 0} \frac{\mathcal{E}[\text{number of transitions from } i \text{ to } j \text{ during } (t, t+h]]}{h} = q_i p_{ij} \qquad (11.3.1)$$

We stated previously that the transition matrix $\mathbf{P}$ and the rates q_is completely define the chain. Now we see that it can also be completely specified if all transition rates q_{ij}s ($i \neq j$) are known, since these rates govern the development of the chain during any infinitesimal duration. In fact, if we are given all q_{ij}s, we can obtain the rates q_is and the transition matrix $\mathbf{P}$ as follows:

---

[2]For lack of better terms, we call both q_i and q_{ij} the transition rates.

1. $q_i = \sum_{j \neq i} q_{ij}$ for all $i \in S$. This is because the expected number of transitions from state i to all other states during an infinitesimal duration $\lim_{h \to 0}(t, t+h)$ is $\sum_{j \neq i} q_i p_{ij} h = h \sum_{j \neq i} q_{ij}$.
2. $p_{ij} = q_{ij}/q_i$ for all $j \neq i$, from Equation (11.3.1).

11.4 TRANSITION DIAGRAMS

Similar to the discrete-time Markov chains, a continuous-time Markov chain can be represented graphically by a *transition diagram*. The only difference is that, while the transition diagram of the former has the transition probability p_{ij} written along the arrow from state i to state j, that of a continuous-time Markov chain has the transition rate q_{ij}.

11.5 THE Q-MATRIX

The transition rates q_{ij}s can be grouped together in the matrix $\mathbf{Q} \equiv [q_{ij}]_{j,i \in S}$, in which the diagonal elements q_{ii}s are defined as

$$q_{ii} \equiv -q_i = -\sum_{j \neq i} q_{ij} \quad \text{for all } i \in S$$

We call this matrix the chain's *Q-matrix.*[3]

It can be verified that

$$\mathbf{Q} = \mathbf{Q}_d(\mathbf{I} - \mathbf{P})$$

where $\mathbf{Q}_d$ is a diagonal matrix in which the (i, i) element is $-q_i$ and all nondiagonal elements are zeros.

Note that *each row of the Q-matrix adds to* 0, not 1. Also, if state i is absorbing, $q_{ij} = 0$ for all $j \neq i$, and all elements of the i-row therefore are zeros.

Continuous-Time Markov Chains in Action: A Geologic Profile. Along the 15,000-feet Seabrook Power Station discharge tunnel, rock types can be classified as: (1) schist, (2) metaquartzite, (3) diorite, and (4) quartzite. To model the rock type changes as a continuous-time Markov chain, Ioannou (1987) obtained the transition matrix of the embedded chain as

$$\mathbf{P} = \begin{array}{c} \\ 1 \\ 2 \\ 3 \\ 4 \end{array} \begin{array}{c} \begin{array}{cccc} 1 & 2 & 3 & 4 \end{array} \\ \begin{pmatrix} 0 & .02 & .23 & .75 \\ .02 & 0 & .50 & .48 \\ .02 & .20 & 0 & .78 \\ .23 & .17 & .60 & 0 \end{pmatrix} \end{array}$$

The transition rates are $q_1 = 138 \times 10^{-5}$ ft^{-1}, $q_2 = 822 \times 10^{-5}$ ft^{-1}, $q_3 = 262 \times 10^{-5}$ ft^{-1}, and $q_4 = 250 \times 10^{-5}$ ft^{-1}.

[3] Some authors call it the *infinitesimal generator*.

This information can now be combined in the following Q-matrix (with unit 10^{-5} ft^{-1}), which describes the chain completely:

$$\mathbf{Q} = \begin{array}{c} \\ 1 \\ 2 \\ 3 \\ 4 \end{array} \begin{array}{c} \begin{array}{cccc} \quad 1 \quad & \quad 2 \quad & \quad 3 \quad & \quad 4 \quad \end{array} \\ \begin{pmatrix} -138 & (.02)(138) & (.23)(138) & (.75)(138) \\ (.02)(822) & -822 & (.50)(822) & (.48)(822) \\ (.02)(262) & (.20)(262) & -262 & (.78)(262) \\ (.23)(250) & (.17)(250) & (.60)(250) & -250 \end{pmatrix} \end{array} \quad \blacksquare$$

11.6 EQUALLY SPACED DISCRETE-TIME CHAINS

In §2.18, we said that one way to study a continuous-time chain is to observe it at equally spaced epochs. In this case, the sojourn times are assumed to have a geometric distribution (§2.19). We now present some reasons as to why it would be more appropriate in most cases to *let the spacing of the observations degenerate to zero*. The sojourn times would then have an exponential distribution, and the chain becomes the continuous-time Markov chain.

1. One difficulty of the equally spaced chains is the determination of the spacing of the observations.
   Continuous-Time Markov Chains in Action: Spacing of Social Processes. According to Singer and Spilerman (1976),

   > [B]ecause most social processes evolve continuously, there usually isn't a compelling reason for preferring one specification of the unit time interval to another. (For instance, in studying *intra*generational occupational mobility, should the unit time interval be five years or three years or six months?) Yet this is a question of great consequence, because an empirically determined matrix (estimated, let us say for this illustration, from observations ten years apart) may be consistent with a discrete-time Markov structure *for some choices of the unit time interval but not for other choices*. . . . Where no substantive meaning can be attached to a particular interval length, this does not imply that the unit time interval can be specified at the convenience of the researcher. . . . Rather, it suggests that the appropriate mathematical structure is a continuous-time formulation. ■

2. If the spacing of the observations is large, many transitions might take place between two consecutive observations; thus, they are not accounted for. When the spacing of a chain degenerates to zero, normally only zero or one transition can occur between two consecutive observations.
3. In many practical applications where the states can be organized in a certain order, *natura non facit saltum*,[4] and we can expect only transitions between neighboring states to take place. This property would be lost if wide spacing is used because many transitions might take place between the observations.

[4]Nature makes no leaps.

Continuous-Time Markov Chains in Action: Bring in the Cloud. Madsen et al. (1985) analyzed 15 years of hourly observations of cloud cover at an airport near Copenhagen, Denmark. Cloud cover is measured as an integer between 0 and 9, where 0 corresponds to completely clear sky, 8 to completely overcast sky, and 9 indicates that the cloud cover is unobservable, which usually is the case in foggy weather or heavy snowfall.

> The number of parameters required for a model in continuous time is considerably reduced as compared with the discrete time model. In a short time interval the only possible changes in cloud cover are transitions to neighboring cloud cover states. The cloud cover does not change momentarily from, e.g., state 2 to state 5. Therefore, basically, the only parameters necessary for the continuous time model are parameters describing transitions between neighboring cloud cover states. There are, however, two exceptions to this principle. . . . [W]e observe that the transition probability from state 0 to state 9 is higher than the transition probability from state 0 to, e.g., state 8. Likewise, the probability of transition from state 9 to state 0 is higher than the probability of transition from state 9 to state 1. This can be explained physically by the fact that foggy weather (state 9) can replace a clear sky (state 0) after sunset due to radiative cooling of the lower air stratum. Transitions from state 9 to state 0 are possible due to radiative heating after sunrise. ■

If a continuous-time chain is Markovian only when observed with a specific spacing, it must be studied as a semi-Markov chain having this spacing as the constant holding times. However, if it is Markovian with any spacing, it is better to let the spacing degenerate to zero and study it as a continuous-time Markov chain.

11.7 WHY THIS MODEL?

We were able to obtain some very significant results for the semi-Markov chain in the previous chapter. So why do we want to impose additional limitations? Is there anything we can do with the continuous-time Markov chains that we could not do earlier with the semi-Markov chains?

1. To start, we have seen that many important results related to a semi-Markov model are dependent only on the expected value of the holding times, not on their actual distributions; thus, in many practical applications, it does not matter whether the holding times are exponentially distributed or not.
2. The memoryless property of the continuous-time Markov chains makes the mathematics easier. For example, to study a semi-Markov chain, we have to study its embedded chain first. This is not necessary for the continuous-time Markov chains, as we shall see later. Also, even within the scope of this book, some *transient results* can be obtained (Problems 11.36 to 11.40).
3. Finally, there are many practical situations in which both the holding time distributions and the transition probabilities *are derived from some common and more fundamental assumptions*. These derivations must be recognized, explored, explained, and

exploited. For most semi-Markov chains, their holding time distribution and transition probabilities are obtained empirically, directly, and separately, as with Kao's (1947a) data for the coronary patients in §10.7. This makes the relationship between the holding times or the transition matrix with the more fundamental assumptions, if there is any, hard to detect. In the next section, we present a large class of continuous-time Markov chains that can be constructed from a more fundamental set of assumptions.

11.8 COMPETITIVE CHAINS

Consider the following scenario. Immediately after a chain moves to state i, all other states compete with each other for its next visit. To compete, *each state is given a stream of events*. The chain will move to whichever state has the first event occurrence. In other words, the chain will leave state i at the first occurrence of the *superposition* of all the competing streams and will move to state j if this first occurrence is associated with state j. After the chain moves to this new state j, it forgets its past and starts a new competition in which all states $k \neq j$ can participate.

If the competing streams are not Markovian, then neither is their superposition. In this case, the holding time h_i at state i may be dependent on the competing streams in a complicated way.

Suppose now that, given the chain is in state i, the stream assigned to state $j \neq i$ is *Poisson with rate* q_{ij}. In this case, from §8.14, we know that the superposition of all competing streams is also Markovian with rate $q_i \equiv \sum_{j \neq i} q_{ij}$. The holding time at state i is therefore exponentially distributed with rate q_i. Also, given the chain is leaving state i, Equation (8.10.1) gives the probability of the chain's moving to state $k \neq i$ as $p_{ik} = q_{ik}/q_i$. We now have a continuous-time Markov chain, which is defined completely by all transition rates q_{ij}s.

A large class of practical problems fits this type of chain, which we refer to as the *competitive* continuous-time Markov chain. In fact, the majority of problems in this chapter and the next asks us to model them as such.

11.9 THE MACHINE-REPAIR EXAMPLE

As an example of a competitive continuous-time Markov chain, consider a company having four machines and three servicepeople. Each machine operates for an exponential duration with rate λ before breaking down. Once a machine is down, if a serviceperson is available, it is repaired immediately; if not, it has to wait until a serviceperson is available. It takes a serviceperson an exponential duration with rate μ to repair a down machine. We assume that the time to break down and the repair times are mutually independent.

We want to study the chain $\{\delta(t)\}_{t \in [0,\infty)}$, where $\delta(t)$ is the number of down machines at time t. Its state space is $S \equiv \{0, 1, 2, 3, 4\}$.

At time t, assume there are i down machines, or $\delta(t) = i$ for $0 \leq i \leq 4$. Hence, there are $4 - i$ working machines.

1. Each of the $4 - i$ working machines can fail with rate λ. Thus, the time until the next machine failure is exponentially distributed with rate $(4 - i)\lambda$. If this happens, the number of down machines increases from i to $i + 1$. We therefore give state $i + 1$ a *Poisson stream of machine failures* with rate $q_{i,i+1} = (4 - i)\lambda$.
2. With only three servicepeople, given i down machines, the number of machines that can be repaired simultaneously is

$$Y_i \equiv \min(3, i)$$

Since each machine can be repaired with rate μ, the duration until the first service completion is therefore exponentially distributed with rate $Y_i\mu$. If this happens, the number of down machines decreases from i to $i - 1$. We therefore give state $i - 1$ a *Poisson stream of service completions* with rate $q_{i,i-1} = Y_i\mu$.
3. The preceding two streams compete with each other. If a service completion occurs first, the chain moves from state i to state $i - 1$; if a machine failure occurs first, it moves from state i to state $i + 1$.

Chain $\{\delta(t)\}_{t\in[0,\infty)}$ is a continuous-time Markov chain because it satisfies our assumptions:

1. The chain's holding time h_i is the duration until either the next machine failure or the next service completion, whichever comes first. This is the duration until the first occurrence in the superposition of the stream of machine failures and the stream of service completions. It is exponentially distributed with rate $q_i = (4 - i)\lambda + Y_i\mu$.
2. The chain will move from state $i < 4$ to state $i + 1$ if a failure occurs before a service completion. From Equation (8.10.1), the probability that this happens is

$$p_{i,i+1} = \frac{(4 - i)\lambda}{(4 - i)\lambda + Y_i\mu} \qquad \text{for all } 0 \le i < 4$$

Similarly, it will move from state $i > 0$ to state $i - 1$ if a service completion occurs before a failure. The probability that this happens is

$$p_{i,i-1} = \frac{Y_i\mu}{(4 - i)\lambda + Y_i\mu} \qquad \text{for all } 1 < i \le 4$$

The embedded chain is thus Markovian with transition matrix:

$$\mathbf{P} = \begin{array}{c} \\ 0 \\ 1 \\ 2 \\ 3 \\ 4 \end{array} \begin{array}{c} \begin{array}{ccccc} 0 & 1 & 2 & 3 & 4 \end{array} \\ \begin{pmatrix} 0 & 1 & 0 & 0 & 0 \\ \frac{\mu}{3\lambda + \mu} & 0 & \frac{3\lambda}{3\lambda + \mu} & 0 & 0 \\ 0 & \frac{2\mu}{2\lambda + 2\mu} & 0 & \frac{2\lambda}{2\lambda + 2\mu} & 0 \\ 0 & 0 & \frac{3\mu}{\lambda + 3\mu} & 0 & \frac{\lambda}{\lambda + 3\mu} \\ 0 & 0 & 0 & 1 & 0 \end{pmatrix} \end{array}$$

The chain's Q-matrix is

$$\mathbf{Q} = \begin{array}{c} \\ 0 \\ 1 \\ 2 \\ 3 \\ 4 \end{array} \begin{array}{c} \begin{array}{ccccc} 0 & 1 & 2 & 3 & 4 \end{array} \\ \begin{pmatrix} -4\lambda & 4\lambda & 0 & 0 & 0 \\ \mu & -(\mu+3\lambda) & 3\lambda & 0 & 0 \\ 0 & 2\mu & -(2\mu+2\lambda) & 2\lambda & 0 \\ 0 & 0 & 3\mu & -(3\mu+\lambda) & \lambda \\ 0 & 0 & 0 & 3\mu & -3\mu \end{pmatrix} \end{array}$$

Its transition diagram is as in Figure 11.9.1.

We can pause here and observe the modeling advantage of the competitive continuous-time Markov chains in this example. There is no empirical inference for the holding times and the transition probabilities. They are explained and expressed in terms of the two *more fundamental parameters* λ and μ. This would not be possible without the Markovian assumptions of the time-to-failures and the repair times.

As this is a special form of a semi-Markov chain, the chain's limiting results can be calculated as in the previous chapter. First, Equations (6.3.1) and (6.3.2) give the long-term visiting rates (which are also the limiting probabilities) of the embedded chain as (Problem 6.41):

$$g_0 = \frac{1}{2(1+\rho)^3} \qquad g_1 = \frac{1+3\rho}{2(1+\rho)^3} \qquad g_2 = \frac{3\rho(1+\rho)}{2(1+\rho)^3}$$

$$g_3 = \frac{\rho^3+3\rho^2}{2(1+\rho)^3} \qquad g_4 = \frac{\rho^3}{2(1+\rho)^3}$$

where

$$\rho \equiv \frac{\lambda}{\mu}$$

The long-term average spacing of the transitions, as defined in (10.10.2), can now be calculated as

$$h \equiv \sum_{i\in S} g_i h_i = \sum_{i\in S} \frac{g_i}{q_i} = \frac{\Delta}{2(1+\rho)^3 4\lambda}$$

where

$$\Delta \equiv 1 + 4\rho + 6\rho^2 + 4\rho^3 + \frac{4}{3}\rho^4 \tag{11.9.1}$$

FIGURE 11.9.1 Transition diagram for the machine-repair problem.

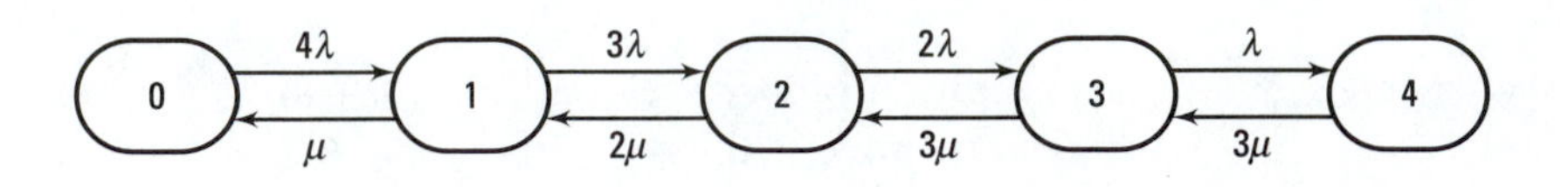

Equation (10.11.1) now gives the long-term visiting rates (and the limiting probabilities) of the continuous-time Markov chain $\{\delta(t)\}_{t\in[0,\infty)}$ as $\gamma_i = g_i/(hq_i)$, or

$$\gamma_0 = \frac{1}{\Delta} \qquad \gamma_1 = \frac{4\rho}{\Delta} \qquad \gamma_2 = \frac{6\rho^2}{\Delta} \qquad \gamma_3 = \frac{4\rho^3}{\Delta} \qquad \gamma_4 = \frac{(4/3)\rho^4}{\Delta} \tag{11.9.2}$$

Observe that these results are not dependent on the actual values of λ or μ, but only their *relative value* $\rho \equiv \lambda/\mu$.

We now present a method to calculate these results directly without having to resort to the embedded chain.

11.10 THE BALANCE EQUATIONS

Consider now an *positive-recurrent irreducible* continuous-time Markov chain, which is not necessarily a competitive one. Since the holding times are exponentially distributed, they are spread out. The chain therefore converges in distribution, and its limiting probabilities are the same as its long-term visiting rates.

Since it is also semi-Markovian, Equation (10.11.1) gives its long-term visiting rates as $\gamma_i = g_i/(hq_i)$. Let us now multiply both sides of this equation by $p_{ij}q_i$ and sum over all $i \in S$:

$$\sum_{i\in S} p_{ij}q_i\gamma_i = \frac{1}{h}\sum_{i\in S} g_i p_{ij}$$

The left-hand side of this equation can be written as $\sum_{i\in S} q_{ij}\gamma_i$ because $p_{ij}q_i = q_{ij}$. Its right-hand side can be written as g_j/h because $\sum_{i\in S} g_i p_{ij} = g_j$, as in Equation (7.5.1). Applying the relation $\gamma_j = g_j/(hq_j)$, or $g_j = \gamma_j hq_j$, again to the right-hand side, we now have

$$q_j\gamma_j = \sum_{i\in S} q_{ij}\gamma_i \quad \text{for all } j \in S \tag{11.10.1}$$

This equation is equivalent to Equation (6.3.6) for the discrete-time Markov chains. It is called the *balance equation* because of the following interpretations:

1. γ_j is the limiting probability that the chain visits state j. Given that it is in state j, q_j is the rate the chain moves out of state j. Thus, $\gamma_j q_j$ is the limiting unconditional rate the chain leaves state j.
2. γ_i is the limiting probability that the chain visits state i. Given that it is in state i, q_{ij} is the rate it moves from state i into state $j \neq i$. Thus, $\gamma_i q_{ij}$ is the limiting unconditional rate the chain moves into state j from state i, and $\sum_{i\in S}\gamma_i q_{ij}$ is the limiting unconditional rate the chain enters state j from all other states.
3. For stability, the limiting unconditional rate the chain leaves each state must be balanced with the limiting unconditional rate the chain enters it.

For chain $\{\delta(t)\}_{t\in[0,\infty)}$ in the machine-repair example, Equation (11.10.1) gives the following set of simultaneous equations:

State	Rate of Leaving = Rate of Entering
0	$(4\rho)\gamma_0 = \gamma_1$
1	$(3\rho+1)\gamma_1 = 4\rho\gamma_0 + 2\gamma_2$
2	$(2\rho+2)\gamma_2 = 3\rho\gamma_1 + 3\gamma_3$
3	$(\rho+3)\gamma_3 = 2\rho\gamma_2 + 3\gamma_4$
4	$(3)\gamma_4 = \rho\gamma_3$

(11.10.2)

Actually, the balance equations can easily be obtained from the transition diagrams. For example, consider state 2 in Figure 11.9.1:

1. There are two arrows coming out of this state, giving the total rate of leaving state 2 as $2\lambda + 2\mu$. The limiting unconditional rate of leaving state 2 is therefore $(2\lambda+2\mu)\gamma_2$.
2. There are also two arrows ending at state 2: one from state 1 with rate 3λ and the other from state 3 with rate 3μ. The limiting unconditional rate of the chain's entering state 2 is therefore $(3\lambda)\gamma_1 + (3\mu)\gamma_3$.
3. Thus, $(2\lambda + 2\mu)\gamma_2 = 3\lambda\gamma_1 + 3\mu\gamma_3$, or $(2\rho+2)\gamma_2 = 3\rho\gamma_1 + 3\gamma_3$.

In matrix form, Equation (11.10.1) can be written as:

$$\boldsymbol{\gamma}\mathbf{Q} = \mathbf{0} \tag{11.10.3}$$

where $\boldsymbol{\gamma} = [\gamma_i]_{i\in S}$ and $\mathbf{0}$ is a row in which all elements are equal to 0. This equation is the continuous-time version of Equation (6.3.1) or (6.4.4).

Like Equation (6.3.1) or (6.4.4), the balance equations in (11.10.1) or (11.10.3) do not form a system of independent equations for us to solve for the unique values of γ_is. For example, in the machine-repair example, we can obtain the balance equation for state 4 by adding those for states 0, 1, 2, and 3 in (11.10.2) side by side. To have a set of independent equations, we must arbitrarily delete one of those balance equations and substitute the following *normalizing equation*:

$$\sum_{i\in S} \gamma_i = 1 \tag{11.10.4}$$

For the machine-repair example, Equations (11.10.2) and (11.10.4) together should give the same results as those obtained previously. (Note that we obtain them without the help of the embedded chain.)

For a *finite* chain having N distinct states, Equation (11.10.4) can be written as $\boldsymbol{\gamma}\mathbf{E} = \mathbf{e}$, where $\mathbf{E}$ and $\mathbf{e}$ are the $(N \times N)$ and $(1 \times N)$ matrices, respectively, with all elements equal to 1. Adding this equation with (11.10.3) yields $\boldsymbol{\gamma}(\mathbf{Q}+\mathbf{E}) = \mathbf{e}$, and hence the following formula (Resnick, 1992b), which is very useful for numerical calculations and is equivalent to Equation (6.3.5):

$$\boldsymbol{\gamma} = \mathbf{e}(\mathbf{Q}+\mathbf{E})^{-1} \tag{11.10.5}$$

Continuous-Time Markov Chains in Action: Effects of Policy Changes in Health Services. Assume a health service system has the following four components of care: (0) no use of service, (1) ambulatory care, (2) hospital care, and (3) domiciliary care. To model the patient movements among these states as a continuous-time Markov chain, let us assume the following hypothetical data:

$$\mathbf{P} = \begin{array}{c} \\ 0 \\ 1 \\ 2 \\ 3 \end{array} \begin{array}{c} \begin{array}{cccc} 0 & 1 & 2 & 3 \end{array} \\ \begin{pmatrix} 0 & .8 & .1 & .1 \\ .7 & 0 & .2 & .1 \\ .1 & .8 & 0 & .1 \\ .4 & .4 & .2 & 0 \end{pmatrix} \end{array}$$

with $\mathcal{E}[h_0] = 200$ days, $\mathcal{E}[h_1] = 40$ days, $\mathcal{E}[h_2] = 10$ days, and $\mathcal{E}[h_3] = 5$ days.

The limiting probabilities can be calculated as: $\gamma_0 = .7869$, $\gamma_1 = .1923$, $\gamma_2 = .0156$, and $\gamma_3 = .0052$ (Problem 11.5, and Problem 10.12 in which we treat the chain as semi-Markovian).

As discussed by Schach and Schach (1972),

> Perhaps the most interesting feature of the model is that it allows a quantitative investigation of changes in the behavior of the health services system. Since individuals move between components of the system, the demands for the services of different components are interrelated. Therefore, a policy change in one component has an impact on all others. Even beyond that, it has an indirect influence on itself, in the sense that the ultimate effect of a policy change may differ considerably from the change as initially instituted.
>
> To illustrate this, let us assume that, on the average, hospital stays are reduced by 20 per cent as the result of a new government policy. This increases . . . [q_2] to 0.12 and decreases . . . [γ_2] to 0.0130. For a population of size 100,000 this means that the average number of hospital patients is reduced from 1561 to 1305, a reduction of 16.4 per cent. Thus the ultimate effect of a 20 per cent reduction in average length of stay results in a mere 16.4 per cent decrease in the hospital population. Without the help of a model which takes into account explicitly all the interrelationships between components, such estimates are difficult to obtain. ■

11.11 BIRTH-AND-DEATH PROCESSES

The machine-repair chain belongs to a very important class of competitive continuous-time Markov chains called the *birth-and-death processes*. In Chapter 7, we introduced the *simple random walk* in which the chain may only move from state i to its *neighboring states* $i-1$ or $i+1$. A birth-and-death process is a competitive continuous-time Markov chain *having a simple random walk as its embedded chain*. Its state space is the set of positive integers $\{0, 1, 2, \ldots\}$.

More specifically, if a birth-and-death process is visiting state $i > 0$, state $i+1$ is given a Poisson stream of births having the *birthrate* $q_{i,i+1} \equiv \lambda_i$ and state $i-1$ is given a Poisson stream of deaths having the *death rate* $q_{i,i-1} \equiv \mu_i$. Birth and death now compete with each other: The chain will move from state i to state $i+1$ if a birth occurs before a death; otherwise, it will move from state i to state $i-1$. We also assume that, if the process is visiting state 0, the death rate $q_{0,-1} \equiv \mu_0$ is 0. (The process is said to have a *barrier* at 0 because it cannot visit a negative state.)

The embedded chain of this process is a simple random walk, as defined in (7.1.1), with $r_i = 0$, $s_i = \lambda_i/(\lambda_i + \mu_i)$, and $q_i = \mu_i/(\lambda_i + \mu_i)$. The holding time at state i is exponentially distributed with rate $\lambda_i + \mu_i$.

Its Q-matrix is

$$\mathbf{Q} = \begin{array}{c} \\ 0 \\ 1 \\ 2 \\ 3 \\ 4 \\ \vdots \end{array} \begin{array}{c} \begin{array}{cccccc} 0 & 1 & 2 & 3 & 4 & \cdots \end{array} \\ \begin{pmatrix} -\lambda_0 & \lambda_0 & 0 & 0 & 0 & \cdots \\ \mu_1 & -(\mu_1+\lambda_1) & \lambda_1 & 0 & 0 & \cdots \\ 0 & \mu_2 & -(\mu_2+\lambda_2) & \lambda_2 & 0 & \cdots \\ 0 & 0 & \mu_3 & -(\mu_3+\lambda_3) & \lambda_3 & \cdots \\ 0 & 0 & 0 & \mu_4 & -(\mu_4+\lambda_4) & \cdots \\ \vdots & \vdots & \vdots & \vdots & \vdots & \ddots \end{pmatrix} \end{array}$$

Its transition diagram is in Figure 11.11.1. Its balance equations are:

State	Rate of Leaving = Rate of Entering
0	$\lambda_0\gamma_0 = \mu_1\gamma_1$
$i \geq 1$	$(\lambda_i + \mu_i)\gamma_i = \lambda_{i-1}\gamma_{i-1} + \mu_{i+1}\gamma_{i+1}$

We have seen how important the random walk models are. As a continuous-time counterpart of this model, the birth-and-death processes are equally important, if not more.

For a birth-and-death model with $\lambda_0 \neq 0$ (hence, state 0 is not absorbing and the process is said to have a *reflective barrier* at 0), we can obtain its limiting distribution in a manner similar to that in §7.5:

Step 1. For all $i > 0$, we express γ_i in terms of γ_0. For notational advantage, we define $\delta^{(0)} \equiv 1$ and

$$\delta^{(i)} \equiv \frac{\lambda_0}{\mu_1}\frac{\lambda_1}{\mu_2}\cdots\frac{\lambda_{i-1}}{\mu_i} \quad \text{for all } i = 1, 2, 3, \ldots \tag{11.11.1}$$

a. From the balance equation for state 0, we obtain:

$$\gamma_1 = \delta^{(1)}\gamma_0$$

FIGURE 11.11.1 Transition diagram for a birth-and-death process.

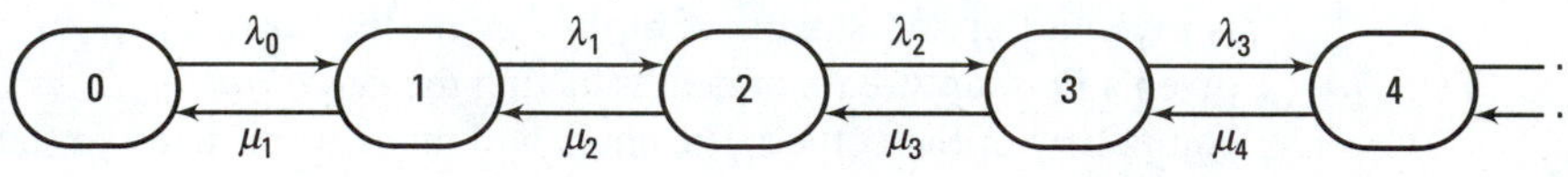

b. Adding the balance equations for states 0 and 1 yields:

$$\lambda_1\gamma_1 = \mu_2\gamma_2 \quad \rightarrow \quad \gamma_2 = \frac{\lambda_1}{\mu_2}\gamma_1 = \delta^{(2)}\gamma_0$$

c. Adding the balance equations for states 0, 1, and 2 yields:

$$\lambda_2\gamma_2 = \mu_3\gamma_3 \quad \rightarrow \quad \gamma_3 = \frac{\lambda_2}{\mu_3}\gamma_2 = \delta^{(3)}\gamma_0$$

d. Continuing in this manner, adding the balance equations from states 0 to state i, we obtain $\lambda_i\gamma_i = \mu_{i+1}\gamma_{i+1}$. Hence,

$$\gamma_i = \delta^{(i)}\gamma_0 \quad \text{for all } i = 0, 1, 2, \ldots \tag{11.11.2}$$

Step 2. We now can solve for γ_0 by applying (11.10.4):

$$\sum_{i=0}^{\infty}\gamma_i = \gamma_0\sum_{i=0}^{\infty}\delta^{(i)} = 1$$

Thus,

$$\gamma_0 = \frac{1}{\Delta} \tag{11.11.3}$$

where

$$\Delta \equiv \sum_{i=0}^{\infty}\delta^{(i)} \tag{11.11.4}$$

Note that γ_0 is nonzero and the process is positive-recurrent if and only if $\Delta < \infty$.

Let us now apply the results to some special cases.

Case 1. The machine-repair model is a special case of the birth-and-death process in which

$$\lambda_i = \begin{cases} (4-i)\lambda & \text{for all } i < 4 \\ 0 & \text{for all } i \geq 4 \end{cases}$$

and

$$\mu_i = Y_i\mu \equiv \min(3, i)\mu \quad \text{for all } i \geq 0$$

Thus, writing $\rho \equiv \lambda/\mu$, we have

$$\delta^{(1)} = \frac{4\rho}{1} \qquad \delta^{(2)} = \left(\frac{4\rho}{1}\right)\left(\frac{3\rho}{2}\right)$$

$$\delta^{(3)} = \left(\frac{4\rho}{1}\right)\left(\frac{3\rho}{2}\right)\left(\frac{2\rho}{3}\right) \qquad \delta^{(4)} = \left(\frac{4\rho}{1}\right)\left(\frac{3\rho}{2}\right)\left(\frac{2\rho}{3}\right)\left(\frac{1\rho}{3}\right)$$

Equation (11.11.4) now gives the same quantity as in Definition (11.9.1):

$$\Delta = 1 + \delta^{(1)} + \delta^{(2)} + \delta^{(3)} + \delta^{(4)} = 1 + 4\rho + 6\rho^2 + 4\rho^3 + \frac{4\rho^4}{3}$$

and the limiting results (11.9.2) can be obtained again.

Case 2. Simple birth-and-death processes: This model assumes

1. a constant birthrate; that is, $\lambda_i = \lambda$ for all $i = 0, 1, 2, \ldots$
2. a constant death rate; that is, $\mu_i = \mu$ for all $i = 1, 2, 3, \ldots$

Let $\rho \equiv \lambda/\mu$. Then $\delta^{(i)} = \rho^i$ for all $i \geq 0$. If $\rho < 1$, then Equation (11.11.4) gives:

$$\Delta \equiv \sum_{i=0}^{\infty} \rho^i = \frac{1}{1-\rho}$$

Thus,

$$\gamma_0 = 1 - \rho \tag{11.11.5}$$

The limiting distribution therefore is *geometric*, or

$$\gamma_i = (1-\rho)\rho^i \quad \text{for all } i = 0, 1, 2, \ldots$$

These limiting probabilities only exist when $\rho < 1$, or when the birthrate λ is *strictly less than* the death rate μ. (When $p = 1$, the process is null-recurrent.) Recall that the embedded chain of this process is a simple random walk with $s = \lambda/(\lambda + \mu)$ and $q = \mu/(\lambda + \mu)$. The condition for positive-recurrence established in §7.5 that $q > s$ is consistent with the condition that $\rho < 1$ established here.

Case 3. The simple-birth-and-linear-death processes: This model assumes

1. a constant birthrate; that is, $\lambda_i = \lambda$ for all $i = 0, 1, 2, 3, \ldots$
2. a linear death rate; that is, $\mu_i = i\mu$ for all $i = 1, 2, 3, \ldots$

Like before, we write $\rho \equiv \lambda/\mu$. Then $\delta^{(i)} = \rho^i/i!$ for all $i \geq 0$. With the help of Appendix (A.2), Equation (11.11.4) now becomes:

$$\Delta = \sum_{i=0}^{\infty} \frac{1}{i!}\rho^i = e^{\rho} < \infty$$

Thus, the limiting distribution exists for all values of λ and μ and is *Poisson* with rate ρ, or

$$\gamma_i = \frac{\rho^i}{i!}e^{-\rho} \quad \text{for all } i = 0, 1, 2, \ldots$$

Continuous-Time Markov Chains in Action: Alliance Formation in International Politics. According to Midlarsky (1983),

> The question of the existence of memory is of great importance in international politics and foreign policymaking, for issues of consequence often rest on it. The rapprochement between Egypt and Israel, for example, relied heavily on the notion of memorylessness,

or at least its institutionalization in the form of forgetting, in encouraging the Egyptians and Israelis to forget past grievances and concentrate on present self-interest. An entire alliance system, the nineteenth-century balance-of-power system, has been asserted to be without memory. Indeed, many have claimed that Henry Kissinger relied heavily on this nineteenth-century model in laying the foundation for the Camp David accords.

From this memeoryless assumption, Midlarsky assumed that the stream of alliance formation is Poisson, the rate of which is independent of the number of alliances already in the system—that is, $\lambda_n = \lambda$. He also assumed that each alliance duration is exponentially distributed with rate μ. Given n alliances in existence at time t, the death rate is thus $n\mu$. This results in a simple-birth-and-linear-death processes in which the number of alliances in existence at any time point is Poisson distributed with rate ρ. This is "precisely as was posited theoretically and confirmed empirically for the nineteenth century in prior research" (Midlarsky, 1981). ■

11.12 SUMMARY

1. A continuous-time Markov chain is a semi-Markov chain having exponentially distributed holding times. It is memoryless at all time points and can be completely specified by the instantaneous transition rates q_{ij}s.
2. If a competitive Markov chain is in state i, then state j $(j \neq i)$ is given a Poisson stream with rate q_{ij}. The chain will move to whichever state has the first event occurrence.
3. The limiting probabilities of a continuous-time Markov chain can be derived directly from the balance Equations (11.10.1) and the normalizing Equation (11.10.4).
4. A birth-and-death process is a competitive continuous-time Markov chain having the random walk as its embedded chain.

PROBLEMS

11.1 Consider an absorbing continuous-time Markov chain. The transition matrix of its embedded chain takes the form (5.1.1).

a. Show that its Q-matrix can be written as

$$\mathbf{Q} = \begin{matrix} \\ (a \text{ rows}) \\ (t \text{ rows}) \end{matrix} \begin{pmatrix} (a \text{ columns}) & (t \text{ columns}) \\ \mathbf{0} & \mathbf{0} \\ -\mathbf{W}_d\mathbf{S} & \mathbf{W}_d(\mathbf{I}-\mathbf{T}) \end{pmatrix}$$

where $\mathbf{W}_d$ is the $(t \times t)$ transient-to-transient portion of matrix $\mathbf{Q}_d$ defined in §11.5.

b. Hence, use Equation (10.7.2) to obtain the following *continuous-time fundamental matrix*:

$$\boldsymbol{\Psi} = -[\mathbf{W}_d(\mathbf{I}-\mathbf{T})]^{-1}$$

11.2 Return to the rat-maze example $\{R(t)\}_{t\in[0,\infty)}$ having transition matrix (5.3.1). Suppose the durations the rat stays in rooms 2, 3, 4, and 5 are exponentially distributed with mean 2 minutes, 7/3 minutes, 8/3 minutes, and 3 minutes, respectively.

a. Write the chain's Q-matrix.
b. Hence, use Problem 11.1 to find the matrix $\mathbf{\Psi}$.
c. Compare with the results in (10.7.4).

11.3 Consider the marketing example in §10.3. Assume now that the durations the company uses brand X, Y, or Z between two consecutive orders are exponentially distributed with means 2.3 weeks, 3.1 weeks, and 3.4 weeks, respectively.
a. Find the Q-matrix for the chain $\{B(t)\}_{t\in[0,\infty)}$ where $B(t)$ is the brand the company uses at time t.
b. Hence, use the balance equations in (11.10.5) to find the proportion of time the company uses each brand. Compare with the results in (10.11.2).

11.4 Calculate the limiting probabilities of rock types modeled by Ioannou (1987) in §11.5.

11.5 Verify the limiting probabilities in the health service example calculated by Schach and Schach (1972) in §11.10. Verify also that when q_2 increases to .12 and p_2 to .013, the expected number of hospital patients is reduced from 1561 to 1305 for a population of size 100,000.

11.6 Hutchinson (1990) observed the midwinter rainfall pattern in Spokane, Washington, for over 25 years. He used the following three states: (0) dry spell, (1) transition spell, and (2) wet spell. The transition probabilities were obtained as $p_{10} = .265$, $p_{01} = p_{21} = 1$ (that is, the transition spells always follow the dry and the wet spells). The average durations of the dry spells, the transition spells, and the wet spells are $\mathcal{E}[h_0] = 28.8$ hours, $\mathcal{E}[h_1] = 1.83$ hours, and $\mathcal{E}[h_2] = 0.84$ hours, respectively.
a. Assume a continuous-time Markov chain and find its Q-matrix.
b. Find the limiting distribution of the chain using Equation (11.10.5).
c. Write the balance equations.
d. Find the limiting distribution of the chain from the balance equations.

11.7 Molecules arrive at a surface of a bacterium as a Poisson stream with rate λ. Only a proportion α of them is acceptable. If a molecule is acceptable, it will stay at the surface for an exponential period with rate μ_A; if not, it will stay for an exponential period with rate μ_U. Find the proportion of time that the surface is unoccupied
a. by studying a regenerative process
b. by studying a semi-Markov chain
c. by studying a continuous-time Markov chain

11.8 A machine has k components, and the life of each is exponentially distributed with rate λ. It only uses one component at a time. If a component fails, it will be replaced immediately by a component that has not failed. The working component that is not in use cannot fail. Once there is no component left, the machine crashes. In this case, it takes an exponential duration with rate μ to repair all failed components simultaneously. Find the proportion of time that there are i working components
a. by studying a regenerative process
b. by studying a semi-Markov chain
c. by studying a continuous-time Markov chain

11.9 A machine works for an exponential duration with rate λ. When it fails, it is repaired immediately. Each repair requires k operations to be done *sequentially*. The duration of operation i is exponentially distributed with rate μ_i. Find the proportion of time that the machine is undergoing operation i

a. by studying a regenerative process
b. by studying a semi-Markov chain
c. by studying a continuous-time Markov chain

11.10 A machine works for an exponential duration with rate λ. When it fails, it is repaired immediately. Each repair requires two operations, A and B, to be done *concurrently*. The machine is up again after both operations are completed. The duration of operation i ($i = A, B$) is exponentially distributed with rate μ_i. Find the proportion of time that the machine is working

a. by studying a regenerative process. *Hint:* Use Problem 8.26.
b. by studying a semi-Markov chain.
c. by studying a continuous-time Markov chain.

11.11 Consider a system having N components *in series*; that is, it only functions when all components function. We assume that each component operates independently. The life of component i is exponentially distributed with rate λ_i, and the duration to repair component i is exponentially distributed with rate μ_i. The system is *interruptive*—that is, when component i fails, the system is shut down and all other components stop operating, and thus cannot fail while component i is being repaired. Find the proportion of time that the system functions

a. by studying a regenerative process
b. by studying a semi-Markov chain
c. by studying a continuous-time Markov chain

11.12 There are two machines, A and B, and two repairpeople. Machine i ($i = A, B$) operates for an exponential duration with rate λ_i before breaking down. If machine i is down, it takes an exponential duration with rate μ_i to repair it. Assume $\lambda_A = \lambda_B = \lambda$ and $\mu_A = \mu_B = \mu$. Find the proportion of time that both machines are up

a. by studying a regenerative process. *Hint:* Use Problem 8.34.
b. by studying a semi-Markov chain.
c. by studying a continuous-time Markov chain.

11.13 Refer to Problem 11.12. Assume now that there is only one repairperson. If a machine breaks down while the repairperson is fixing the other machine, it has to wait. Assume $\lambda_A = \lambda_B = \lambda$ and $\mu_A = \mu_B = \mu$. Find the proportion of time that both machines are up

a. by studying a regenerative process. *Hint:* Use Problem 8.33.
b. by studying a semi-Markov chain.
c. by studying a continuous-time Markov chain.

11.14 Refer to Problem 11.13. Assume now that machine A has a *higher priority* than machine B. If machine A breaks down when the repairperson is fixing machine B, she will leave machine B to work on machine A and resume working on B only when she finishes with A. If machine B is down when she is fixing machine A, machine B will have to wait until she finishes with A. Assume $\lambda_A = \lambda_B = \lambda$ and $\mu_A = \mu_B = \mu$. Find the proportion of time that both machines are up

a. by studying a regenerative process. *Hint:* Use Problem 8.36.
b. by studying a semi-Markov chain.
c. by studying a continuous-time Markov chain.

11.15 Refer to Problem 11.12. Let us now assume that only one machine is needed at a time. If machine A is working, machine B is used as a *standby redundant* and thus cannot fail. Only when machine A is under repair is machine B put in use and thus can fail. Assume $\lambda_A = \lambda_B = \lambda$ and $\mu_A = \mu_B = \mu$. Since we only need one machine at a time, it is more important to find the proportion of time that both machines are down than the proportion of time that both machines are up. Find both proportions

a. by studying a regenerative process. *Hint:* Use Problem 8.37.
b. by studying a semi-Markov chain.
c. by studying a continuous-time Markov chain.

11.16 Let $m(t)$ and $f(t)$ be the numbers of male and female members in a population, respectively. Let λ be the instantaneous rate that a male member mates with a female. Each time they mate, an offspring is produced, equally likely to be male or female. Represent the vector chain $\{m(t), f(t)\}_{t\in[0,\infty)}$ as a continuous-time Markov chain. Obtain its transition diagram.

11.17 An organism can be either of type A or type B. A type A organism can split into two type B organisms with rate λ_A. A type B organism can change into a type A organism with rate λ_B. Let $a(t)$ and $b(t)$ be the numbers of type A and type B organisms at time t, respectively. Represent the vector chain $\{a(t), b(t)\}_{t\in[0,\infty)}$ as a continuous-time Markov chain. Obtain its transition diagram.

11.18 Consider an elementary chemical reaction of the form $A \leftrightarrow B$. Here, each molecule of A can randomly react with a catalyst and be converted into a molecule of B with rate λ_A; similarly, each molecule of B can randomly react with a catalyst and be converted into a molecule of A with rate λ_B. Assume the total number N of molecules A and B remains constant throughout the process. Let $X(t)$ be the number of molecules A at time t. Show that chain $\{X(t)\}_{t\in[0,\infty)}$ is a finite birth-and-death process and its limiting distribution is binomial with parameters $(N, \lambda_B/(\lambda_A + \lambda_B))$.

11.19 (Holgate, 1967) Elephants tend to group themselves into herds. Two herds may meet and amalgamate; a herd may split into two smaller ones. For a population of N elephants, let $X(t)$ be the number of herds at time t. We assume that, if $X(t) = i$, the duration until two herds meet and amalgamate is exponentially distributed with rate $\mu_i = (i-1)\mu$ and the duration until a herd splits into two is exponentially distributed with rate $\lambda_i = (N-i)\lambda$. (Why?)

a. Show that $\{X(t)\}_{t\in[0,\infty)}$ is a birth-and-death process.
b. Show that the limiting distribution of the number of herds is one plus a binomial variable; that is,

$$\gamma_i = \binom{N-1}{i-1}\left(\frac{\lambda}{\lambda+\mu}\right)^{i-1}\left(\frac{\mu}{\lambda+\mu}\right)^{N-i} \qquad \text{for all } 1 \le i \le N$$

c. When there are k herds, the average herd size is N/k. Show that the long-term average herd size is $[(\lambda+\mu)^N - \mu^N]/\lambda(\lambda+\mu)^{N-1}$.

11.20 Consider a general machine-repair problem with M machines and S servicepeople ($S \leq M$). Each machine operates for an exponential duration with rate λ before breaking down. It takes a serviceperson an exponential duration with rate μ to repair a down machine. Let $\delta(t)$ be the number of down machines at time t.

a. Show that chain $\{\delta(t)\}_{t\in[0,\infty)}$ is a birth-and-death process having limiting distribution

$$\gamma_i = \begin{cases} \binom{M}{i} \rho^i \gamma_0 & \text{for } 0 \leq i \leq S \\ \binom{M}{i} \dfrac{i!}{S!S^{i-S}} \rho^i \gamma_0 & \text{for } S < i \leq M \end{cases}$$

b. When $M = 2$ and $S = 1$, compare with the result obtained in Problem 11.13.

c. When $M = 4$ and $S = 3$, compare with the result obtained in §11.9.

d. Assume that $S = M$. (This situation arises when a machine's operator is also its serviceperson.) Show that the limiting distribution is binomial. Explain heuristically why.

11.21 A system has three reparable machines and three servicepeople. It is called a *two-out-of-three* system because it works only when at least two machines work. The system's *availability* $A(t)$ is the probability that it functions at time t regardless of whether it has failed before t. Find the limiting availability of this system

a. from the result obtained in Problem 11.20

b. from the result obtained in Problem 8.32

11.22 Find the limiting distribution of a finite birth-and-death process having state space $\{1, 2, \ldots, N\}$ with constant death rates μ and birthrate

$$\lambda_i = \begin{cases} (N-i)\lambda & \text{for all } 1 \leq i < N \\ 0 & \text{for all } N \leq i \end{cases}$$

11.23 Consider a finite birth-and-death process having state space $\{1, 2, \ldots, N\}$ with constant death rates μ and birthrate

$$\lambda_i = \begin{cases} \dfrac{N-i}{N(i+1)}\lambda & \text{for all } 0 \leq i < N \\ 0 & \text{for all } N \leq i \end{cases}$$

Show that its limiting distribution is binomial.

11.24 Consider $\{\ell, \ell+1, \ldots, u-1, u\}$, a finite birth-and-death process having state space with

$$\lambda_i = \begin{cases} i(u-i)\lambda & \text{for all } \ell \leq i \leq u \\ 0 & \text{otherwise} \end{cases}$$

and

$$\mu_i = \begin{cases} i(i-\ell)\mu & \text{for all } \ell \le i \le u \\ 0 & \text{otherwise} \end{cases}$$

Show that

$$\gamma_i = \binom{u-\ell}{i-\ell}\left(\frac{\lambda}{\mu}\right)^{i-\ell}\frac{\ell}{i}\gamma_\ell \quad \text{for all } \ell \le i \le u$$

11.25 Consider an infinite birth-and-death process $\{X(t)\}_{t\in[0,\infty)}$ having $\lambda_i = (i+1)\lambda$ and $\mu_i = i^2\mu$ for all $i = 0, 1, 2, \ldots$. Show that its limiting distribution is Poisson with parameter λ/μ. *Hint:* Use Appendix (A.2).

11.26 Consider an infinite birth-and-death process $\{X(t)\}_{t\in[0,\infty)}$ having $\lambda_i = \lambda/(i+1)$ and $\mu_i = \mu$ for all $i \ge 0$. Show that its limiting distribution is Poisson with parameter λ/μ. *Hint:* Use Appendix (A.2).

11.27 Consider an infinite birth-and-death process $\{X(t)\}_{t\in[0,\infty)}$ having $\lambda_i = \alpha^i\lambda$ and $\mu_i = \mu$ for all $i \ge 0$.

a. Find the condition for positive-recurrence.

b. If the chain is positive-recurrent, show that

$$\gamma_i = \rho^i\alpha^{(i-1)i/2}\gamma_0 \quad \text{for all } i = 1, 2, 3, \ldots$$

11.28 Consider an infinite birth-and-death process having constant death rate $\mu_i = \mu$ and birth rate

$$\lambda_i = \frac{N+i}{N(i+1)}\lambda \quad \text{for all } i \ge 0$$

Show that its limiting distribution is negative binomial. Compare with Problem 11.23. *Hint:* Use Equation (1.14.2) and Appendix (A.14).

11.29 Consider an infinite birth-and-death process $\{X(t)\}_{t\in[0,\infty)}$ having $\lambda_i = (i+2)\lambda$ and $\mu_i = i\mu$ for all $i \ge 0$.

a. Find the condition for positive-recurrence. *Hint:* Use Appendix (A.10).

b. If the chain is positive-recurrent, find its limiting distribution. *Hint:* Use Problem 1.8.

11.30 Consider an infinite birth-and-death process having $\lambda_i = (i+1)\lambda$ and $\mu_i = (i+1)\mu$ for all $i \ge 0$.

a. Find the condition for positive-recurrence. *Hint:* Use Appendix (A.5).

b. If the chain is positive-recurrent, find its limiting distribution.

11.31 Consider an infinite birth-and-death process having *linear growth with immigration* and linear death; that is, $\lambda_i = (i\lambda + \theta)$ and $\mu_i = i\mu$ for all $i \ge 0$. Apply the *ratio test*[5] for the convergence of a series to Equation (11.11.4) to find the condition

[5] $\sum_{n=0}^{\infty} a_n$ is convergent if and only if $\lim_{n\to\infty} \frac{a_{n+1}}{a_n} < 1$.

of positive-recurrence for this chain. Explain why this condition is independent of the value of θ.

11.32 In §7.3, we noted that an infinite irreducible chain having an absorbing state might never be absorbed. Consider a general infinite birth-and-death process having birthrate λ_i and death rate μ_i. Let state 0 be absorbing; that is, $\lambda_0 = 0$ and $\mu_0 = 0$.

a. Similar to Problem 7.7a, use first-step analysis to show that the probability that the chain will eventually be absorbed into state 0 is

$$f_{i0} = \begin{cases} \dfrac{\Omega_i}{1+\Omega_1} & \text{if } \Omega_1 < \infty \\ 1 & \text{if } \Omega_1 = \infty \end{cases}$$

where $\omega^{(i)} \equiv \prod_{k=1}^{i} \mu_k/\lambda_k$ and $\Omega_i \equiv \sum_{k=i}^{\infty} \omega^{(k)}$ for all $i = 1, 2, 3, \ldots$.

b. Let t_{i0} be the duration before the chain is absorbed into state 0 from state i. Use first-step analysis and the fact that $\mathcal{E}[t_{i0}] = \mathcal{E}[t_{i,i-1}] + \mathcal{E}[t_{i-1,0}]$ to show that $\mu_i \mathcal{E}[t_{i,i-1}] = 1 + \lambda_i \mathcal{E}[t_{i+1,i}]$.

c. Hence, for all $i = 0, 1, 2, \ldots$,

$$\mu_i \mathcal{E}[t_{i,i-1}] = 1 + \frac{\lambda_i}{\mu_{i+1}} + \frac{\lambda_i}{\mu_{i+1}}\frac{\lambda_{i+1}}{\mu_{i+2}} + \frac{\lambda_i}{\mu_{i+1}}\frac{\lambda_{i+1}}{\mu_{i+2}}\frac{\lambda_{i+2}}{\mu_{i+3}} + \cdots$$

Compare with Problem 7.7f.

11.33 Consider a *simple* birth-and-death process in which $\lambda_i = \lambda$ and $\mu_i = \mu$. Let $\rho \equiv \lambda/\mu$.

a. Show that the probability the chain will ever visit state 0 from state i is

$$f_{i0} = \begin{cases} \left(\dfrac{1}{\rho}\right)^i & \text{if } \rho > 1 \\ 1 & \text{if } \rho \leq 1 \end{cases}$$

i. by using Problem 11.32

ii. by showing that $f_{i0} = (f_{10})^i$ and then using first-step analysis

b. Hence, obtain the condition for transience and recurrence. Compare the results with those in §7.3.

c. Assume for the rest of the problem that the chain is recurrent. Explain why $\mathcal{E}[t_{i0}] = i\mathcal{E}[t_{10}]$. Hence, show that the expected duration to reach state 0 from state i is

$$\mathcal{E}[t_{i0}] = \frac{i}{\mu - \lambda} \quad \text{for all } i = 0, 1, 2, \ldots$$

i. by using Problem 11.32

ii. by using first-step analysis

d. Hence, obtain the condition for positive-recurrence and null-recurrence.

e. Assume for the rest of the problem that the chain is positive-recurrent. Show that the chain is regenerative at each entrance into state 0 and that the expected regenerative cycle is

$$\mathcal{E}[t_{00}] = \frac{\mu}{\lambda(\mu - \lambda)}$$

f. Use the foregoing result and the regenerative argument to obtain Equation (11.11.5).

11.34 Consider a general infinite *linear* birth-and-death process in which $\lambda_i = i\lambda$ and $\mu_i = i\mu$. (This applies to a population in which each individual behaves independently and each has a birthrate of λ and a death rate of μ.) Let $\rho \equiv \lambda/\mu$.

a. Explain why the classification of this chain is the same as that for the simple birth-and-death processes in Problem 11.33. *Hint:* Use Problem 11.32a.

b. Use Problem 11.32c to show that the expected duration to reach state 0 from state i is

$$\mathcal{E}[t_{10}] = \begin{cases} \infty & \text{if } \rho \geq 1 \\ -\dfrac{\ln(1-\rho)}{\lambda} & \text{if } \rho < 1 \end{cases}$$

Hint: Use Appendix (A.5).

11.35 Consider a general infinite birth-and-death process having birthrate λ_i and death rate μ_i. Assume $\mu_0 = 0$.

a. Let $t_{i,i+1}$ be the time for the chain to move from state i to state $i+1$. Use first-step analysis to find $\mathcal{E}[t_{i,i+1}]$. Compare with Problem 7.8a.

b. Hence, find $\mathcal{E}[t_{i,i+1}]$ when $\lambda_i = \lambda$ and $\mu_i = \mu$ for all $i \in S$.

11.36 For a continuous-time Markov chain $\{X(t)\}_{t\in[0,\infty)}$ with Q-matrix $\mathbf{Q}$, it can be shown that the *transient results* equivalent to Equation (3.4.1) for the discrete-time chain can be obtained as

$$\mathbf{P}(t) \equiv [\Pr\{X(t) = j | X(0) = i\}]_{i,j\in S} = \exp\{\mathbf{Q}t\}$$

where, similar to Appendix (A.2), the exponential function of a matrix $\mathbf{A}$ is defined as

$$\exp\{\mathbf{A}\} \equiv \mathbf{I} + \sum_{n=1}^{\infty} \frac{1}{n!}\mathbf{A}^n$$

a. A Poisson process $\{m(t)\}_{t\in[0,\infty)}$ introduced in §8.11 is a continuous-time Markov chain. Obtain its Q-matrix and show that

$$[\mathbf{Q}^n]_{ij} = \begin{cases} \dbinom{n}{j-i}(-1)^{n+j-i}\lambda^n & \text{for all } 0 \leq j-i \leq n \\ 0 & \text{otherwise} \end{cases}$$

b. Hence, use the foregoing transient result to find $\mathbf{P}(t)$. Compare with Equation (8.11.1).

11.37 Suppose the Q-matrix of a continuous-time Markov chain can be *decomposed* into

$$\mathbf{Q} = \mathbf{HJH}^{-1}$$

where $\mathbf{J}$ is a diagonal matrix, the (i, i)-element of which is j_{ii}.

a. Show that

$$\mathbf{P}(t) = \mathbf{H}\exp\{\mathbf{J}t\}\mathbf{H}^{-1}$$

where $\exp\{\mathbf{J}t\}$ is a diagonal matrix, the (i, i)-element of which is $\exp\{j_{ii}t\}$. Compare with Problem 3.23.

b. For a two-state chain, verify that

$$Q \equiv \begin{pmatrix} -\alpha & \alpha \\ \beta & -\beta \end{pmatrix} = \frac{1}{\alpha+\beta}\begin{pmatrix} 1 & \alpha \\ 1 & -\beta \end{pmatrix}\begin{pmatrix} 0 & 0 \\ 0 & -(\alpha+\beta) \end{pmatrix}\begin{pmatrix} \beta & \alpha \\ 1 & -1 \end{pmatrix}$$

Hence, show that

$$P(t) = \frac{1}{\alpha+\beta}\begin{pmatrix} \beta & \alpha \\ \beta & \alpha \end{pmatrix} + \frac{\exp\{-(\alpha+\beta)t\}}{\alpha+\beta}\begin{pmatrix} \alpha & -\alpha \\ -\beta & \beta \end{pmatrix}$$

Compare with Problem 3.24.

11.38 Consider the marketing example in Problem 11.3.

a. Verify that the Q-matrix can be written as $\mathbf{HJH}^{-1}$, with

$$\mathbf{H} = \begin{pmatrix} \frac{23}{136} & \frac{\sqrt{6073}}{8{,}259{,}280}\left(-263+565\sqrt{6073}\right) & \frac{\sqrt{6073}}{8{,}259{,}280}\left(263+565\sqrt{6073}\right) \\ \frac{23}{136} & -\frac{23\sqrt{6073}}{8{,}259{,}280}\left(857+5\sqrt{6073}\right) & -\frac{23\sqrt{6073}}{8{,}259{,}280}\left(-857+5\sqrt{6073}\right) \\ \frac{23}{136} & -\frac{23\sqrt{6073}}{8{,}259{,}280}\left(-1047+5\sqrt{6073}\right) & -\frac{23\sqrt{6073}}{8{,}259{,}280}\left(1047+5\sqrt{6073}\right) \end{pmatrix}$$

$$\mathbf{J} = \begin{pmatrix} 0 & 0 & 0 \\ 0 & -\frac{3169}{12{,}121}+\frac{5\sqrt{6073}}{12{,}121} & 0 \\ 0 & 0 & -\frac{3169}{12{,}121}-\frac{5\sqrt{6073}}{12{,}121} \end{pmatrix}$$

$$\mathbf{H}^{-1} = \begin{pmatrix} 1 & \frac{62}{23} & \frac{51}{23} \\ 1 & -\frac{179}{322}-\frac{5\sqrt{6073}}{322} & -\frac{143}{322}+\frac{5\sqrt{6073}}{322} \\ 1 & -\frac{179}{322}+\frac{5\sqrt{6073}}{322} & -\frac{143}{322}-\frac{5\sqrt{6073}}{322} \end{pmatrix}$$

b. Hence, use Problem 11.37 to find the distribution of the brand the company uses at time $t = 2$ weeks.

c. Find the limiting distribution of the brand the company uses from the foregoing diagonalized form. Compare with the result obtained in Problem 11.3.

11.39 Consider an absorbing continuous-time Markov chain having the Q-matrix as in Problem 11.1. In this problem, we are especially interested in the unconditional duration ψ until the chain is absorbed, starting with an initial distribution $\boldsymbol{\pi}_T^{(0)} \equiv [\pi_i^{(0)}]_{i \in T}$. Without loss of generality, let us lump all the absorbing states together into one state and label it 0.

a. Use the transient result in Problem 11.36 to show that

$$F(x) \equiv \Pr\{\psi < x\} = 1 - \boldsymbol{\pi}_T^{(0)} \exp\{\boldsymbol{\Omega} t\}\mathbf{e}$$

where $\boldsymbol{\Omega} \equiv \mathbf{W}_d(\mathbf{I} - \mathbf{T})$ as defined in Problem 11.1 and the exponential function of a matrix was defined in Problem 11.36. The distribution $F(x)$ is known in the literature as the *phase-type* distribution. The pair $(\boldsymbol{\pi}_T^{(0)}, \boldsymbol{\Omega})$ is called the *representation* of $F(x)$.

b. Consider the hyperexponential distribution (8.8.5). Let $k = 3$. Represent this distribution as a special case of the phase-type distribution.

c. Explain why the Erlang distribution with parameters $(3, \lambda/3)$ is a special phase-type distribution. Hence, obtain Equation (8.12.1). *Hint:*

$$\begin{pmatrix} 1 & 0 & 0 \\ -1 & 1 & 0 \\ 0 & -1 & 1 \end{pmatrix}^n = \begin{pmatrix} \binom{n}{0} & 0 & 0 \\ -\binom{n}{1} & \binom{n}{0} & 0 \\ \binom{n}{2} & -\binom{n}{1} & \binom{n}{0} \end{pmatrix} \quad \text{for all } n \geq 2$$

11.40 A *Yule process* $\{X(t)\}_{t \in [0,\infty)}$ is a birth-and-death process having a linear birthrate (that is, $\lambda_i = i\lambda$ for all $i \geq 0$) and no death (that is, $\mu_i = 0$ for all $i \geq 0$).

a. Let $t_{i,j}$ be the time it takes the chain to move from state i to state j for all $0 < i < j$. Use Problem 8.27c to find $\Pr\{t_{1,i+1} \leq t\}$.

b. Hence, show that

$$p_{1i}(t) = e^{-\lambda t}(1 - e^{-\lambda t})^{i-1}$$

c. What is $\mathcal{E}[X(t) \mid X(0) = 1]$?

CHAPTER 12

Markovian Queues

12.1 QUEUING THEORY

Living in this crowded world, we have to wait all the time: To go to a supermarket, we slowly edge our car of the future along the freeway of the past; at the supermarket, we patiently line up at a checkout counter. Queuing theory is a branch of applied probability that seeks to understand these phenomena.

Beginning more than 80 years ago with the study of waiting times in telephone systems, queuing theory now finds important applications in many other fields such as computer networking, manufacturing, and traffic management. Not only that, it also shares the same basic mathematical structure with other problems such as inventory, dams, and insurance risks. Understanding queuing models will lead to the understanding of these problems.

12.2 QUEUING SYSTEMS

Generally, a *queuing system* is any facility at which the *customers* arrive and then stay for a certain duration before departing.[1]

Let the *interarrival time* t_n be the duration between the arrivals of the nth and the $(n+1)$th customers. We assume that all t_ns ($n \geq 1$) are mutually independent and identically distributed. The arrivals of the customers thus form a renewal stream, with the renewal rate, or the *arrival rate*, $\lambda \equiv 1/\mathcal{E}[t_1]$. We call the process $\{t_n\}_{n=1,2,\ldots}$ the *arrival process*.

---

[1]Queuing theory vocabulary should be interpreted abstractly and applied generally. For example, although we will refer to a customer as "he" or "she," this does not necessarily imply a human being, let alone any gender preference. It could be a down machine arriving at a repairing facility or a computer program input to a computer.

FIGURE 12.2.1 A realization of a queuing model.

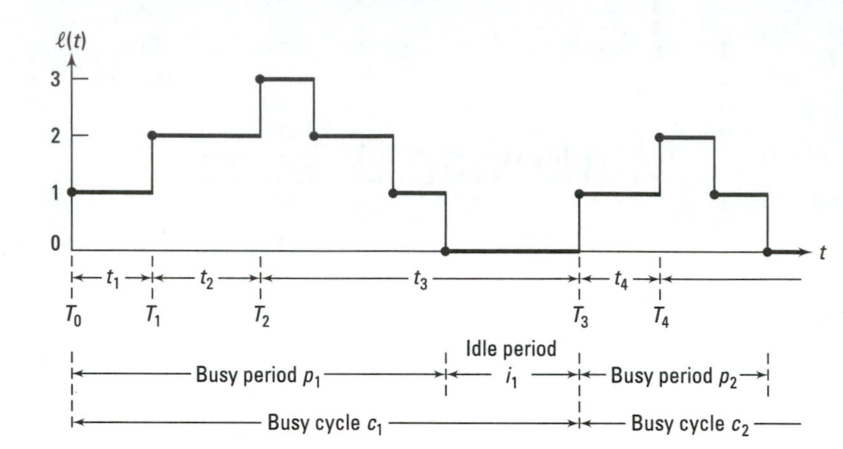

The customers arrive at the system because they need to be served by the system's *servers*.[2] We assume that the nth customer requires a *service time* s_n ($n \geq 1$) and that all s_ns are mutually independent and identically distributed. The process $\{s_n\}_{n=1,2,\ldots}$ is called the *service process*. We also assume that all interarrival times are independent of all service times.

These arrival and service processes are not sufficient to describe a queuing system completely. To study its behavior, we need further information. In this chapter, we assume that there is only one waiting line. We also assume the *service discipline* is *first-come/first-served*; that is, if an arriving customer finds all servers busy, and if he is willing to wait, then he will wait at the end of the waiting line. (This does not mean that he has to physically stand at the end of the waiting line, but rather that he will be the last to receive service among those currently waiting.)

One major property of a queuing system is that it is *regenerative*: It starts anew whenever an arriving customer finds an empty system. We call its nth regenerative cycle the *busy cycle* c_n. We assume that the first customer arrives at time $T_0 = 0$; hence, the first cycle is independent and identically distributed with all later cycles (Figure 12.2.1).

12.3 KENDALL NOTATION

Kendall (1951) proposed a notational system to describe the basic parameters of a queuing system. Its abridged form is $A/B/k$, where A describes the type of interarrival time distribution, B the type of service time distribution, and k the number of servers.

1. If we make no assumption about the distribution, A or B takes the notation G (*G*eneral).
2. If the interarrival times or the service times are exponentially distributed, A or B is replaced by M (*M*arkovian).

---

[2]We will refer to a server as "it."

3. If the interarrival times or the service times have the Erlang distribution, we write A or B as E_k.
4. If the interarrival times or the service times are constant, A or B becomes D (*D*eterministic).

Some service facilities have a *finite capacity* and thus can only accommodate up to N customers, including those being served. If an arriving customer finds N other customers already in the system, she will be lost, or denied entry, and will immediately depart from the system. In this case, we extend our notation and denote the system by $A/B/k/N$.

12.4 PERFORMANCE MEASURES

Having described a queuing system, we now identify the quantities that we wish to calculate. Normally referred to collectively as the system's *performance measures*, they are:

1. L_q, the long-term time-average of the continuous-time chain $\{\ell_q(t)\}_{t\in[0,\infty)}$, where $\ell_q(t)$ is the number of customers waiting in the queue at time t.
2. L, the long-term time-average of the continuous-time chain $\{\ell(t)\}_{t\in[0,\infty)}$, where $\ell(t)$ is the number of customers in the system at time t; that is, both waiting in the queue and being served.
3. W_q, the long-term customer-average of the discrete-time process $\{q_n\}_{n=1,2,\ldots}$, where q_n is the duration the nth customer waits in the queue before receiving service.
4. $W = W_q + \mathcal{E}[s_1]$, the long-term customer-average of the discrete-time process $\{w_n\}_{n=1,2,\ldots}$, where $w_n \equiv q_n + s_n$ is the duration the nth customer stays in the system; that is, both waiting in the queue and being served.

Queuing Theory in Action: Linear Costs for Waiting Times. The main concern of queuing theory (and in this book especially) is the long-term averages. We should note that using the average of a certain quantity in decision making implies that the corresponding cost, or disutility, is *linearly* proportional to its value. However, according to Larson (1987), this is normally not the case.

> [T]he effectiveness of urban emergency response systems depends in a nonlinear way on system response time.
>
> . . . In emergency medical services the report of a person having suffered a myocardial infarction (i.e., heart attack) indicates that on-scene professional emergency medical services should start within 5 minutes after the infarction or the probability of death is almost 1.0. In considering the heart attack victim's personal "disutility" of a 5-minute response delay versus a 2.5-minute response delay, it seems clear that the 5-minute delay is "more than twice as bad" than the 2.5-minute delay. ■

12.5 LITTLE'S RESULT

There is a very elegant equation in the literature, which is normally referred to as *Little's result*. This equation is applicable to all queuing systems in which the customers arrive, remain for some time, and then leave. It simply states that[3]

$$L = \Lambda W \tag{12.5.1}$$

where Λ is the *(long-term) arrival rate* of the customers to the system, defined as

$$\Lambda \equiv \lim_{t\to\infty} \frac{\mathcal{E}\left[\text{number of arrivals during } (0, t]\right]}{t} \tag{12.5.2}$$

If the stream of arrivals is renewal, then Λ is defined as in §9.3. If it is Markovian, it is also the instantaneous rate as defined in §8.5.

It is important to note that, at the same service facility, we can define the system differently and thus obtain different forms of Little's result. As an example, consider a service facility:

1. We can consider the *waiting line alone* as a system. In that case, Little's result yields:

$$L_q = \Lambda W_q \tag{12.5.3}$$

2. We can consider the *servers alone* as a system. In that case, Little's result yields:

$$b = \Lambda_s \mathcal{E}[s_1] \tag{12.5.4}$$

where

a. b is the long-term time-average of the number of customers being served simultaneously, or the number of busy servers.

b. Λ_s is the long-term rate of customers entering into a service, or the rate of *service beginnings*.

c. $\mathcal{E}[s_1]$ is the long-term average of the durations the customers spend with the servers, which is the average service time of all customers.

As another example, consider a servicing facility having a finite capacity N in which an arriving customer is denied entry if $\ell(t) = N$ at the time of his arrival.

1. If we define a system as including those denied customers, Equation (12.5.1) becomes $L = \lambda W_1$, where λ is the arrival rate of all customers and W_1 is the average time in the system of all customers, including the zero durations of the denied customers.

2. If we exclude the denied customers from our system, Equation (12.5.1) becomes $L = \Lambda W_2$, where Λ is the arrival rate of the admitted customers only (which is smaller than λ) and W_2 is the average duration in the system of the admitted customers only (which is larger than W_1). Note that, unless stated otherwise, our default definition of a *(queuing) system* excludes the denied customers, who are said to depart *from the system* immediately at their arrivals.

[3]For a simple proof of this result, see Eilon (1969).

12.6 IDENTIFYING A CONTINUOUS-TIME MARKOV CHAIN

For the rest of this chapter, we shall study queuing systems in which *the arrival stream is Poisson and all the service times are exponentially distributed.* We call them the *Markovian* queues. In Kendall's notation, they are the $M/M/k$ models. In the next chapter, we shall study some non-Markovian queuing models.

We can say generally that most Markovian queues can be studied by analyzing some suitably defined continuous-time Markov chains. Normally, a good candidate is the chain $\{\ell(t)\}_{t\in[0,\infty)}$, where $\ell(t)$ is the number of customers in the system at time t. This chain changes state whenever there is an arrival at, or a departure from, the system. With the Markovian assumption of the arrival and service processes in this chapter, it is very likely that we have here a competitive continuous-time Markov chain. If not, we might be able to make it Markovian by modifying the state space or introducing some suitable supplementary variables.

Analyzing chain $\{\ell(t)\}_{t\in[0,\infty)}$ would allow us to find L. We then can use Little's result to calculate other quantities such as L_q, W, and W_q.

In this chapter, we do not dwell on any particular model to discuss its detailed results. We look instead at as many models as possible, mainly to illustrate *how the appropriate continuous-time Markov chain can be identified.* Once we have done that, we move on to a different model, leaving the reader to obtain the balance equations and hence the performance measures in the problem section if appropriate.[4]

12.7 $M/M/k/N$ QUEUES

Let us first study an $M/M/k/N$ queuing system. Here, customers arrive as a Poisson stream with rate λ. If an arriving customer finds N other customers already in the system, he immediately departs from the system. Otherwise, he remains there and eventually is served with an exponentially distributed service time having rate μ.

This is an example in which the chain $\{\ell(t)\}_{t\in[0,\infty)}$, with state space $\{0, 1, 2, \ldots, N\}$, is a *finite birth-and-death process*, as studied in §11.11. Here, the chain can only move from state i to state $i+1$ when there is an arrival or move from state i to state $i-1$ when there is a service completion. The birthrates are constant; that is, $\lambda_i = \lambda$ for all $0 \leq i < N$. The death rates are $\mu_i = \min(k, i)\mu$ for all $0 \leq i$ because the system will serve all i customers if possible, using up to k servers.

We leave it to the reader to obtain the limiting distribution of $\{\ell(t)\}_{t\in[0,\infty)}$ from the general results of the birth-and-death processes in §11.11 (Problem 12.9). It turns out that this distribution is dependent on $\rho \equiv \lambda/\mu$ alone, which is the *relative strength* of λ and μ, not on their absolute values.

Queuing Theory in Action: Calling L. L. Bean. L. L. Bean, Inc. is known for retailing outdoor goods and apparel. According to Quinn et al. (1991),

> With annual sales of \$580 million in 1988, L. L. Bean conservatively estimated that it lost \$10 million of profit because it allocated telemarketing resources suboptimally.

[4] When the chain $\{\ell(t)\}_{t\in[0,\infty)}$ is infinite, it might not be easy to obtain the performance measures from the balance equations.

Customer-service levels had become clearly unacceptable: in some half hours, 80 percent of the calls dialed received a busy signal because the trunks [that is, the telephone lines] were saturated; those customers who got through might have waited 10 minutes for an available agent. . . . On exceptionally busy days, the total orders lost because of trunk "busies" (incoming calls not finding an idle trunk) and caller abandonment (after waiting for an agent) were estimated—based on conservative retry probabilities—to approach $500,000 in gross revenues. When annualized, based on call volume, the accumulated penalty cost of these allocations of resources amounted to $10 million in lost profits in 1988.

. . . The $(M/M/s/s)$ model . . . and the general finite-queue model $(M/M/s/K)$ are used to estimate the operating characteristics [that is, the performance measures] for the trunks and agents, respectively. . . . These operating characteristics are then used to assess the economic impact of blocked calls, abandoned calls, and queue times.

. . . Although the . . . analysis indicates conservatively-estimated annual profit gains with the model of approximately $10 and $9.2 million for 1988 and 1989, respectively, the intangible, long-term benefits to L. L. Bean may well out-weight these tangible profit gains. As a result of our management science project, estimated to have cost only $40,000, telemarketing management now has a tool that allows it to simultaneously optimize all of the queuing resources over which it has control.

. . . Managers also report benefits in operations, indicating that L. L. Bean's reputation in the eyes of calling customers has improved and that the number of problems that agents now experience with callers has been drastically reduced. ■

12.8 *M/M/k* QUEUES

Let us now impose no limit on the foregoing system's capacity; that is, we let $N \to \infty$. Hence, no arriving customer is lost. We now have an $M/M/k$ queuing system. With state space $\{0, 1, 2, \ldots\}$, the chain $\{\ell(t)\}_{t\in[0,\infty)}$ becomes an *infinite* birth-and-death process having birthrates $\lambda_i = \lambda$ and death rates $\mu_i = \min(k, i)\mu$.

When $\lambda < k\mu$, or $\rho \equiv \lambda/\mu < k$, the reader can verify from §11.11 that

$$\gamma_i = \begin{cases} \dfrac{\rho^i}{i!}\gamma_0 & \text{for all } 0 \le i \le k \\ \dfrac{\rho^i}{k^{i-k}k!}\gamma_0 & \text{for all } i > k \end{cases} \tag{12.8.1}$$

where

$$\gamma_0 = \left[\frac{\rho^k}{k!}\left(\frac{k}{k-\rho}\right) + \sum_{i=0}^{k-1}\frac{\rho^i}{i!}\right]^{-1} \tag{12.8.2}$$

When $k = 1$, this gives a *geometric* distribution for the limiting number of customers in the *single-server* queue; that is,

$$\gamma_i = (1-\rho)\rho^i \quad \text{for all } i \ge 0 \tag{12.8.3}$$

Some important observations about the $M/M/k$ queues are in order here:

1. As chain $\{\ell(t)\}_{t\in[0,\infty)}$ is infinite, we must be careful with its classification. As discussed in §11.11,

a. when $\lambda > k\mu$, or $\rho > k$, this chain is *transient* and the system is said to be *unstable*. As the arrival rate is higher than the maximum total service rate, the system has more and more *service backlog*, and the number of customers in the system will drift toward infinity.

b. when $\lambda = k\mu$, or $\rho = k$, this chain is *null-recurrent* and the system is also said to be *unstable*. The number of customers in the system does not drift in any direction, but the expected return time to any state is infinite.

c. when $\lambda < k\mu$, or $\rho < k$, this chain is *positive-recurrent* and the system is said to be *stable*. As the arrival rate is less than the maximum total service rate, the number of customers in the system tends to drift toward zero, but then is reflected back to one whenever a customer arrives at an empty system.

2. Recall that we denote the long-term time-average of the number of busy servers as b and the number of servers as k. Let us call the ratio b/k the *utilization factor* of the queuing system.

a. For a unstable system in which $\lambda/\mu \geq k$, Equation (12.5.4) yields $b = k$ because $\Lambda_s = k/\mathcal{E}[s_1]$. (Why?) Thus, the servers are always busy, and the utilization factor is 1.

b. For a stable system in which $\lambda/\mu < k$, Equation (12.5.4) yields $b = \lambda/\mu \equiv \rho$ because $\Lambda_s = 1/\mathcal{E}[t_1]$. (Why?) Thus, $b < k$ or the servers are idle occasionally, and the utilization factor ρ/k is strictly less than 1.

3. In *heavy traffic*, when $\lambda \to k\mu$, or when the utilization factor $\rho/k \to 1$, both the limiting expected number of customers L and the limiting expected time in system W *increase very rapidly to infinity*. This is evident in Figure 12.8.1 for the $M/M/1$ queue, in which Equations (12.8.3) and (1.12.7) give

$$L = \frac{\rho}{1-\rho} \tag{12.8.4}$$

FIGURE 12.8.1 *L* **versus the utilization factor in an** $M/M/1$ **queue.**

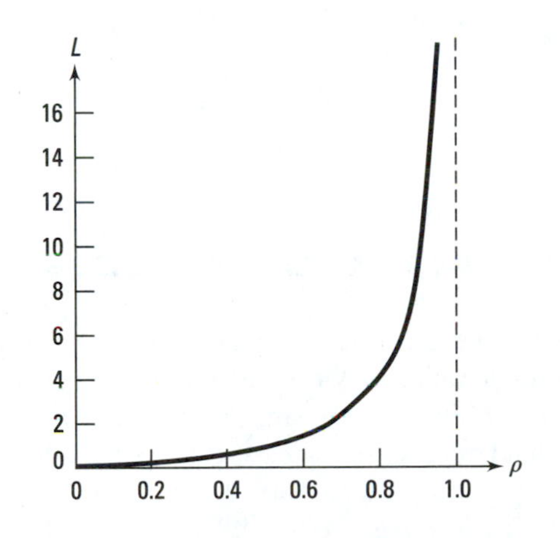

Queuing Theory in Action: Maximizing the Utilization Factor. According to Suri et al. (1995),

> Traditional views of manufacturing differ from modern views of manufacturing. ... Since companies usually based performance measures on equipment utilizations, managers had an incentive to schedule factories at or over 100 percent capacity to ensure full utilization [that is, they try to push the utilization factor closer to 1]. Using actual data on facilities and products, we are able to demonstrate to managers how this emphasis on high utilization results in long lead times for their product. ■

4. At any time t, let $d(t)$ be the duration until the next departure of a customer from the system. *Burke's theorem* states that, as $t \to \infty$, $d(t)$ is exponentially distributed with rate λ and is independent of the number of customers in the system at time t. In other words, the limiting stream of customer departures from an $M/M/k$ queue is Poisson with rate λ (Burke, 1976).

It is easy to see that the departure rate must be the same as the arrival rate λ for the system to remain stable. However, it is hard to accept intuitively the fact that the limiting duration until the next departure is independent of the number of customers in the system. At least we would expect that $d(t)$ is dependent on whether the system is empty or not. For example, consider the $M/M/1$ queue:

a. Given the system is empty, or $\ell(t) = 0$, then $d(t)$ is the sum of the duration until the next arrival (which is exponentially distributed with rate λ) and the service time of the arriving customer (which is exponentially distributed with rate μ). Hence, its expected value is $\mathcal{E}[d(t) \mid \ell(t) = 0] = 1/\lambda + 1/\mu$.

b. Given the system is not empty, or $\ell(t) > 0$, then $d(t)$ is the remaining service time of the customer being served (which is exponentially distributed with rate μ). Hence, its expected value is $\mathcal{E}[d(t) \mid \ell(t) > 0] = 1/\mu$.

c. Thus, the distribution of the duration until the next departure is a *mixture* of both:

$$\Pr\{d(t) < \tau\} = \Pr\{d(t) < \tau \mid \ell(t) = 0\} \Pr\{\ell(t) = 0\} + \Pr\{d(t) < \tau \mid \ell(t) > 0\} \Pr\{\ell(t) > 0\}$$

It turns out that the *limiting emptiness probability* $\lim_{t\to\infty} \Pr\{\ell(t) = 0\} \equiv \gamma_0$ of an $M/M/k$ queue provides the right mixture such that $d(t)$ becomes exponentially distributed with rate λ when $t \to \infty$. For the $M/M/1$ queue, from Equation (12.8.3), we have $\gamma_0 = 1 - \lambda/\mu$; hence,

$$\lim_{t\to\infty} \mathcal{E}[d(t)] = \left(\frac{1}{\lambda} + \frac{1}{\mu}\right)\left(1 - \frac{\lambda}{\mu}\right) + \left(\frac{1}{\mu}\right)\left(\frac{\lambda}{\mu}\right) = \frac{1}{\lambda}$$

12.9 $M/M/k$ QUEUES WITH FEEDBACK

Consider an $M/M/k$ queuing system having arrival rate λ, service rate μ, and an infinite capacity. We assume further that, after a customer has received service, there is a probability α that she returns, or is *fedback*, to the end of the queue and waits for another round of service. For each customer, the total number of rounds of service she receives has a geometric distribution with mean $1/(1-\alpha)$, and her expected total service time is therefore $1/(\mu(1-\alpha))$.

Let us first define the system as including the fedback customers (system A in Figure 12.9.1). As the outsiders of this system, who are concerned with the total number of customers in it, it does not matter to us whether a customer, after receiving a round of service, returns to the end of the queue and waits or to the front of the queue and immediately receives another round of service.

If a customer immediately receives another round, her total service time will be exponentially distributed with an *effective service rate* $\mu(1-\alpha)$. (Why?) We therefore can study this system as a normal $M/M/k$ queue. All the results relating to the limiting behavior of the chain $\{\ell(t)\}_{t\in[0,\infty)}$ are still applicable, and the stream of customer departures from this system is Poisson with rate λ as $t \to \infty$.

However, the problem becomes harder if we exclude the fedback customers from our system (system B in Figure 12.9.1). This view forces us to distinguish between the stream of *exogenous arrivals* (which is Poisson with rate λ) and the stream of *total arrivals* (which is the superposition of the stream of exogenous arrivals and the stream of feedbacks). We also have to distinguish between the stream of *departures* (which is Poisson with rate λ) and the stream of *service completions* (which can be decomposed into the stream of departures and the stream of feedbacks).

It is important for our future discussion to note that the feedback stream is not Poisson. It is dependent on the number of customers in the system in a complicated way. An extreme example is when the queue has a very short expected service time and a very long expected interarrival time. Here, a customer loops rapidly through the system and then departs; a long duration later, another customer arrives and loops rapidly and so on. The stream of feedbacks thus has *clusters*, and therefore is not Poisson.

Note that the stream of feedbacks is a random decomposition of the stream of service completions, not of the stream of departures. So although the latter is Poisson, the stream of feedback is not.

Not only is the stream of total arrivals non-Poisson, but its long-term rate, as defined in (12.5.2), is also not λ, but is $\Lambda = \lambda/(1-\alpha)$. This is because $\Lambda = \lambda + \alpha\Lambda$. (Why?)

FIGURE 12.9.1 ***M/M/k* queues with feedback.**

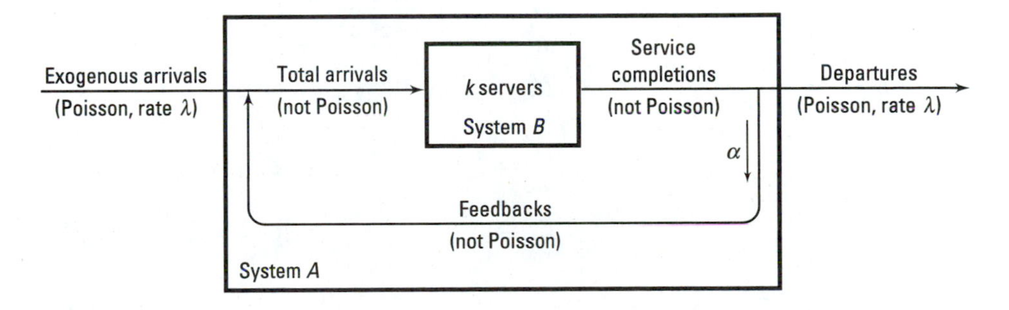

12.10 $M^{[G]}/M/1$ QUEUES

So far, we have seen queuing systems in which the chain $\{\ell(t)\}_{t\in[0,\infty)}$ is a birth-and-death process. Let us now look at one example in which it is not, although it is still a continuous-time Markov chain. This happens when the customers arrive in *groups* (or *batches*) as a *compound Poisson stream* (§8.16); that is, the groups arrive at the system as a Poisson stream with rate λ, and the size of each group is a random variable G. Assuming one server and no limit on system capacity, we denote this system by $M^{[G]}/M/1$.

For illustration, assume that the group size G can only be one or two, or

$$\Pr\{G = 1\} \equiv g_1 \qquad \Pr\{G = 2\} \equiv g_2 = 1 - g_1 \qquad \Pr\{G > 2\} = 0$$

The Poisson stream of group arrivals can now be decomposed into two independent Poisson streams: one comprising all single arrivals (with rate λg_1) and the other all double arrivals (with rate λg_2).

The process $\{\ell(t)\}_{t\in[0,\infty)}$ is a competitive continuous-time Markov chain. If $\ell(t) = n > 0$, there are three Poisson streams competing with each other: one with rate λg_2 assigned to state $n+2$, one with rate λg_1 to state $n+1$, and one with rate μ to state $n-1$. The chain will move to whichever state has the first event occurrence. If $\ell(t) = 0$, only states 1 and 2 compete. The transition diagram of this chain is shown in Figure 12.10.1.

Because its value can increase by more than one, $\{\ell(t)\}_{t\in[0,\infty)}$ is not a birth-and-death process, so we don't have any existing results to adapt to. However, the method to obtain its limiting distribution is very similar to that of the birth-and-death processes in §11.11:

Step 0. We write the balance equations as:

State	Rate of Leaving = Rate of Entering
0	$\lambda\gamma_0 = \mu\gamma_1$
1	$(\lambda+\mu)\gamma_1 = \mu\gamma_2 + \lambda g_1\gamma_0$
$i > 1$	$(\lambda+\mu)\gamma_i = \mu\gamma_{i+1} + \lambda g_1\gamma_{i-1} + \lambda g_2\gamma_{i-2}$

Step 1. For all $i > 0$, we express all γ_i in terms of γ_0 and $\rho \equiv \lambda/\mu$:

$$\begin{aligned}
\frac{\gamma_1}{\gamma_0} &= \rho \\
\frac{\gamma_2}{\gamma_0} &= g_2\rho + \rho^2 \\
\frac{\gamma_3}{\gamma_0} &= 2g_2\rho^2 + \rho^3 \\
\frac{\gamma_4}{\gamma_0} &= g_2^2\rho^2 + 3g_2\rho^3 + \rho^4 \\
\frac{\gamma_5}{\gamma_0} &= 3g_2^2\rho^3 + 4g_2\rho^4 + \rho^5 \\
\frac{\gamma_6}{\gamma_0} &= g_2^3\rho^3 + 6g_2^2\rho^4 + 5g_2\rho^5 + \rho^6 \\
&\vdots
\end{aligned}$$

FIGURE 12.10.1 Transition diagram of chain $\{\ell(t)\}_{t\in[0,\infty)}$ in an $M^{[G]}/M/1$ queue.

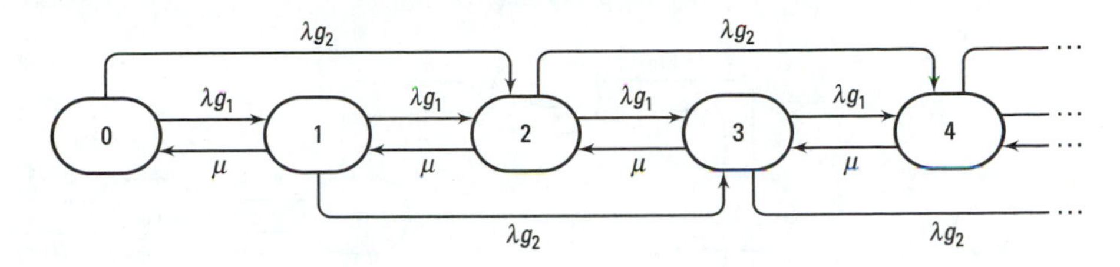

Generally, it can be shown that

$$\frac{\gamma_i}{\gamma_0} = \sum_{k=[i/2]}^{i} \binom{k}{i-k} (g_2)^{i-k} \rho^k \quad \text{for all } i \geq 2$$

where $[i/2]$ is the smallest integer that is larger than, or equal to, $i/2$.

Step 2. Adding all these equations side by side and then applying Equation (11.10.4) yields:

$$\frac{1}{\gamma_0} = 1 + (1+g_2)\rho + (1+g_2)^2\rho^2 + (1+g_2)^3\rho^3 + (1+g_2)^4\rho^4 + \cdots$$

$$= \sum_{i=0}^{\infty} (\rho\mathcal{E}[G])^i$$

where the *average group size* is $\mathcal{E}[G] = (g_1 + 2g_2) = (1 + g_2)$.

Thus,

$$\gamma_0 = \begin{cases} 1 - \rho\mathcal{E}[G] & \text{if } \rho\mathcal{E}[G] < 1 \\ 0 & \text{if } \rho\mathcal{E}[G] \geq 1 \end{cases}$$

While λ is the arrival rate of the groups, $\lambda\mathcal{E}[G]$ is the *arrival rate of the customers*. Again, the stability condition $\rho\mathcal{E}[G] < 1$ simply requires that the arrival rate $\lambda\mathcal{E}[G]$ of the customers is strictly less than the system's service rate μ.

12.11 *M/M/2* QUEUES WITH SERVER PREFERENCE

So far, we have only seen examples in which $\{\ell(t)\}_{t\in[0,\infty)}$ is a continuous-time Markov chain. We now give an example in which it is not, but can be made one by modifying its state space.

Assume that an $M/M/2$ queue has two servers A and B. When an arriving customer sees an empty system, she will go to server A with probability α and to server B with probability $1-\alpha$. Assume A serves with rate μ_A and B with rate μ_B.

The information that $\ell(t) = 1$ alone does not help us predict the future because we do not know which server is serving the only customer or which service rate this customer is receiving. The process $\{\ell(t)\}_{t\in[0,\infty)}$, with state space $\{0, 1, 2, \ldots\}$ is therefore not Markovian. To make it Markovian, we need to split state 1 into states A (indicating

FIGURE 12.11.1 **Transition diagram of chain $\{\ell(t)\}_{t\in\infty}$ in an $M/M/2$ queue with server preference.**

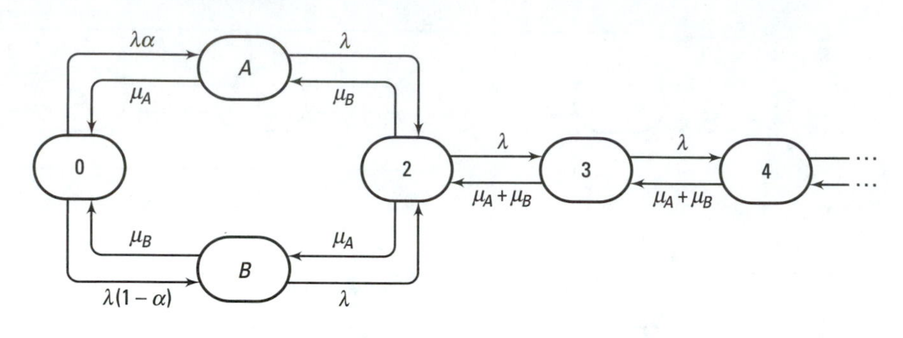

that the customer is being served by server A) and B (indicating that the customer is being served by server B). Now we have a continuous-time Markov chain. Its transition diagram is shown in Figure 12.11.1. The balance equations and the limiting probabilities can now be obtained in the normal manner (Problem 12.25).

12.12 $M/M/1$ QUEUES WITH N-POLICY

Here is an example in which $\{\ell(t)\}_{t\in[0,\infty)}$ is not a continuous-time Markov chain, but can be made one by introducing a supplementary variable. Consider an $M/M/1$ queuing system in which, every time the server is empty, it waits until there are N customers in the system before serving again. However, once the server is working, it continues to serve even when the number of customers in the system falls below N. It only stops when the system becomes empty.

Here, the information that $\ell(t) = i < N$ is not sufficient for us to predict the future. We need to know whether the server is still waiting for another $N - i$ customers to begin service or the service has already begun. To have a continuous-time Markov chain, we need to introduce the following supplementary variable when $0 < \ell(t) < N$:

$$I(t) = \begin{cases} 0 & \text{if the server is waiting at time } t \\ 1 & \text{if the server is serving at time } t \end{cases}$$

Now the vector chain $\{\ell(t), I(t)\}_{t\in[0,\infty)}$ is a continuous-time Markov chain. Its state space is $\{0, (1, 0), (1, 1), \ldots, (N-1, 0), (N-1, 1), N, (N+1), \ldots\}$.[5]

If $N = 3$, the transition diagram of this chain is shown in Figure 12.12.1. We leave it to the reader to derive the balance equations and the limiting probabilities for this system (Problem 12.32).

---

[5]Sometimes, the distinction between the method of introducing supplementary variables and the method of refining the state space is not clear. In this example, we can say that state $i < N$ was split into two states $(i, 0)$ and $(i, 1)$.

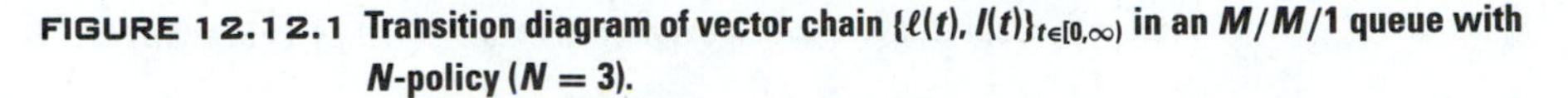

FIGURE 12.12.1 Transition diagram of vector chain $\{\ell(t), I(t)\}_{t\in[0,\infty)}$ in an $M/M/1$ queue with N-policy ($N = 3$).

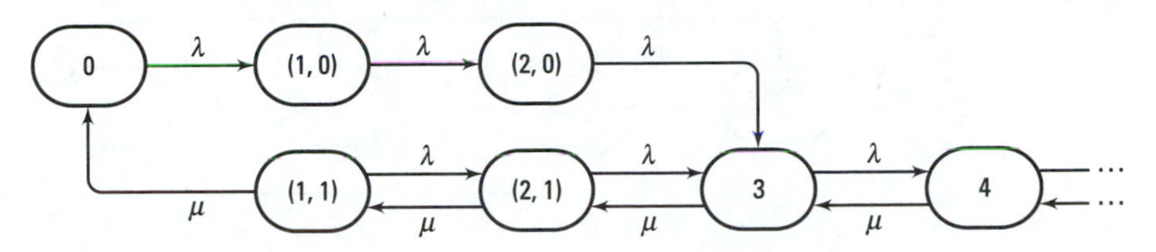

12.13 $M/M^{[G]}/1$ QUEUES

In the following queuing system, we demonstrate the advantages of *lumping* the states (§2.17). Here, the server in an $M/M/1$ queue is able to serve *up to two customers simultaneously*. After a service completion, if the server sees at least two customers waiting, it will serve two customers simultaneously with rate μ, and they will depart from the system at the same time. If it sees only one customer waiting, it will serve this customer alone with the same rate μ. One example might be a double-chair ski lift.

To study the chain $\{\ell(t)\}_{t\in[0,\infty)}$, we need to include a supplementary variable

$$\sigma(t) = \begin{cases} 0 & \text{if the server is idle at time } t \\ 1 & \text{if the server is serving one customer at time } t \\ 2 & \text{if the server is serving two customers at time } t \end{cases}$$

Figure 12.13.1 gives the transition diagram of the vector chain $\{\ell(t), \sigma(t)\}_{t\in[0,\infty)}$.

A simpler chain would be obtained if we lump states $(i, 1)$ and $(i+1, 2)$ together for all $i \geq 1$. This yields the chain $\{\ell_q(t)\}_{t\in[0,\infty)}$, where $\ell_q(t)$ is the number of customers

FIGURE 12.13.1 Transition diagram of the vector chain $\{\ell(t), \sigma(t)\}_{t\in[0,\infty)}$ in an $M/M^{[G]}/1$ queue.

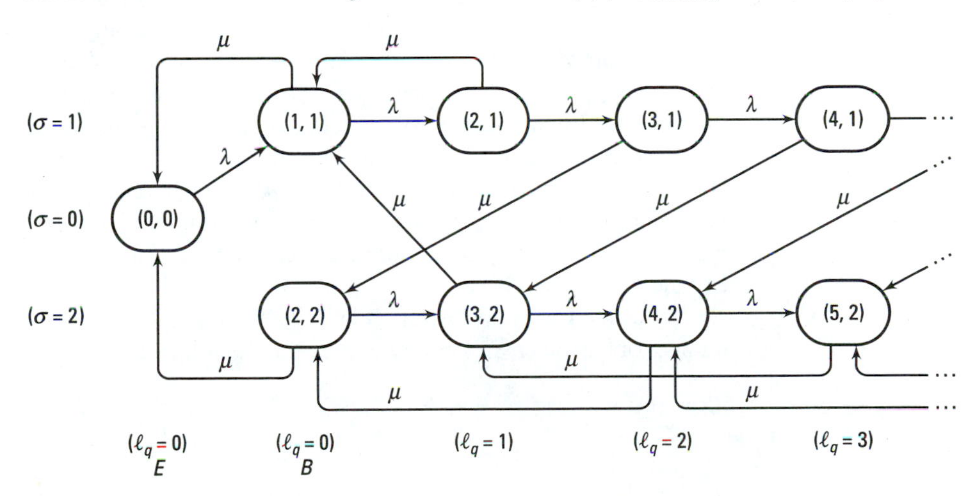

FIGURE 12.13.2 Transition diagram of chain $\{\ell_q(t)\}_{t\in[0,\infty)}$ in an $M/M^{[G]}/1$ queue.

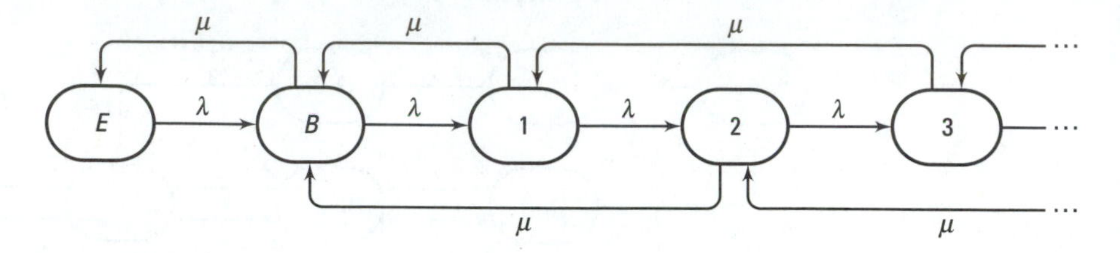

waiting for service at time t. Its state space is $\{E, B, 1, 2, \ldots\}$, where state $E \equiv (0, 0)$ represents the empty system, and state B is state $(1, 1)$ or state $(2, 2)$ representing the empty queue together with one or two customers being served. The transition matrix of this chain is shown in Figure 12.13.2.

In lumping the states, we lose the information about the individual states. If we are looking for the limiting distribution of $\{\ell(t)\}_{t\in[0,\infty)}$, then chain $\{\ell_q(t)\}_{t\in[0,\infty)}$, although simpler than chain $\{\ell(t)\}_{t\in[0,\infty)}$, is of no use. In this particular example, however,

$$\lim_{t\to\infty} \Pr\{\ell(t) = 0\} = \lim_{t\to\infty} \Pr\{\ell_q(t) = E\}$$

Furthermore, we can obtain L of chain $\{\ell(t)\}_{t\in[0,\infty)}$ as follows. First, we obtain L_q from chain $\{\ell_q(t)\}_{t\in[0,\infty)}$. We then calculate $W_q = L_q/\lambda$ using Little's result (12.5.3). From that, $W = W_q + 1/\mu$. Finally, $L = \lambda W$ as in Equation (12.5.1) (Problem 12.33).

12.14 QUEUES IN SERIES WITH BLOCKING

So far, we have considered queuing systems in which there is only one waiting line. We now depart from this type and move toward the *queuing networks*, having more than one *station*, and each station has its own waiting line. Departing from one station, a customer may proceed to another station to receive additional service before leaving the network.

We start by studying one of the simplest forms of queuing networks, in which there are two single-server stations 1 and 2 *in series*.[6] Both have *no waiting room*. The customers arrive at station 1 in a random manner with rate λ. If an arriving customer finds server 1 busy, she will be turned away from the system; otherwise, she will be served immediately by the single server in station 1, with an exponential rate μ_1. After being served in station 1, she moves immediately to station 2 if server 2 is free; otherwise, since 2 has no waiting room, she is *blocked* and must remain at station 1 until station 2 is vacant, thus preventing station 1 from accepting any new customer. This customer will depart from the system after being served by the single server in station 2, with an exponential rate μ_2.

Let us start with the vector chain $\{\ell_1(t), \ell_2(t)\}_{t\in[0,\infty)}$, where $\ell_i(t)$ $(i = 1, 2)$ is the number of customers in station i at time t. Its state space is $\{(0, 0), (1, 0), (0, 1), (1, 1)\}$.

---

[6]Some authors say they are *in tandem*.

FIGURE 12.14.1 Transition diagram of vector chain $\{\ell_1(t), \ell_2(t)\}_{t\in[0,\infty)}$ in two $M/M/1$ queues in series with blocking.

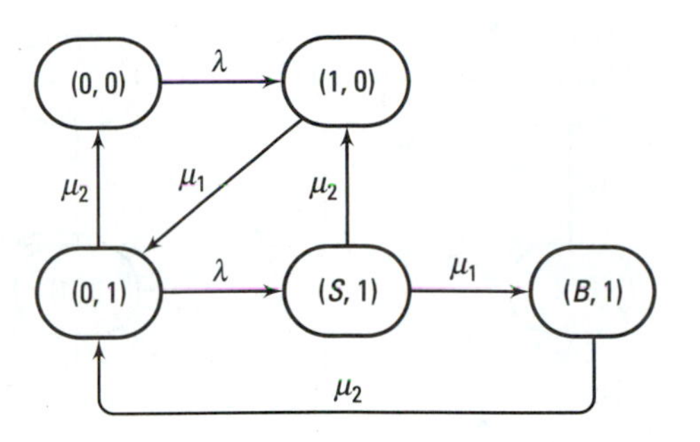

This chain is not Markovian because state (1,1) does not indicate whether the customer in station 1 is being served or blocked. To make it Markovian, we must split state (1, 1) into two states $(S, 1)$ and $(B, 1)$, where $(S, 1)$ indicates that the customer at station 1 is being served and $(B, 1)$ is being blocked. The transition diagram is in Figure 12.14.1.

12.15 QUEUES IN SERIES WITHOUT BLOCKING

We now study the same model as in the previous section but give infinite capacities to both stations 1 and 2. Here, the customers are never turned away when server 1 is busy or blocked when server 2 is busy. Immediately after being served by server 1, he moves to station 2 and waits there if necessary.

Now the infinite vector chain $\{\ell_1(t), \ell_2(t)\}_{t\in[0,\infty)}$, with state space $\{(0, 0), (1, 0), (0, 1), (1, 1), \ldots\}$, is a continuous-time Markov chain. Its transition diagram is shown in Figure 12.15.1.

For all $i, j > 0$, we can write the balance equations as

State	Rate of Leaving = Rate of Entering
$(0, 0)$	$\lambda\gamma_{0,0} = \mu_2\gamma_{0,1}$
$(i, 0)$	$(\lambda + \mu_1)\gamma_{i,0} = \mu_2\gamma_{i,1} + \lambda\gamma_{i-1,0}$
$(0, j)$	$(\lambda + \mu_2)\gamma_{0,j} = \mu_2\gamma_{0,j+1} + \mu_1\gamma_{1,j-1}$
(i, j)	$(\lambda + \mu_1 + \mu_2)\gamma_{i,j} = \mu_2\gamma_{i,j+1} + \mu_1\gamma_{i+1,j-1} + \lambda\gamma_{i-1,j}$

Assuming $\rho_1 \equiv \lambda/\mu_1 < 1$ and $\rho_2 \equiv \lambda/\mu_2 < 1$, the reader can verify the following limiting *joint probabilities*:

$$\begin{aligned}\gamma_{ij} &\equiv \lim_{t\to\infty} \Pr\{\ell_1(t) = i, \ell_2(t) = j\} \\ &= (1 - \rho_1)(\rho_1)^i(1 - \rho_2)(\rho_2)^j \qquad \text{for all } i, j \geq 0 \qquad (12.15.1)\end{aligned}$$

FIGURE 12.15.1 Transition diagram of chain $\{\ell_1(t), \ell_2(t)\}_{t\in[0,\infty)}$ in two $M/M/1$ queues in series without blocking.

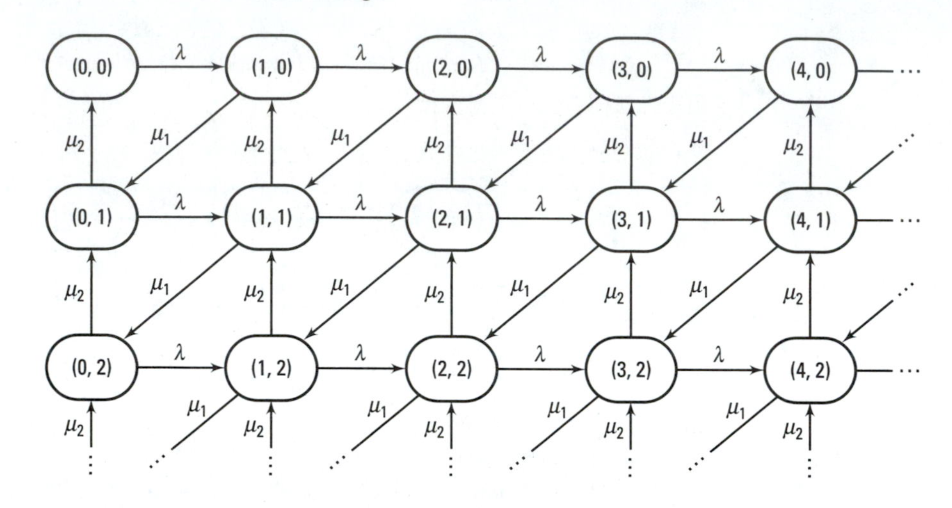

Should we be surprised by this equation? Not if we remember from §12.8 that, as $t \to \infty$, the departure stream from station 1 is Poisson with rate λ, independent of the number of customers in station 1. This departure stream in turn is the arrival stream to station 2. Consequently, station 2 is independent of station 1 and can be studied as an independent $M/M/1$ queue, or

$$\gamma_{ij} = \lim_{t\to\infty} \Pr\{\ell_1(t) = i\} \lim_{t\to\infty} \Pr\{\ell_2(t) = j\}$$

Equation (12.15.1) can now be obtained from Equation (12.8.3). Note that, for stability, we need $\mu_1, \mu_2 > \lambda$.

This *product-form result* can be generalized to a network of N stations in series in which station i ($i = 1, \ldots, N$) has an infinite capacity and has k_i servers, all serving with the same rate μ_i. The customers arrive at station 1 first from outside the network as a Poisson stream with rate λ. They then have to go through all stations in *sequential* order, waiting at each if necessary, before leaving the system.

Here, as $t \to \infty$, each station develops independently of each other. All have the same arrival rate λ. If $\lambda < k_i\mu_i$ for all $i = 1, \ldots, N$, then the product-form result gives the joint probability of the state of the network simply as the product of the *marginal probabilities* $\lim_{t\to\infty} \Pr\{\ell_i(t) = j\}$.

12.16 ACYCLIC OPEN JACKSON NETWORKS

A generalization of queues in series is a queuing network in which the customers may move among the stations in any order. The general form of a *Jackson network* is as follows. There are N stations labeled $1, 2, \ldots, N$. Each station has an infinite capacity and thus can admit all arriving customers and allow them to wait for service if necessary. Station i has k_i servers; all serve with the same exponential rate μ_i. After being served at station i, there is a probability p_{ij} that the customer moves to station j for an additional round of service.

A Jackson network is *open* when station i may receive customers directly from outside the network as a Poisson stream with rate λ_i. If the number of stations is finite, then for system stability, each customer must leave the network eventually. The probability that a customer departs from the network immediately after receiving service at station i is $1 - \sum_{j=1}^{N} p_{ij}$.

In this section, we restrict ourselves to the *acyclic* open Jackson networks, in which the customers may not return to the same station. This requires that they cannot be fedback to station i instantaneously, or $p_{ii} = 0$. They also cannot be fedback to station i via other stations.

If we allow the customers to return to the same station, we know from §12.9 that the stream of the total arrivals to this station, including the feedback stream, is non-Poisson. The acyclic requirement assures that the feedback streams do not exist.

If the stream of total arrivals to station i is Poisson having rate Λ_i, then as $t \to \infty$, the stream of departures from station i is also Poisson with the same rate, independent of the numbers of customers at all stations. The stream of customers moving from station i to station j is therefore also Poisson with rate $p_{ij}\Lambda_i$.

The stream of total arrivals to station j is the superposition of the Poisson arrivals from outside the system (with rate λ_j) and all Poisson arrivals from other stations (with rate $\sum_{k=1}^{N} p_{kj}\Lambda_j$). For stability, its total arrival rate must be equal to its departure rate, or

$$\Lambda_j = \lambda_j + \sum_{k=1}^{N} p_{kj}\Lambda_k \quad \text{for all } j = 1, 2, \ldots, N \tag{12.16.1}$$

This set of *flow-balance equations* allows us to obtain the total arrival rate at each station.

For illustration, consider an acyclic open Jackson network having four stations 1, 2, 3, and 4, as depicted in Figure 12.16.1. Let $\lambda_1 = 3$ arrivals/hour, $\lambda_2 = 2$ arrivals/hour, $\lambda_3 = 4$ arrivals/hour, and $\lambda_4 = 5$ arrivals/hour. Also let

$$[p_{ij}] = \begin{array}{c} \\ 1 \\ 2 \\ 3 \\ 4 \end{array} \begin{array}{c} \begin{array}{ccccc} 1 & 2 & 3 & 4 & \text{leave} \end{array} \\ \begin{pmatrix} 0 & .3 & .4 & .1 & .2 \\ 0 & 0 & 0 & .6 & .4 \\ 0 & 0 & 0 & .3 & .7 \\ 0 & 0 & 0 & 0 & 1 \end{pmatrix} \end{array}$$

From Equation (12.16.1),

$$\begin{cases} \Lambda_1 = 3 \\ \Lambda_2 = 2 + .3\Lambda_1 \\ \Lambda_3 = 4 + .4\Lambda_1 \\ \Lambda_4 = 5 + .1\Lambda_1 + .6\Lambda_2 + .3\Lambda_3 \end{cases}$$

FIGURE 12.16.1 An acyclic open Jackson network.

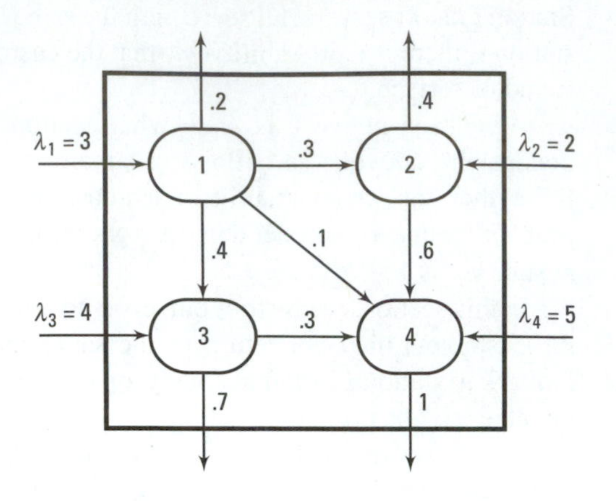

Thus, $\Lambda_1 = 3$ arrivals/hour, $\Lambda_2 = 2.9$ arrivals/hour, $\Lambda_3 = 5.2$ arrivals/hour, and $\Lambda_4 = 8.6$ arrivals/hour.

As $t \to \infty$, each station in an acyclic network behaves *independently* as an $M/M/k$ queue. If $\Lambda_i < k_i\mu_i$ for all $1 \le i \le N$, the limiting joint probabilities are simply the product of the limiting marginal probabilities.

Suppose each station in this network has a single server. Let the service rates be $\mu_1 =$ 5 services/hour, $\mu_2 = 4$ services/hour, $\mu_3 = 7$ services/hour, and $\mu_4 = 10$ services/hour. Then the limiting joint probability of seeing three customers in station 1, two in station 2, one in station 3, and four in station 4 is simply

$$\left(1-\frac{3}{5}\right)\left(\frac{3}{5}\right)^3\left(1-\frac{2.9}{4}\right)\left(\frac{2.9}{4}\right)^2\left(1-\frac{5.2}{7}\right)\left(\frac{5.2}{7}\right)\left(1-\frac{8.6}{10}\right)\left(\frac{8.6}{10}\right)^4$$

12.17 GENERAL OPEN JACKSON NETWORKS

Let us now consider an open Jackson network in which the customers may visit the same station more than once. The stream of total arrivals at each station now may include the stream of feedbacks and therefore *may not be Poisson.*

Does this mean that, before analyzing any open network, we have to check first to see whether it is acyclic or not? Fortunately, the answer is no. The *Jackson theorem* shows that, as far as the limiting joint probabilities of the number of customers in each station are concerned, we can treat any open network *as an acyclic network* and apply the *product-form result*. In other words, we can regard station i *as if it were an independent* $M/M/k_i$ *queue*, the total arrival rate Λ_i of which satisfies the flow-balance equations (12.16.1).

The proof of the Jackson theorem is cumbersome but not difficult (Jackson, 1957). All we need do is obtain the balance equations for the continuous-time Markov vector-chain $\{\ell_1(t), \ell_2(t), \ldots, \ell_N(t)\}_{t\in[0,\infty)}$ and then verify that the product-form result satisfies these balance equations and the normalization equation (11.10.4).

We need to remember that, for stability, we must have,

$$\Lambda_i < k_i \mu_i \quad \text{for all } 1 \le i \le N$$

For example, consider the network having three stations 1, 2, and 3. Let the arrival rates from outside the network be $\lambda_1 = 3$ arrivals/hour, $\lambda_2 = 2$ arrivals/hour, and $\lambda_3 = 4$ arrivals/hour. Also let

$$[p_{ij}] = \begin{array}{c} \\ 1 \\ 2 \\ 3 \end{array} \begin{array}{c} \begin{array}{cccc} 1 & 2 & 3 & \text{leave} \end{array} \\ \begin{pmatrix} 0 & 1/2 & 0 & 1/2 \\ 1/4 & 1/4 & 1/4 & 1/4 \\ 1/3 & 1/3 & 1/3 & 0 \end{pmatrix} \end{array}$$

Equation (12.16.1) gives:

$$\begin{cases} \Lambda_1 = 3 \qquad\qquad\qquad (1/4)\Lambda_2 + (1/3)\Lambda_3 \\ \Lambda_2 = 2 + (1/2)\Lambda_1 + (1/4)\Lambda_2 + (1/3)\Lambda_3 \\ \Lambda_3 = 4 + \qquad\quad + (1/4)\Lambda_2 + (1/3)\Lambda_3 \end{cases}$$

This yields $\Lambda_1 = 74/7$ arrivals/hour, $\Lambda_2 = 104/7$ arrivals/hour, and $\Lambda_3 = 81/7$ arrivals/hour.

Assume each station has one server, and the service rates are $\mu_1 = 14$ services/hour, $\mu_2 = 15$ services/hour, and $\mu_3 = 12$ services/hour. Then $\rho_1 = (74/7)/14$, $\rho_2 = (104/7)/15$, and $\rho_3 = (81/7)/12$.

The limiting joint probability of seeing two customers in station 1, four in station 2, and three in station 3 is simply:

$$\lim_{t\to\infty} \Pr\{\ell_1(t) = 2, \ell_2(t) = 4, \ell_3(t) = 3\} = (1-\rho_1)(\rho_1)^2(1-\rho_2)(\rho_2)^4(1-\rho_3)(\rho_3)^3$$

Equation (12.8.4) gives the limiting expected number of customers in the whole network as

$$L = \frac{\rho_1}{1-\rho_1} + \frac{\rho_2}{1-\rho_2} + \frac{\rho_3}{1-\rho_3}$$

Little's result now gives the limiting expected duration a customer spends in the network as $W = L/(\lambda_1 + \lambda_2 + \lambda_3)$.

12.18 SUMMARY

In this chapter,

1. we identified the basic parameters to describe a queuing system.
2. we presented Little's result (12.5.1) and its variations, which are applicable to many queuing models.
3. we studied queuing systems in which the arrival stream is Poisson and the service times are exponentially distributed. We demonstrated how they can be studied by constructing suitable continuous-time Markov chains.

4. we showed how the limiting joint probabilities of an open Jackson network can be obtained using the product-form result, in which each station is treated as being independent.

PROBLEMS

12.1 Consider an $M/M/k$ queuing system.

a. Obtain Equation (12.8.1) from Equation (11.11.2).

b. Use Equations (12.8.1) and (12.8.2) to show that the limiting probability of an arriving customer having to wait is

$$p_w = \frac{k\rho^k}{k!(k-\rho)}\gamma_0$$

c. Use Equation (12.8.1) and Appendix (A.10) to show that

$$L_q = \frac{\rho^{k+1}}{(k-1)!(k-\rho)^2}\gamma_0$$

d. Use Little's results to obtain a similar expression for L, W, and W_q.

e. Find γ_0, γ_i, L, L_q, W, and W_q when $k = 1$ and $k = 2$.

12.2 Consider an $M/M/\infty$ queuing system. Show that the limiting distribution of the chain $\{\ell(t)\}_{t\in[0,\infty)}$ is Poisson with parameter λ/μ.

12.3 Consider an $M/M/1$ queuing system.

a. Explain why the chain $\{\ell_q(t)\}_{t\in[0,\infty)}$ is not Markovian.

b. Obtain Equation (12.8.3) from Equation (12.8.1).

c. By conditioning on the number of customers in the system at the time of an arrival, show that

$$\lim_{n\to\infty} \Pr\{w_n \geq t\} = \exp\{-(\mu-\lambda)t\} \quad \text{for all } t \geq 0$$

Hint: Use the Erlang distribution (8.12.1) and Appendix (A.2).

d. Hence, find W. *Hint:* Use Problem 1.51.

e. By conditioning on the number of customers in the system at the time of an arrival, show that

$$\lim_{n\to\infty} \Pr\{q_n \geq t\} = \rho\exp\{-(\mu-\lambda)t\} \quad \text{for all } t > 0$$

f. Hence, find W_q.

g. Let $\$b\mu$ be the cost of the server per hour, idle or not, and $\$c$ be the cost of each customer per hour in the system. Find the service rate that minimizes the total cost.

12.4 In an $M/M/1$ queuing system, let d_n be the number of customers in the system immediately after the departure of the nth customer.

a. Find the transition matrix of $\{d_n\}_{n=0,1,2,\ldots}$. Show that it has the form of Problem 7.12a. *Hint:* Use Problem 8.48d.

b. Hence, use Problem 7.15b to show that the limiting distribution of $\{d_n\}_{n=0,1,2,\ldots}$ is the same as that of $\{\ell(t)\}_{t\in[0,\infty)}$.

12.5 Consider a system having two independent $M/M/1$ queuing stations, each with its own waiting line. Let the arrival rate to each station be $\lambda/2$ and the service rate of each server be μ. We now *pool* the two stations together and thus form an $M/M/2$ queuing system with one waiting line. The arrival rate to the pooled system is therefore λ, and the service rate of each server remains at μ. Compare the two systems in terms of L, L_q, W, and W_q. Explain why pooling improves the system's performance.

12.6 Refer to Problem 12.5. It is now suggested that, instead of having two servers in the pooled system, each serving with rate μ, we replace them with a single server working with rate 2μ. The arrival rate of this new $M/M/1$ system is still λ. Compare the two systems in terms of L, L_q, W, and W_q.

12.7 For an $M/M/1$ queuing system, use Problem 8.52 to show that the limiting time until the next departure has an exponential distribution with rate λ.

12.8 Consider an $M/M/2$ queuing system.

a. Given $\ell(t) = 0$, find the expected duration until the next departure after t. *Hint:* Use Problem 8.21.

b. Given $\ell(t) = 1$, find the expected duration until the next departure after t.

c. Given $\ell(t) > 1$, find the expected duration until the next departure after t.

d. Hence, use the results obtained in Problem 12.1e to find the expected duration until the next departure after t as $t \to \infty$.

12.9 Consider an $M/M/k/N$ queuing system.

a. Show that

$$\gamma_i = \begin{cases} \dfrac{\rho^i}{i!}\gamma_0 & \text{for all } 0 \le i \le k \\ \dfrac{\rho^i}{k^{i-k}k!}\gamma_0 & \text{for all } k < i \le N \end{cases}$$

where

$$\gamma_0 = \left[\frac{\rho^k}{k!}\sum_{j=0}^{N-k}\left(\frac{\rho}{k}\right)^j + \sum_{i=0}^{k-1}\frac{\rho^i}{i!}\right]^{-1}$$

b. When $N \to \infty$, obtain Equations (12.8.1) and (12.8.2).

c. When $N = k$, obtain γ_i for all $0 \le i \le k$. This is known as *Erlang's loss formula*. Find all γ_is when $N = k = 2$.

d. In an $M/M/1/N$ queuing system, find L. (Note that the result is different when $\lambda = \mu$.) Hence, find the expected time spent in the system

i. of all arriving customers.

ii. of those who actually entered the queue. *Hint:* Use Appendix (A.9).

e. In an $M/M/1/3$ queuing system, find γ_i for all $i = 1, 2, 3$.

12.10 In an $M/M/1/3$ queuing system, let d_n denote the number of customers left behind at the departure of the nth customer. Find the transition matrix of the discrete-time chain $\{d_n\}_{n=0,1,2,\ldots}$ and its limiting distribution.

12.11 Passengers arrive at a taxi station singly as a Poisson stream with rate λ. Taxis arrive at the same station singly as a Poisson stream with rate μ. The arriving passenger will wait if there is no taxi available; otherwise, she will be taken away immediately by a waiting taxi. The arriving taxi will wait if there is no waiting passenger. Define a continuous-time Markov chain to obtain the limiting distributions of the numbers of passengers and taxis at the station. Draw its transition diagram. In the following cases, find the condition of recurrence and the limiting distributions of the numbers of passengers and taxis at the station:

a. There is no limit on the number of waiting passengers and taxis.
b. Only M taxis can wait, but there is no limit on the number of waiting passengers.
c. There is no limit on the number of waiting taxis, but only N passengers can wait.
d. Only M taxis and N passengers can wait.

12.12 Consider an $M/M/k$ queuing system in which the servers help each other. If there are $i < k$ customers in the system, the servers are divided into i groups and each group serves a customer. If there are $i \geq k$ customers in the system, only k customers can be served by the k servers. Define a continuous-time Markov chain to obtain the limiting distribution of the number of customers in the system.

12.13 Consider an $M/M/1$ queuing system in which the server works at a different rate whenever there are fewer than k customers in the system. More specifically, if there are fewer than k customers in the system, the service rate is μ_1; otherwise, it is μ. Find the limiting distribution of $\{\ell(t)\}_{t\in[0,\infty)}$. Why does the condition for positive-recurrence depend on λ and μ but not on μ_1?

12.14 Consider an $M/M/1$ queuing system with *reneging* in which a waiting customer can leave the waiting line and depart from the system at any time. For a customer who is waiting in the queue at time t, assume that the probability that he leaves the system during $(t, t+h)$ is αh as $h \to 0$. Consider chain $\{\ell(t)\}_{t\in[0,\infty)}$.

a. Draw its transition diagram.
b. Hence, obtain the following recursive relation:

$$(i\alpha + \mu)\gamma_{i+1} = \lambda\gamma_i \quad \text{for all } i = 1, 2, 3, \ldots$$

c. When $\alpha = \mu$, find the limiting distribution. *Hint:* Use Problem 12.2.

12.15 Consider an $M/M/1$ queuing system with *balking* in which an arriving customer might get discouraged by the number of other customers already in the system and thus does not enter. More specifically, given i customers in the system, we assume that the arrival rate is $\lambda_i = \lambda/(i+1)$.

a. Find the limiting distribution of $\{\ell(t)\}_{t\in[0,\infty)}$. Hence, find L. *Hint:* Use Problem 11.26.
b. Find the expected time spent in the system of all arriving customers.
c. Find the expected time spent in the system of those who actually enter the system. *Hint:* Use Appendix (A.3).

12.16 Consider an $M/M/k/k$ queuing system having two types of customers, X and Y. Type i customers arrive as a Poisson stream with rate λ_i ($i = X, Y$). If an arriving type X customer finds all servers busy, she will leave the system without receiving service. If an arriving type Y customer finds more than n busy servers, he will leave the system without receiving service. Let the service rate of each server be μ.

a. Obtain the limiting distribution of chain $\{\ell(t)\}_{t \in [0,\infty)}$.
b. Hence, find the limiting probability that a type X customer does not receive service.
c. Find the limiting probability that a type Y customer does not receive service.

12.17 Consider an $M/M/1/1$ queuing system in which the server may *break down*. Immediately after each breakdown, it is repaired. Let the repair times and service times be exponentially distributed with rates δ and μ, respectively. The server only breaks down when serving. If serving at time t, the probability of its breakdown during $(t, t+h)$ is βh as $h \to 0$. Immediately after each breakdown, the customer being served departs from the system without receiving complete service. No customer is allowed to enter the system when the server is being repaired.
a. Find the proportion of time that the system is empty
i. by studying a continuous-time Markov chain
ii. by studying a semi-Markov chain
iii. by studying a regenerative process
b. Hence, find the probability that an arriving customer receives complete service.

12.18 Refer to Problem 12.17 in which the server in an $M/M/1/1$ queuing system may break down when serving (with rate β). Assume now that, immediately after each breakdown, the customer being served does not depart from the system but stays there to wait for the continuation of service after the server is repaired.
a. Find the proportion of time that the system is empty
i. by studying a continuous-time Markov chain
ii. by studying a semi-Markov chain
iii. by studying a regenerative process
b. Hence, find the probability that an arriving customer receives complete service.

12.19 Refer to Problem 12.17 in which the server in an $M/M/1/1$ queuing system may break down when serving (with rate β). Assume now that one customer is allowed to enter the system to wait when the server is being repaired.
a. Find the proportion of time that the system is empty
i. by studying a continuous-time Markov chain
ii. by studying a semi-Markov chain
iii. by studying a regenerative process
b. Hence, find the probability that an arriving customer receives complete service.

12.20 Refer to Problem 12.17 in which the server in an $M/M/1/1$ queuing system may break down when serving (with rate β). Assume now that it may also break down when idle. If the server is idle at time t, the probability of its breaking down during $(t, t+h)$ is ιh as $h \to 0$. Also, one customer is allowed to enter the system to wait when the server is being repaired. Construct a continuous-time Markov chain to find the probability that an arriving customer will receive complete service. Do not try to solve the balance equations.

12.21 Refer to Problem 12.20 in which the server in an $M/M/1$ queuing system may break down when serving (with rate β) or when idle (with rate ι). Assume now that there is an infinite waiting room. Also, immediately after each breakdown, the customers do not depart from the system but stay there to wait for service, or the continuation of service, after the server is repaired. All arriving customers are allowed to enter the system to wait for service. Construct a continuous-time Markov chain to find the proportion of time that the system is empty. Do not try to solve the balance equations.

12.22 Refer to Problem 12.20 in which the server in an $M/M/1$ queuing system may break down when serving (with rate β) or when idle (with rate ι). Assume now that there is an infinite waiting room. Also, immediately after each breakdown, all customers are lost (as in the case of a computer processor). All arriving customers are allowed to enter the system to wait for service. Construct a continuous-time Markov chain to find the proportion of time that the system is empty. Do not try to solve the balance equations.

12.23 Consider the $M/M/2$ queuing system with server preference studied in §12.11. Assume now that $\mu_A = \mu_B$ and there is only room for two customers. Find the limiting distribution of $\{\ell(t)\}_{t\in[0,\infty)}$

a. for a general value of α. *Hint:* Use Problem 12.9c.

b. when $\alpha = 1$.

c. when $\alpha = .5$.

12.24 Refer to Problem 12.23. Assume now that, besides the customers already described, there is another type of customer who arrives at the system with rate λ_B and can only be served by server B. Upon arrival, if they find server B idle, they will be served by B; otherwise, they will leave the system without receiving service. Define a continuous-time Markov chain to find the limiting distribution of $\{\ell(t)\}_{t\in[0,\infty)}$. Write the balance equations but do not solve them.

12.25 Refer to the $M/M/2$ queuing system having an infinite waiting room and with server preference studied in §12.11. Assume now that $\mu_A = \mu_B$.

a. Find the limiting distribution of $\{\ell(t)\}_{t\in[0,\infty)}$.

b. Compare with Problem 12.23.

12.26 Consider the $M/M/2$ queuing system with server preference studied in §12.11. Let $\alpha = 1$. Also, assume that whenever server A becomes idle, the customer being served by B immediately switches to A. Find the condition for positive-recurrence for the chain $\{\ell(t)\}_{t\in[0,\infty)}$ and its limiting distribution if it exists.

12.27 Consider an $M/M/1$ queuing system in which the server needs a *setup* time; that is, when an arriving customer finds the server idle, she has to wait for it to be set up before receiving service. The setup times are exponentially distributed with rate σ. Define a continuous-time Markov chain to obtain the limiting distribution of the number of customers in the system. Draw its transition diagram. Do not try to solve the balance equations.

12.28 Consider an $M/M/1$ queuing system in which the server goes on vacation immediately when the system becomes empty. Upon returning from a vacation, if it sees the system empty, it will immediately take another vacation. Assume the duration of each vacation is exponentially distributed with rate υ. Define a continuous-time Markov chain to obtain the limiting distribution of the number of customers in the system. Draw its transition diagram. Do not try to solve the balance equations.

12.29 Refer to Problem 12.28 in which the server goes on vacation. Assume now that it takes a vacation only once each time. Upon returning from a vacation, if it sees the system empty, it will stay and wait for the next arriving customer. Define a continuous-time Markov chain to obtain the limiting distribution of the number of customers in the system. Draw its transition diagram. Do not try to solve the balance equations.

12.30 A fax machine can only receive one incoming fax at a time. Assume faxes arrive as a Poisson stream with rate λ. If an incoming fax finds the machine idle, it will be processed by the machine immediately. The process time is exponentially distributed with rate μ. If an incoming fax finds the machine busy, it will be sent again later, after an exponential duration with rate θ. Construct a continuous-time Markov chain to find the limiting distribution of the number of faxes waiting to be re-sent at time t. Draw its transition diagram. Do not try to solve the balance equation.

12.31 Consider a queuing system in which customers arrive as a Poisson stream with rate λ. At the beginning of her service, independent of anything else, a customer must decide whether she wants type I service (having rate μ_α) or type II service (having rate μ_β). Let the probability she picks the former be α. Construct a continuous-time Markov chain to find the limiting distribution of the total number of customers in the system. Draw its transition diagram. Do not try to solve the balance equations.

12.32 Consider the $M/M/1$ queuing system with N-policy studied in §12.12. Let $N = 2$.

- **a.** Find the condition for positive-recurrence for the chain $\{\ell(t)\}_{t\in[0,\infty)}$.
- **b.** For the rest of this problem, assume that this chain is positive-recurrent. Find its limiting distribution.
- **c.** Hence, find L. *Hint:* Use Appendix (A.10).

12.33 Consider the $M/M^{[G]}/1$ queuing system studied in §12.13.

- **a.** Find the condition for positive-recurrence for the chain $\{\ell_q(t)\}_{t\in[0,\infty)}$. *Hint:* Try $\gamma_i = c\alpha^i \gamma_E$. Then determine c and α.
- **b.** For the rest of this problem, assume that this chain is positive-recurrent. Find its limiting distribution.
- **c.** Hence, find L_q.
- **d.** Show that the limiting probability that the server serves a single customer is $(\lambda/\mu)\gamma_E + \gamma_1$, where $\gamma_1 \equiv \lim_{t\to\infty} \Pr\{\ell_q(t) = 1\}$. *Hint:* Consider chain $\{\sigma(t)\}_{t\in[0,\infty)}$, where $\sigma(t)$ is the number of customers being served concurrently at time t.

12.34 An $M/M/1$ queuing system serves two types of customers, X and Y. A type X customer has *priority* over a type Y customer: If a type X customer arrives when the server is serving a type Y customer, the server will discontinue its service to the latter and serve the former immediately. (This is called a *preemptive* system.) In such a case, the type Y customer will remain in the system to receive additional service later. If an arriving type X customer finds the server serving another type X customer, he will wait for his turn ahead of other waiting type Y customers. If an arriving type Y customer finds the server busy, she will wait for her turn at the end of the queue. For $i = X, Y$, let the interarrival times and the service times of type i customers be exponentially distributed with rates λ_i and μ_i, respectively; let $\ell_i(t)$ be the numbers of type i customers in the system at time t.

a. Obtain the limiting distribution for the chain $\{\ell_X(t)\}_{t\in[0,\infty)}$.

b. Obtain the balance equations for the vector chain $\{\ell_X(t), \ell_Y(t)\}_{t\in[0,\infty)}$.

c. Show by induction that, for all $j \geq 0$,

$$\lambda_Y \lim_{t\to\infty} \sum_{i=0}^{\infty} \Pr\{\ell_Y(t) = j\} = \mu_Y \lim_{t\to\infty} \Pr\{\ell_X(t) = 0, \ell_Y(t) = j+1\}$$

d. Hence, show that

$$\lim_{t\to\infty} \Pr\{\ell_X(t) = 0, \ell_Y(t) = 0\} = 1 - \lambda_X/\mu_X - \lambda_Y/\mu_Y$$

12.35 Consider a queuing system with preemptive priority as in Problem 12.34. However, assume now that only one customer is allowed to wait. Thus, we have an $M/M/1/2$ queuing system. Let $\lambda_X = 5$, $\mu_X = 7$, $\lambda_Y = 6$, and $\mu_Y = 3$. Find the limiting distributions of the numbers of type X and type Y customers in the system.

12.36 An $M/M/1/3$ queuing system serves two types of customers, X and Y. A type X customer has preemptive priority over a type Y customer. If a type Y customer is preempted, she will return to the end of the queue to receive additional service later. Waiting type X customers will be served before waiting type Y customers. If an arriving type X customer finds three customers in the system, one waiting type Y customer will have to leave, if there is any; if not, this type X customer will leave. For $i = X, Y$, let the interarrival times and the service times of type i customers be exponentially distributed with rates λ_i and μ_i, respectively. Define a continuous-time Markov chain to obtain the limiting distributions of the numbers of type X and type Y customers. Draw its transition diagram. Do not try to solve the balance equations.

12.37 Consider an $M/M/\infty$ queuing system having two types of customers, X and Y. For $i = X, Y$, let the interarrival times of type i customers be exponentially distributed with rate λ_i, the service times of type i customers be exponentially distributed with rate μ_i, and the numbers of type i customers in the system at time t be $\ell_i(t)$.

a. Obtain the balance equations for the continuous-time Markov vector chain $\{\ell_X(t), \ell_Y(t)\}_{t\in[0,\infty)}$.

b. Verify that the following product-form satisfies all balance equations:

$$\gamma_{ij} \equiv \lim_{t\to\infty} \Pr\{\ell_X(t) = i, \ell_Y(t) = j\} = c\left(\frac{1}{i!}\rho_X^i\right)\left(\frac{1}{j!}\rho_Y^j\right) \qquad \text{for all } i, j \geq 0$$

where $\rho_X \equiv \lambda_X/\mu_X$ and $\rho_Y \equiv \lambda_Y/\mu_Y$.

c. Let $\ell(t) \equiv \ell_X(t) + \ell_Y(t)$. Obtain $\gamma_n \equiv \lim_{t\to\infty} \Pr\{\ell(t) = n\}$. Hence, show that

$$c = e^{-\rho_X - \rho_Y}$$

d. Explain why the limiting joint probability γ_{ij} is the product of the limiting marginal probabilities.

12.38 Consider the same queuing system as in Problem 12.37. However, the number of servers is now reduced to k, and system capacity is also limited to k. Thus, we have an $M/M/k/k$ system.

a. Verify that the product-form in Problem 12.37b still satisfies all balance equations.
b. Find the expression for c.
c. Compare with the Erlang loss formula in Problem 12.9c.

12.39 Consider a queuing system having five servers and two types of customers, X and Y. For $i = X, Y$, let the arrival rates of type i customers be λ_i. A type X customer cannot wait: If she finds all servers busy, she will leave the system immediately; otherwise, she will be served by one server with an exponential rate μ_X. A type Y customer has to be served concurrently by two servers, and his service time is exponentially distributed with rate μ_Y. He also cannot wait: If he finds fewer than two idle servers upon arrival, he will leave the system immediately.

a. Draw the transition diagram for the vector chain $\{\ell_X(t), \ell_Y(t)\}_{t\in[0,\infty)}$, where $\ell_i(t)$ is the number of type i customers in the system at time t ($i = X, Y$).
b. Verify that the following product-form satisfies all balance equations:

$$\gamma_{ij} \equiv \lim_{t\to\infty} \Pr\{\ell_X(t) = i, \ell_Y(t) = j\} = \gamma_{00}\left(\frac{1}{i!}\rho_X^i\right)\left(\frac{1}{j!}\rho_Y^j\right) \qquad \text{for all } i, j \geq 0$$

where $\rho_X \equiv \lambda_X/\mu_X$ and $\rho_Y \equiv \lambda_Y/\mu_Y$.
c. When the number of servers becomes infinite, what do you expect the expression of γ_{ij} to be in terms of λ_X, λ_Y, μ_X, and μ_Y?

12.40 Consider a *closed, cyclic* queue in which M customers circulate between two stations. They join the queue in front of the first station to be served with rate μ_1. After that, they join the queue in front of the second station to be served with rate μ_2. They then return to the first station and repeat the process. Draw the transition diagram and find the limiting distribution of the number of customers in station 1.

12.41 Refer to Problem 8.69. Actually, C is a single attraction in an amusement park having three separate rides. The duration of each ride is exponentially distributed with mean 30 minutes. D is the ticket booth for this attraction, having two windows. The service time of each is exponentially distributed with mean 5 minutes.

a. What is the limiting joint distribution of the number of people at C and D?
b. What is the expected total duration a person spends at this amusement park—that is, at C or D?

12.42 Refer to Problem 8.70. Actually, A is a gas station having three pumps; each serves with an exponentially distributed duration with mean 10 minutes. B is a restaurant having ten tables; the duration of each meal is exponentially distributed with mean 30 minutes. C is a supermarket having four checkout counters; the duration of each checkout is exponentially distributed with mean 20 minutes.

a. What is the limiting joint distribution of the number of cars at A, B, and C?
b. What is the expected total duration a car spends at this city?

12.43 For two queues in series without blocking studied in §12.15, verify that the expressions in (12.15.1) satisfy all balance equations.

12.44 Consider a queuing network having two single-server stations 1 and 2, serving with rates μ_1 and μ_2, respectively. Customers arrive as a Poisson stream with rate λ. Upon arrival, they pick the station having a *shorter waiting line*. If both lines are the same, they pick station 1 with probability α. Once at a station, they cannot switch to the other. Draw the transition diagram for this system.

12.45 Consider a network having two $M/M/1/1$ stations 1 and 2, serving with rates μ_1 and μ_2, respectively. If an arriving customer finds station 1 idle, she will be served by this station; otherwise, she will leave the system without receiving service. After being served by station 1, if she finds station 2 idle, she will be served by this station; otherwise, she will leave the system. Define a continuous-time Markov chain to obtain

a. the limiting probability that station 1 is busy
b. the limiting probability that station 2 is busy
c. the limiting average number of customers in the system
d. the limiting average duration a customer spends in the system

12.46 Consider an $M/M/1$ queuing system with feedback as studied in §12.9. Assume now that it takes a returning customer an exponential duration with rate δ to reach the end of the waiting line. Let $\ell(t)$ be the number of customers both waiting in line and being served at time t and let $r(t)$ be the number of customers on the process of returning to the end in the queue.

a. Explain why $\gamma_{i,j} \equiv \lim_{t\to\infty} \Pr\{\ell(t) = i, r(t) = j\}$ has a product-form. *Hint:* Use Problem 12.2.
b. Check that this form satisfies all balance equations.

CHAPTER 13

General Single-Server Queues: Simulation and Other Techniques

13.1 $G/G/1$ QUEUES

During the long journey we have made so far in this book, we have restricted ourselves mainly inside the cozy Markovian world. It is now time for us to venture out and see what it is like having to carry baggage full of memory to advance to the future.

We have just studied the Markovian queues, so let us stick to this subject. However, we now remove the Markovian assumption and study the *general* queuing models, where the interarrival times t_ns are not necessarily exponentially distributed but are independent and identically distributed with a common mean $\mathcal{E}[t_1]$. In addition, the service times s_ns are not necessarily exponentially distributed but are independent and identically distributed with a common mean $\mathcal{E}[s_1]$.

It turns out that general queuing systems are extremely difficult to analyze. So let us concentrate only on the *single-server* queue having an infinite capacity, in which customers are served in order of their arrivals. In Kendall notation, they are the $G/G/1$ models. (Sometimes they are also referred to as the $GI/G/1$ models, where I stands for independence.)

Now the process $\{\ell(t)\}_{t\in[0,\infty)}$ is no longer a continuous-time Markov chain or even a semi-Markov chain. To make it Markovian, we must introduce two supplementary variables: One is the duration until the next departure; the other is the duration until the next arrival. (Why?) Unfortunately, this method of inclusion of supplementary variables is beyond the scope of this book.

We shall approach the subject using regenerative arguments instead. But before doing so, we must emphasize that the main purpose of this chapter is not to analyze the $G/G/1$ queuing models per se. It just happens to be an excellent example demonstrating many approaches one can take in studying non-Markovian processes. We hope that, even if you cannot see yourselves ever having to analyze queuing systems, this chapter is still useful, and you can apply the underlying principles to whatever is your application.

13.2 BUSY CYCLES

Recall that we denoted the long-term time-average number of busy servers as b in §12.5. For the $G/G/1$ queue, if $b < 1$, the single server must occasionally be idle, and the system is said to be *stable*. Little's result (12.5.4) states that this happens when $b = \mathcal{E}[s_1]/\mathcal{E}[t_1] < 1$ or when the expected service time $\mathcal{E}[s_1]$ is strictly less than the expected interarrival time $\mathcal{E}[t_1]$, which we shall henceforth assume.

One major property of a stable $G/G/1$ queuing system is that it is *regenerative* whenever an arriving customer finds an empty system. We call its nth regenerative cycle the *busy cycle* c_n. Because the system must be empty occasionally, $\mathcal{E}[c_n] < \infty$. The busy cycle c_n has two parts: the *busy period* p_n, in which the server is busy, and the *idle period* i_n, in which the server stays idle (Figure 12.2.1).

Let us assume that there is a regenerative point at time $t = 0$; that is, a customer arrives at time $t = 0$ and finds an empty system. In this case, the system's behavior within any busy cycle is independent and probabilistically identical to that within the first cycle, starting from time $t = 0$. Thus, all the information about the system's behavior, whether transient or limiting, can be related to that within the first busy cycle (Minh, 1980).

We assume further that the durations of the busy cycles are spread out. This can be satisfied if the interarrival times, or the service times, are. In this case, from §9.10, the system's pointwise limits exist and are equal to its long-term time-averages.

Let us now see how we can express the limiting results in terms of the quantities within the first busy cycle.

13.3 WAITING TIMES AND IDLE TIMES

We shall now derive the following relationship:

$$W_q = \frac{\mathcal{E}\left[(u_1)^2\right]}{-2\mathcal{E}[u_1]} - \frac{\mathcal{E}\left[(i_1)^2\right]}{2\mathcal{E}[i_1]} \tag{13.3.1}$$

where

1. $u_1 \equiv s_1 - t_1$, which is independent and identically distributed with $u_n \equiv s_n - t_n$ for all $n > 1$. (Note that we assume $\mathcal{E}[u_1] < 0$.)
2. i_1 is the duration of the first idle period, or the *idle time*, which is independent and identically distributed with the nth idle time i_n for all $n > 1$.

To prove this, let us first denote by q_n $(n \geq 1)$ the duration the nth customer waits in queue before receiving service. If she arrives at the system at time T_n, she will depart at time $T_n + q_n + s_n$. The $(n+1)$th customer, who will arrive at time $T_n + t_n$, therefore has to wait for a duration of

$$\begin{aligned} q_{n+1} &= \max((T_n + q_n + s_n) - (T_n + t_n), 0) \\ &= \max(q_n + s_n - t_n, 0) \\ &= \max(q_n + u_n, 0) \end{aligned} \tag{13.3.2}$$

Let us denote the other half of this equation as

$$y_n \equiv -\min(q_n + u_n, 0) \quad \text{for all } n \geq 1 \tag{13.3.3}$$

Note that $y_n \geq 0$. Furthermore, given the $(n+1)$th customer arrives after the departure of the nth customer, $y_n = (T_n + t_n) - (T_n + q_n + s_n) > 0$. In this case, y_n is the duration of the idle period before the arrival of the $(n+1)$th customer, who does not have to wait. Because all idle periods are independent and identically distributed with common first and second moments $\mathcal{E}[i_1]$ and $\mathcal{E}[(i_1)^2]$, respectively, for all $n \geq 1$, we can write

$$\frac{\mathcal{E}\left[(y_n)^2\right]}{\mathcal{E}\left[y_n\right]} = \frac{\mathcal{E}\left[(y_n)^2 | y_n > 0\right] \Pr\{y_n > 0\}}{\mathcal{E}\left[y_n | y_n > 0\right] \Pr\{y_n > 0\}} = \frac{\mathcal{E}\left[(i_1)^2\right]}{\mathcal{E}\left[i_1\right]} \tag{13.3.4}$$

Adding Equations (13.3.2) and (13.3.3) side by side yields

$$q_{n+1} - y_n = q_n + u_n \quad \text{for all } n \geq 1 \tag{13.3.5}$$

Taking expectations on both sides of (13.3.5) and letting $n \to \infty$ (keeping in mind that $\lim_{n\to\infty} \mathcal{E}[q_{n+1}] = \lim_{n\to\infty} \mathcal{E}[q_n] \equiv W_q$), we obtain

$$\lim_{n\to\infty} \mathcal{E}[y_n] = -\lim_{n\to\infty} \mathcal{E}[u_n] = -\mathcal{E}[u_1] \tag{13.3.6}$$

If we now square both sides of (13.3.5) before taking expectations (remembering that $q_{n+1} y_n = 0$ for all $n \geq 1$) and then let $n \to \infty$, we will also obtain

$$\lim_{n\to\infty} \mathcal{E}\left[(y_n)^2\right] = 2W_q \mathcal{E}\left[u_1\right] + \mathcal{E}\left[(u_1)^2\right] \tag{13.3.7}$$

Hence, from Equation (13.3.4), we obtain

$$\frac{\mathcal{E}\left[(y_n)^2\right]}{\mathcal{E}\left[y_n\right]} = \frac{2W_q \mathcal{E}\left[u_1\right] + \mathcal{E}\left[(u_1)^2\right]}{-\mathcal{E}\left[u_1\right]} = \frac{\mathcal{E}\left[(i_1)^2\right]}{\mathcal{E}\left[i_1\right]}$$

which yields Equation (13.3.1).

If we know how to calculate the first and second moments of the idle time i_1, Equation (13.3.1) would allow us to find the expression for the limiting expected waiting time W_q. Little's results then can relate this quantity with other performance measures such as L, L_q, and W.

Unfortunately, this is where our formal analysis of the $G/G/1$ queues must stop. We stop not because of the scope of this book but because *closed-form* formulas of $\mathcal{E}[i_1]$ and $\mathcal{E}[(i_1)^2]$, and hence of all other quantities of interest, are *not yet known* for the $G/G/1$ queues.

So how do we advance from here? Let us explore a few available options.

13.4 MODEL APPROXIMATION: $M/G/1$ QUEUES

One way of continuing our analysis is to introduce *additional assumptions* into the model, making it tractable. This method normally requires us to retreat into the Markovian world. Fortunately, for the single-server queues, there is no need for us to withdraw into it completely: We must now assume that *the interarrival times are exponentially distributed* with rate λ, but we can still allow the distribution of the service times to remain general.

From §8.9, the Poisson assumption of the arrival stream now yields

$$\mathcal{E}[t_1] = \frac{1}{\lambda} \qquad \mathcal{E}\left[(t_1)^2\right] = \frac{2}{\lambda^2}$$

Hence, with $\rho \equiv \mathcal{E}[s_1]/\mathcal{E}[t_1] = \lambda\mathcal{E}[s_1]$,

$$-\mathcal{E}[u_1] = \mathcal{E}[t_1 - s_1] = \frac{1}{\lambda} - \frac{\rho}{\lambda} = \frac{1-\rho}{\lambda} > 0$$

and

$$\mathcal{E}\left[(u_1)^2\right] = \mathcal{E}\left[(s_1 - t_1)^2\right] = \mathcal{E}\left[(s_1)^2\right] - 2\mathcal{E}\left[s_1\right]\frac{1}{\lambda} + \frac{2}{\lambda^2} = \mathcal{E}\left[(s_1)^2\right] + \frac{2(1-\rho)}{\lambda^2}$$

Furthermore, because the arrival stream is Poisson, the duration from any time point to the next customer's arrival is always exponentially distributed. Being the duration from the time a departing customer leaves an empty system to the next customer's arrival, *an idle time is also exponentially distributed with rate* λ, or

$$\mathcal{E}[i_1] = \frac{1}{\lambda} \qquad \mathcal{E}\left[(i_1)^2\right] = \frac{2}{\lambda^2}$$

These results now enable us to derive from Equation (13.3.1) the following *Pollaczek-Khinchin (P-K)* formula for the $M/G/1$ queues:

$$W_q = \frac{\lambda\mathcal{E}\left[(s_1)^2\right]}{2(1-\rho)} \tag{13.4.1}$$

Generally, we must be careful when adding assumptions to our model to make it tractable. Plausible arguments must be provided to justify these assumptions; otherwise, the model would be too contrived to be useful. Remember that there are many practical applications in which the Markovian assumptions cannot be justified.

On the other hand, we also should not be too strict with our models. Even when the Markovian assumption appears to be rather unrealistic, having some results with it is better than having no result at all. Actually, this is the position that we have been relying on throughout the book—the very models themselves are approximations of the real-world problems. We just have to remember that, in many applications, the introduction of the Markovian assumption makes the $M/G/1$ model an *approximation* to the intractable $G/G/1$ model.

13.5 ANALYTICAL APPROXIMATIONS

Instead of introducing additional assumptions and then obtaining the exact results for the approximating model, we can leave the model as it is and try to obtain the *analytical approximations* of the results or their *upper bounds* and *lower bounds*. While we do not have any nice lower bound for the performance measures of the $G/G/1$ queues, here are some upper bounds and approximations:

1. Note that the term $\mathcal{E}[(i_1)^2]/2\mathcal{E}[i_1]$ in Equation (13.3.1) is nonnegative. Ignoring it, we obtain an upper bound for the limiting expected waiting time in the $G/G/1$ queues as

$$W_q \le \frac{\mathcal{E}\left[(u_1)^2\right]}{-2\mathcal{E}[u_1]} \tag{13.5.1}$$

2. From Equations (1.10.6), (13.3.6), and (13.3.7), we have

$$\begin{aligned}
-2W_q\mathcal{E}[u_1] &= \mathcal{E}\left[(u_1)^2\right] - \lim_{n\to\infty} \mathcal{E}\left[(y_n)^2\right] \\
&= \mathcal{E}\left[(u_1)^2\right] - \lim_{n\to\infty} \text{Var}[y_n] - \lim_{n\to\infty} \mathcal{E}^2[y_n] \\
&= \mathcal{E}\left[(u_1)^2\right] - \lim_{n\to\infty} \text{Var}[y_n] - \mathcal{E}^2[u_1] \\
&= \text{Var}[u_1] - \lim_{n\to\infty} \text{Var}[y_n] \\
&\le \text{Var}[u_1]
\end{aligned}$$

Because $-2\mathcal{E}[u_1] > 0$, we now have another upper bound for W_q as

$$W_q \le \frac{\text{Var}[u_1]}{-2\mathcal{E}[u_1]} \tag{13.5.2}$$

This upper bound is *sharper* than the upper bound (13.5.1) because $\text{Var}[u_1] \le \mathcal{E}[(u_1)^2]$.

3. It turns out that, as the traffic becomes heavy, or $\rho \to 1$, the idle time i_1 approaches 0. This makes $\mathcal{E}[(i_1)^2]$ approach 0 faster than $\mathcal{E}[i_1]$, and hence $\mathcal{E}[(i_1)^2]/\mathcal{E}[i_1] \to 0$. Thus, the upper bound (13.5.1) becomes closer and closer to W_q as $\rho \to 1$. That is why the upper bound (13.5.2), being sharper than (13.5.1), becomes known as the *Kingsman heavy traffic approximation* for W_q, or

$$\lim_{\rho\to 1} W_q \approx \frac{\text{Var}[u_1]}{-2\mathcal{E}[u_1]} = \frac{\text{Var}[t_1] + \text{Var}[s_1]}{-2\mathcal{E}[u_1]} = \frac{\lambda(\text{Var}[t_1] + \text{Var}[s_1])}{2(1-\rho)} \tag{13.5.3}$$

4. When the approximation (13.5.3) is applied to an $M/G/1$ queue, we obtain

$$\lim_{\rho\to 1} W_q \approx \frac{\lambda\text{Var}[s_1] + 1/\lambda}{2(1-\rho)}$$

which is greater than the exact *P-K* formula (13.4.1). For this reason, Marchal (1976) proposed that (13.5.3) should be scaled down to

$$\lim_{\rho \to 1} W_q \approx \frac{\mathrm{Var}[s_1] + \mathcal{E}^2[s_1]}{\mathrm{Var}[s_1] + 1/\lambda^2} \times \frac{\lambda(\mathrm{Var}[t_1] + \mathrm{Var}[s_1])}{2(1-\rho)}$$

At least, this approximation is exact for the $M/G/1$ queues.[1]

13.6 NUMERICAL METHODS

Like many other branches of applied mathematics, early study of stochastic processes focused mainly on the *analytic* closed-form solutions, such as the *P-K* formula for the $M/G/1$ queues or the heavy traffic approximations. With these solutions, one can obtain the numerical results with a few keystrokes on a calculator or analyze them further to obtain the optimal system design.

Overconcern with the analytic solutions has one subtle consequence, namely, the underconcern with the interpretation and understanding of the intermediate results. In the traditional formal analysis of many queuing systems, the system's structure is ignored after it yields a set of *difference-differential equations*. After that, the analysis mainly involves looking for a mathematical procedure to solve this set of equations. Normally, the method of solution requires manipulating the *Laplace transforms* using complex variables, and hence has scant relevance to, and provides little interpretive value for, the model under study.

Also, most final results in queuing theory, including those for the $G/G/1$ queues, are left in terms of the Laplace transforms. Although these solutions have existed for some time, most of them do not make sense to the practitioners, who have long lamented that queuing theory is well hidden behind the Laplace curtain.

This book uses a different approach. It tries to obtain the results using "probabilistic thinkings." It explores the structures of the processes, visualizes their developments, and tries to give intuitive interpretations for the intermediate results. In this way, we were able to derive and explain many results without ever having to write a single differential equation.

The same approach has been an underlying philosophy in an important branch of study called *computational probability*, which uses *numerical methods* for solving stochastic models. Of course, numerical methods can help in the computations of existing analytic solutions, such as in providing efficient algorithms to invert the Laplace transforms and to do other calculations. However, computational probability means much more than just supporting the formal analysis and continuing where it stops. It can—and must—be involved at the very start of the analysis, constantly paying attention to the computational implementation and interpreting intermediate results.

According to Neuts (1981),

> We therefore define computational probability as the study of stochastic models with a genuine added concern for algorithmic feasibility over a wide, realistic range of parameter

[1]See Shanthikumar and Buzacott (1980) for other approximations in the literature.

> values. We have imposed upon our own work and that of our students the requirement that careful and exhaustive computer studies be performed before research results are proposed as the solution to a given problem. The limitations on a specific algorithm are as real as are restrictive conditions on the validity of a theorem. They offer . . . the same challenge and stimulus to further research.
>
> This self-imposed constraint has not been felt as a burden. It has, on the contrary, led us to seek alternatives to certain classical analytic methods that are often difficult and risky in their numerical implementation. Unifying structural properties of Markov chains have been identified, which, through purely probabilistic arguments, lead to highly stable numerical procedures. Rather than proceeding through purely formal manipulations, the analysis of many problems now runs parallel to the steps of the algorithms. The latter then acquire a significance, which is often of independent interest in the interpretation of numerical results.

Computational probability allows the analysis of models more general than the $M/G/1$ queues. The distributions of the interarrival or service times that lend themselves to numerical methods range from a manageable family of Erlang distributions to a more versatile *phase-type distribution* (which, as shown in Problem 11.39, is the distribution of the duration until absorption of an absorbing continuous-time Markov chain). This is a major field in itself, and its coverage is not intended to be within the scope of this book. We only mention it here to encourage readers to pay attention to the computational implementations of all analytical results.

13.7 SIMULATION

In §1.4, we said that one way to obtain the desired probability is by observing the experiment empirically. The major drawback of this method is that it is limited to existing and repeating phenomena. For our queuing problems, this is prohibitive. First, even if the system exists, it might take a long time for it to reach its limiting behavior. Second, we normally are not interested in the behavior of an existing system. We study it because we might want to change its parameters to improve its performance. (For example, we want to add another server.) Thus, besides the existing system, we need an array of other system designs for comparison.

In a nutshell, a *discrete-event simulation* is an *artificial and controlled experiment*. It is artificial because it is done inside a computer. The events are imaginary. It is controlled because it needs a mathematical model having a set of parameters. If these parameters are changed, another experiment can be carried out, and its performance can be compared.

Table 13.7.1 gives a simulation of the development of a $G/G/1$ queuing system in which the interarrival times are uniformly distributed within the interval (2.0, 3.2) hours and the service times are uniformly distributed within the interval (1.8, 3.0) hours. The movements of the first three customers are simulated as follows:

1. For the first customer:

   a. We assume that he arrives at time $T_1 = 0$ (as shown in column 4) and his waiting time is $q_1 = 0$ (as shown in column 5).

Table 13.7.1 Simulating a Queuing System Having Uniformly Distributed Interarrival and Service Times

1 Customer # i	2 R_1	3 Interarrival Time t_{i-1}	4 Arrival Time T_i	5 Waiting Time q_i	6 R_2	7 Service Time s_i
1			0	0	0.833	2.800
2	0.102	2.122	2.122	0.678	0.467	2.361
3	0.758	2.909	5.031	0.129	0.708	2.650
4	0.372	2.447	7.478	0.332	0.228	2.074
5	0.972	3.167	10.645	0	0.481	2.377
6	0.117	2.140	12.785	0.237	0.669	2.603
7	0.879	3.055	15.840	0	0.211	2.053
8	0.780	2.937	18.777	0	0.159	1.991
9	0.180	2.215	20.992	0	0.273	2.128
10	0.296	2.355	23.347	0	0.199	2.039
11	0.354	2.424	25.771	0	0.355	2.226
12	0.198	2.237	28.008	0	0.458	2.350
13	0.359	2.431	30.439	0	0.127	1.953
14	0.218	2.262	32.701	0	0.686	2.623
15	0.764	2.917	35.618	0	0.977	2.972
16	0.503	2.604	38.221	0.369	0.799	2.758
17	0.629	2.755	40.977	0.372	0.720	2.663
18	0.175	2.209	43.186	0.826	0.754	2.705
19	0.273	2.327	45.514	1.203	0.654	2.584
20	0.090	2.108	47.622	1.679	0.548	2.458

b. Using a *random number generator*, we generate a random number—that is, a number *uniformly distributed within the interval* [0, 1) (say, 0.833, as in column 6).

This random number is then *linearly* scaled into a *random variate* uniformly distributed within the interval (1.8, 3.0) to be used as s_1—that is, as in column 7,

$$s_1 = 1.8 + (3.0 - 1.8)(.833) = 2.800 \text{ hours}$$

2. For the second customer:

a. A random number is generated (say, 0.102, as in column 2).

This random number is transformed into a random variate uniformly distributed within the interval (2.0, 3.2) to be used as the interarrival time t_1—that is, as in column 3,

$$t_1 = 2.0 + (3.2 - 2.0)(.102) = 2.122 \text{ hours}$$

Hence, as shown in column 4, the second customer arrives at time

$$T_2 = T_1 + t_1 = 0 + 2.122 = 2.122$$

b. Since the first customer leaves at time $T_1 + s_1$, column 5 shows the second customer has to wait for a duration of

$$q_2 = (T_1 + s_1) - T_2 = (0 + 2.800) - 2.122 = 0.678 \text{ hours}$$

c. A random number is generated (0.467, in column 6) giving his service time as $s_2 = 2.361$ (in column 7).

3. For the third customer,
a. A random number (0.758, in column 2) gives $t_2 = 2.909$ (in column 3); hence, column 4 shows his arrival time as

$$T_3 = T_2 + t_2 = 2.122 + 2.909 = 5.031$$

b. Thus, his waiting time (as shown in column 5) is

$$q_3 = (T_2 + q_2 + s_2) - T_3 = 0.129 \text{ hours}$$

Continuing in this manner, we can generate a substantial number of waiting times in column 5 in a short amount of computer time. Estimates of W, W_q, L, L_q, and all other quantities of interest can now be obtained using the techniques of classical statistical inference. From our simulation of the first 20 waiting times in Table 13.7.1, W_q can be estimated as 0.291.

If the interarrival times or the service times are not uniformly distributed, a suitable *random variate generator* can be used to transform a random number to a number having their distributions.

Since simulation is a very popular method, let us now present it in more detail. We start by discussing how a sequence of random numbers can be generated.

13.8 RANDOM NUMBER GENERATORS

A common method of generating a random number uniformly distributed in the interval [0, 1) by a computer is to use a *formula* to obtain a sequence of numbers $\{U_n\}_{n=0,1,2,\ldots}$ in which the value of U_n is calculated from that of U_{n-1}. The most commonly used formula is the *linear congruential generator*:

$$U_n = (bU_{n-1} + c) \bmod (m) \quad \text{for all } n > 0 \tag{13.8.1}$$

which gives U_n as the remainder when $bU_{n-1} + c$ is divided by m. The quantity b is called the *multiplier*, m the *modulus,* and c the *increment*. The starting value U_0 is called the *seed* of the sequence.

If both constants b and c are not zero, this generator is called the *mixed* congruential generator; if $c = 0$, it is called the *multiplicative* or *pure* congruential generator.

Since U_n is a number in the interval $[0, m)$, once its value is obtained, a random number R in the interval [0, 1) can be delivered as

$$R = \frac{U_n}{m}$$

Some important observations are in order here:

1. For R to be a random number, the sequence $\{U_n\}_{n=0,1,2,\ldots}$ must be mutually independent, or the value of U_{n+1} is independent of all its previous values from U_0 to U_n. The use of a formula as in the foregoing linear congruential generator (and most other generators) clearly violates this property. The number R generated therefore is not really random but only appears to be so. For that reason, it is normally referred to as a *pseudo*-random number.

2. An undesirable property of sequences of the pseudo-random numbers is *identical cycles*. Consider a number U_1, which is obtained by a formula from another number U_0. Sooner or later, the value of U_0 will appear again in the sequence as the value of U_n. Because the same formula is used, the value of U_{n+1} must be the same as that of U_1. In this manner, for a fixed value of i, all U_{kn+i} must be identical for all $k \geq 0$. An infinite sequence $\{U_n\}_{n=0,1,2,\ldots}$ is thus simply one fixed cycle repeating itself indefinitely.

Cycling is unavoidable, and all we can do is try to make the cycle length large. For the linear congruential generator, the cycle length cannot be greater than m (Why?), and the value of m therefore should be large. Since the word size in many computers is 32 bits, with 1 bit reserved for the algebraic sign, 2^{31} seems to be a good choice for m. However, when $c = 0$, number-theoretic results have shown that this choice would give a maximum cycle length of only 2^{29}. Fortunately, it can also be shown that when $m = 2^{31} - 1$, which is a prime number, the maximum cycle length can be $2^{31} - 2$.

3. The speed and efficiency of the generator are also another consideration. When m takes the form 2^k, the modulo operation may be carried out most efficiently (Problem 13.12). However, when $m = 2^{31} - 1$, it can also be shown that the numbers can be generated almost as quickly.

4. A random number R should be uniformly distributed within the interval $[0, 1)$; that is, all numbers in that interval should have an equal chance of appearing. Unfortunately, pseudo-random numbers also violate this property because only numbers that appear in the first cycle keep appearing, while all other numbers cannot. A good pseudo-random number generator should at least provide numbers that are *dense* in the interval $[0, 1)$.[2] For the linear congruential generator, this is another reason for a large value of m.

5. One desirable feature of the pseudo-random numbers is that they are *replicable*; that is, given the same seed, exactly the same sequence can be reproduced. This is very helpful for comparing alternative systems or for debugging.

6. Despite all their shortcomings, well-tested linear congruential generators can produce sequences that can be regarded as genuinely random for all practical purposes. The search for the best generator is still going on. The following generator has been used extensively in the literature and was recommenced as a minimal standard against which other generators should be judged (Park and Miller, 1988):

$$U_n = 7^5 U_{n-1} \bmod \left(2^{31} - 1\right) = 16807 U_{n-1} \bmod \left(2^{31} - 1\right)$$

13.9 RANDOM VARIATE GENERATORS

We now explain and illustrate some algorithms, called the *random variate generators*, to generate the *random variates* having a particular distribution from the random numbers.

---

[2]Similarly, a pair of random numbers should be uniformly distributed and dense in $[0, 1) \times [0, 1)$ and so on.

There has been a very extensive research effort on the subject, and we do not try to cover all generators here. We also do not try to present the best generators, but only those that can illustrate the method under discussion.

Before doing so, we should note that *speed* is normally the most important criterion in choosing a generator. Both the *marginal execution time* to generate one additional random variate and the *setup time* in preparation for the generation need to be minimized. This criterion, however, has been overemphasized in the literature to the detriment of other criteria. In addition to being fast, a good generator should be exact; that is, the variates it generates must have the desired distribution within the limitations of machine accuracy and of the random number generator used. It should work well with all parameter values. It should be easy to understand and implement. Its program should be compact. It should avoid the use of large tables and multiple calls to mathematical functions such as *log*, *sine*, or *cosine*. Furthermore, it should not accentuate the deficiency of the pseudo-random number generator used.

13.10 INVERSE TRANSFORM TECHNIQUE FOR THE DISCRETE DISTRIBUTIONS

Let X be a discrete random variable having probability mass function $p_i \equiv \Pr\{X = x_i\}$, with $\sum_{i=0}^{\infty} p_i = 1$. Let $F_i \equiv \sum_{k=0}^{i} p_k$. To generate a random variate having this distribution, we simply generate a random number R within $[0, 1)$ and then deliver x_i if $F_{i-1} \leq R < F_i$. Since R is uniformly distributed, the probability that the random variate generated is x_i is $F_i - F_{i-1} = p_i$.

Let $F(x)$ be the distribution function of X. Then Figure 13.10.1(a) shows that this method is equivalent to taking the inverse of $F(x)$. That is why it is known as the *inverse transform technique*.

FIGURE 13.10.1 The inverse transform technique.

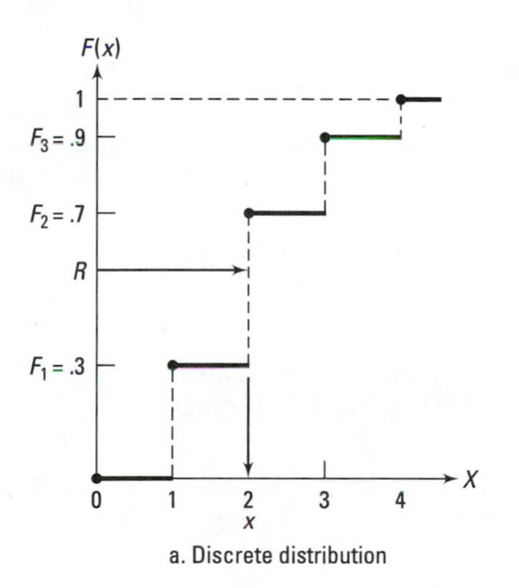

a. Discrete distribution

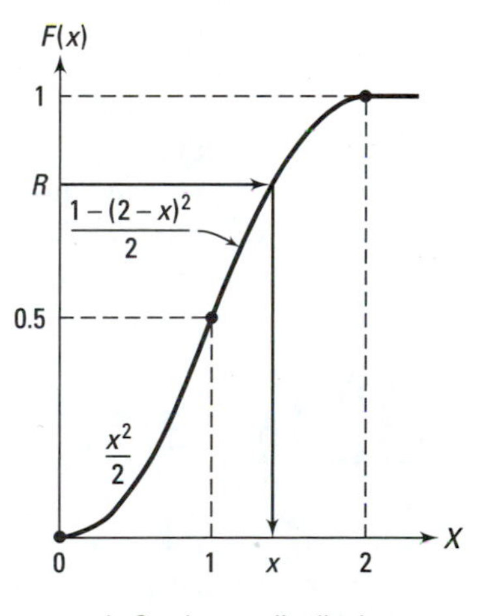

b. Continuous distribution

For illustration, assume $p_1 \equiv \Pr\{X = 1\} = .3$, $p_2 \equiv \Pr\{X = 2\} = .4$, $p_3 \equiv \Pr\{X = 3\} = .2$, and $p_4 \equiv \Pr\{X = 4\} = .1$. Based on the value of the random number R, a variate x can be generated as

$$x = \begin{cases} 1 & \text{if } 0 \le R < 0.3 \\ 2 & \text{if } 0.3 \le R < 0.7 \\ 3 & \text{if } 0.7 \le R < 0.9 \\ 4 & \text{if } 0.9 \le R < 1 \end{cases}$$

Using *table lookup* as in the preceding example would be appropriate when the number of distinct values of X is finite. Otherwise, either we have to *truncate* the value of X or compute F_i only when needed. The latter approach can be used in generating the *Poisson* variates having $p_i = \lambda^i e^{-\lambda}/i!$. Here, after generating a random number R, we first compare it with $F_0 = p_0 = e^{-\lambda}$; if $R \ge F_0$, we then compare it with $F_1 = F_0 + \lambda e^{-\lambda}$; if $R \ge F_1$, we next compare it with $F_2 = F_1 + \lambda^2 e^{-\lambda}/2$; and so on. When needed, F_i is calculated recursively from the relationship $F_{i+1} - F_i = p_{i+1} = \lambda p_i/(i+1)$ for all $i \ge 0$. The algorithm is as follows:

Step 1. Set $i = 0$, $p = e^{-\lambda}$, $F = p$.
Step 2. Generate a random number R in $[0, 1)$.
Step 3. If $R < F$, deliver i, stop.
Step 4. Set $p = \lambda p/(i+1)$, $F = F + p$, and $i = i + 1$. Go to step 3.

If possible, we must go further to make the generator more efficient. For example, consider the *geometric* distribution with parameter p. Let $q \equiv 1 - p$. To generate a geometric variate, the inverse transform technique suggests that we first generate a random number R and then deliver the value i such that $F_{i-1} = 1 - q^{i-1} \le R < F_i = 1 - q^i$ (Equation 1.12.2).

This comparison is equivalent to $q^i < 1 - R \le q^{i-1}$. Also, since R is uniformly distributed within $[0, 1)$, it can take the place of $1 - R$, and we can also deliver the value i such that $q^i < R \le q^{i-1}$.

Note that the variate i is the minimum of all integers k such that $q^k < R$ or such that $\ln q^k = k \ln q < \ln R$ (because logarithm is an increasing function). As $0 \le q \le 1$, we have $\ln q < 0$, and j is thus also the minimum value of all integers k such that $k > \ln R/\ln q$. Thus, after generating R, there is no need for number comparisons, and we simply deliver the integer portion of $(\ln R/\ln q) + 1$.

13.11 INVERSE TRANSFORM TECHNIQUE FOR THE CONTINUOUS DISTRIBUTIONS

As illustrated in Figure 13.10.1(b), the foregoing inverse transform technique can be adapted to a continuous random variable having distribution function $F(x)$ as follows: Let $F^{-1}(y)$ be the inverse function of $F(x)$.

Step 1. Generate a random number R in [0, 1).
Step 2. Deliver $F^{-1}(R)$.

This technique is especially suitable for random variates, the inverse function of which is simple to calculate.

As an example, suppose X has the following *triangular* distribution:

$$f(x) = \begin{cases} x & \text{for all } 0 \le x \le 1 \\ 2 - x & \text{for all } 1 < x \le 2 \\ 0 & \text{otherwise} \end{cases}$$

Then, as shown in Figure 13.10.1(b),

$$F(x) = \begin{cases} 0 & \text{for all } x < 0 \\ \dfrac{x^2}{2} & \text{for all } 0 \le x < 1 \\ 1 - \dfrac{(2-x)^2}{2} & \text{for all } 1 \le x < 2 \\ 1 & \text{for all } 2 \le x \end{cases}$$

Note that $F(0) = 0$, $F(1) = .5$, and $F(2) = 1$. Also, the inverse of $F(x)$ is

$$F^{-1}(y) = \begin{cases} \sqrt{2y} & \text{for } 0 \le y < 0.5 \\ 2 - \sqrt{2(1-y)} & \text{for } 0.5 \le y < 1 \end{cases}$$

This suggests the following algorithm to generate a random variate having this distribution:

Step 1. Generate a random number R in [0, 1).
Step 2. If $R \le 0.5$, deliver $\sqrt{2R}$; else deliver $2 - \sqrt{2(1-R)}$.

An important application of this technique is when X has an *exponential* distribution with rate λ; that is, $F(x) = 1 - e^{-\lambda x}$ for all $0 \le x$. Since the inverse function of $1 - e^{-\lambda x}$ is $-\ln(1-y)/\lambda$, an exponential variate can be obtained as $-\ln R/\lambda$.

13.12 ACCEPTANCE–REJECTION TECHNIQUE FOR THE DISCRETE DISTRIBUTIONS

Suppose we want to generate a discrete random variable X with $\Pr\{X = 0\} = 2/3$ and $\Pr\{X = 1\} = 1/3$; that is, the probability of having 0 is twice that of having 1. The simplest form of the *acceptance–rejection technique* suggests that we first generate a random variate y taking value 0 or 1 with equal probability. If 1 is generated, we accept it with probability .5 and reject it with probability .5. Thus, on the average, half of the 1s generated will be discarded, and the resulting variates would have the same distribution as X.

Table 13.12.1 Illustration of the Acceptance–Rejection Technique

1	2	3	4	5
	R_1	y $p_0 = .5$ $p_1 = .5$	R_2	x $p_0 = 1/3$ $p_1 = 2/3$
1	0.76	1	0.24	(rejected)
2	0.04	0		0
3	0.74	1	0.45	(rejected)
4	0.98	1	0.96	1
5	0.60	1	0.13	(rejected)
6	0.86	1	0.65	1
7	0.40	0		0
8	0.43	0		0
9	0.19	0		0
10	0.72	1	0.74	1
11	0.91	1	0.26	(rejected)
12	0.18	0		0
13	0.27	0		0
14	0.56	1	0.23	(rejected)
15	0.23	0		0

Table 13.12.1 provides a numerical example of this algorithm:

1. Column 2 gives a sequence of random numbers R_1s uniformly distributed in [0, 1).
2. Column 3 gives the variates y having value 0 or 1 with equal probability. (They are obtained by applying the inverse transform technique to R_1; that is, $y = 0$ if $R_1 < 0.5$ and $y = 1$ otherwise.)
3. Column 4 assigns a random number R_2 for each variate 1 in column 3.
4. If $R_2 < 0.5$, we reject the variate 1 in column 3; otherwise, we accept and deliver it to column 5. The variates in column 5 have the same distribution as X, with $\Pr\{X = 0\} = 2\Pr\{X = 1\}$.

The power of this method becomes more apparent when we apply it to the continuous random variables.

13.13 ACCEPTANCE–REJECTION TECHNIQUE FOR THE CONTINUOUS DISTRIBUTIONS

The logic behind the acceptance–rejection technique for the continuous distributions is as follows:

1. Consider a probability density function $f(x)$. Let N be any positive number. If we can generate points uniformly distributed under the curve of $Nf(x)$, the x coordinates of these points have probability density function $f(x)$.

2. The reverse is also true: Assume we have an efficient method to generate a sequence of variates $\{z_n\}_{n=0,1,2,\ldots}$ having probability density function $g(x)$. Consider now those variates generated within an infinitesimal interval $[x, x+h)$ when $h \to 0$—that is, all z_is such that $x \leq z_i < x+h$. We note that their number is linearly proportional to the value of $g(x)$: The higher the value of $g(x)$, the more points generated within the infinitesimal interval $[x, x+h)$. Thus, if each of the variates $\{z_n\}_{n=0,1,2,\ldots}$ is assigned a random number R, the pairs $(z_i, Rg(z_i))$ are *uniformly distributed under the curve* $g(x)$.

If M is any positive number, the pairs $(z_i, RMg(z_i))$ are also uniformly distributed under the curve $Mg(x)$.

3. Thus, to generate a random variable X having probability density function $f(x)$:

a. we first find a random variable Y having probability density function $g(x)$ and a number M such that $f(x) \leq Mg(x)$ for all $-\infty < x < \infty$. In other words, the area under the curve $f(x)$ must fit entirely inside the area under the curve $Mg(x)$. The function $g(x)$ is known as the *envelope*, or the *majorizing function*.

b. we then generate a point uniformly distributed under the curve $Mg(x)$. We accept this point if it is also under the curve $f(x)$ and reject it otherwise. The excepted points are uniformly distributed under the curve $f(x)$, and their x coordinates must have the probability density function $f(x)$.

The algorithm therefore is:

Step 1. Generate z having probability density function $g(x)$.
Step 2. Generate a random number R in [0, 1).
Step 3. If $R > f(z)/Mg(z)$, go to step 1. (A reject)
Step 4. Deliver z. Stop.

Note that the pairs $(z, RMg(z))$ are uniformly distributed under the curve of $Mg(x)$. In step 3, we accept them if they are also under the curve $f(x)$—that is, if $RMg(z) < f(z)$. However, instead of comparing $RMg(z)$ with $f(z)$, we compare R with $f(z)/Mg(z)$ because we normally can simplify the ratio $f(z)/Mg(z)$.

The efficiency of the technique depends on finding an envelope $g(x)$ that matches $f(x)$ well and is easy to generate. We also do not want the value of M to be unnecessarily large. It should be just large enough for $f(x) \leq Mg(x)$. Normally, we pick M as the maximum value of $f(x)/g(x)$.

For example, consider $f(x) \equiv 20x(1-x)^3$ for all $0 < x < 1$. As shown in Figure 13.13.1, this probability density function attains its maximum value when $x = 0.25$, with $f(0.25) = 135/64$. Let us use $g(x) = 1$ as the majorizing function and $M = 135/64$. Now we have $f(x)/Mg(x) = (256/27)x(1-x)^3$ and the following procedure to generate a variate of $f(x)$:

Step 1. Generate a random number R_1 in [0, 1). (The x coordinate)
Step 2. Generate a random number R_2 in [0, 1). (The y coordinate)
Step 3. If $R_2 > (256/27)R_1(1-R_1)^3$, go to step 1. (Reject R_1)
Step 4. Deliver R_1. Stop.

FIGURE 13.13.1 The acceptance–rejection technique for $f(x) = 20x(1-x)^3$.

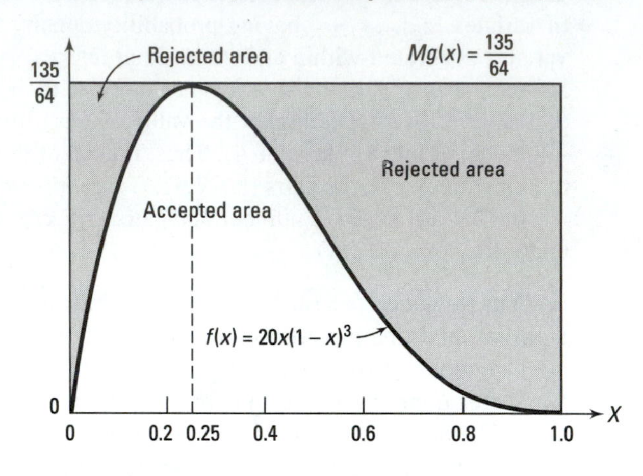

Figure 13.13.2 illustrates how we can generate a variate having the *standard normal* distribution using the exponential distribution $g(x) = e^{-x}$ as the envelope. We first only generate variates from half of this distribution (having positive value) and then randomly assign a positive or negative sign to them. The probability density function of the *half-normal* is (Equation 1.18.1):

$$f(x) = \sqrt{\frac{2}{\pi}} e^{-x^2/2} \quad \text{for all } 0 < x < \infty$$

Since $f(x)/g(x)$ is maximized when $x = 1$, we choose $M = f(1)/g(1) = \sqrt{2e/\pi}$. Then $f(x)/Mg(x) = \exp\{-(x-1)^2/2\}$, and we have the following procedure (Figure 13.13.2):

Step 1. Generate a random number R_1 in $[0, 1)$. Set $z = -\ln R_1$. (z has the exponential distribution.)

FIGURE 13.13.2 Acceptance–rejection technique for the standard normal variates.

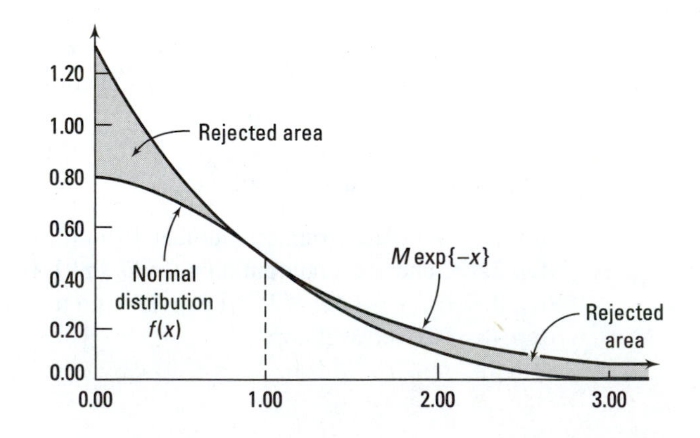

Step 2. Generate a random number R_2 in [0, 1).
Step 3. If $R_2 > \exp\{-(1-z)^2/2\}$, go to step 1. (Reject z)
Step 4. Generate a random number R_3 in [0, 1). If $R_3 < 0.5$, deliver z; else deliver $-z$. (Give a random sign for z.) Stop.

A major drawback of the acceptance–rejection technique is that it can waste a lot of random numbers. One way to improve the technique is to reuse them. As an example, consider the *gamma* distribution. Its probability density function is

$$f(x) = \frac{x^{\alpha-1}\exp(-x)}{\Gamma(\alpha)} \quad \text{for all } 0 \le x < \infty \tag{13.13.1}$$

where the function

$$\Gamma(x) \equiv \int_0^{\infty} t^{x-1}e^{-t}\,dt \quad \text{for all } 0 \le x < \infty$$

is known as the *gamma function* and α is called the *shape parameter* of the distribution. Let $a \equiv \alpha - 1$.

Note that $\Gamma(x+1) = x\Gamma(x)$ for all $x > 0$. Thus, when α is an integer, we have $\Gamma(\alpha) = (\alpha - 1)!$ (Appendix A.17), and this distribution becomes the *Erlang* distribution (8.12.2).

Let $N \equiv \Gamma(\alpha)(e/a)^a$. Then $f(x)$ can be scaled to

$$\begin{aligned} h(x) \equiv Nf(x) &= \left(\frac{xe}{a}\right)^a \exp\{-x\} \\ &= \exp\left\{a\ln\left(\frac{x}{a}\right) + a - x\right\} \qquad \text{for all } 0 \le x < \infty \end{aligned}$$

to avoid evaluating $\Gamma(\alpha)$ and to yield $h(a) = 1$.

In 1988, I proposed a gamma variate generator based on a modification of the acceptance–rejection technique (Minh, 1988). To illustrate the idea, let us restrict our discussion to the generation of a gamma variate within the interval $[a, x_5)$, where $x_5 \equiv a + 2\sqrt{a}$ (Figure 13.13.3).

According to the acceptance–rejection technique, we would generate points uniformly distributed within the rectangular $EGBC$ and then accept only points under the curve $h(x)$. All points within the large area above this curve are rejected.

Now let $x_4 \equiv a + \sqrt{a}$ be the midpoint between a and x_5. The reflection through point J of the curve of $h(x)$ in the interval (a, x_4) is $g(x) \equiv 2h(x_4) - h(2x_4 - x)$ in the interval (x_4, x_5). The area A_1, above the curve $g(x)$ and within the rectangle $DJIC$, is thus identical in shape with the area A_2, under the curve $h(x)$ and within the rectangle $FGHJ$. Also, it can be shown that $g(x) \ge h(x)$ for all $x_4 \le x \le x_5$, or the area A_1 always stays above the curve $h(x)$.

Basically, my technique requires only the generation of points uniformly distributed within the rectangle $EFIC$. Consider now one point in this rectangle having x as its x coordinate.

FIGURE 13.13.3 Generating a gamma variate when $\alpha = 7$.

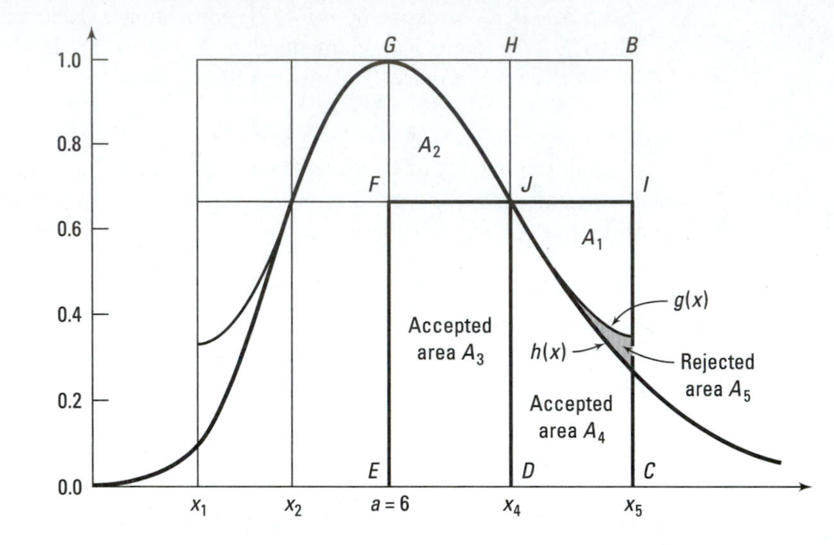

1. If it is within the area A_3, which is the rectangle $EFJD$, we deliver x without having to calculate the value of $h(x)$ for comparison.
2. If it is within the area A_4, which is within the rectangle $DJIC$ and is under the curve $h(x)$, we also deliver x.
3. If it is within the area A_1, which is within the rectangle $DJIC$ and above the curve $g(x)$, we deliver the x coordinate of its reflection through the point J; that is, $x_4 - (x - x_4) = 2x_4 - x$.
4. If it is within the area A_5, which is within the rectangle $DJIC$, above the curve $h(x)$, and under the curve $g(x)$, it is rejected. (This is a small area.)

The same idea is applicable for generating a variate within the interval $[x_1, a)$, where $x_1 \equiv a - 2\sqrt{a} + 1$ for $\alpha > 2$. For a variate within the interval $[0, x_1)$ and the tail $[x_5, \infty)$, the normal acceptance–rejection technique can be used with the exponential distribution as the envelope.

13.14 GENERATING SPECIFIC DISTRIBUTIONS

The inverse transform technique and the acceptance–rejection technique are very general and applicable to many distributions. Therefore, they do not take into consideration the unique features of the random variable under consideration. For a specific distribution, if we explore its unique properties, we might be able to derive a more efficient method to generate its random variates. Here are some examples:

1. As the number of Bernoulli trials until the first success, a *geometric* variate with parameter p can be generated as follows (§1.12):

   Step 1. Set $i = 0$.
   Step 2. Set $i = i + 1$ (updating the number of trials i).

Step 3. Generate a random number R in $[0, 1)$.
Step 4. If $R \geq p$ (a failure), go to step 2 (for another trial); else deliver i.

2. As the number of successes in N Bernoulli trials, a *binomial* variate with parameters (N, p) can be generated as follows (§1.13):

   Step 1. Set $i = 0$ (the number of successes) and $m = 0$ (the number of trials).
   Step 2. Generate a random number R in $[0, 1)$.
   Step 3. If $R < p$ (a success), set $i = i + 1$ (updating the number of successes).
   Step 4. Set $m = m + 1$ (updating the number of trials).
   Step 5. If $m < N$, go to step 2; else deliver i.

3. An *Erlang* variate with parameters (k, λ) can be considered as a sum of k independent exponential variables, each having rate $k\lambda$ (§8.12). Each exponential variable in turn can be generated as $-\ln R_i/(k\lambda)$. Thus, we have the following Erlang generator:

   Step 1. Generate k random numbers $R_1, \ldots, R_k$.
   Step 2. Deliver $-\ln(R_1 \cdots R_k)/(k\lambda)$.

4. Given a sequence of independent and identically distributed random variables $\{X_n\}_{n=0,1,2,\ldots}$, the *central limit theorem* states that, as $k \to \infty$, $\sum_{n=1}^{k} X_n$ approaches the normal distribution with mean $k\mathcal{E}[X_1]$ regardless of the distribution of X_i. Thus, we have the following generator that provides a variate having a distribution close to the standard normal $N[0, 1)$:

   Step 1. Generate 12 random numbers $R_1, \ldots, R_{12}$.
   Step 2. Deliver $R_1 + \cdots + R_{12} - 6$.

13.15 SIMULATING $E_2/E_2/1$ QUEUES

In §13.7, we demonstrated how we can simulate the development of a single server queuing system in which both the interarrival times and service times are uniformly distributed. Now we know how to simulate the behaviors of queuing systems having other distributions. In fact, simulation is so versatile that it can simulate virtually any queuing model. A complicated model would make the programming more tedious but not impossible.

For example, Table 13.15.1 gives the simulation of an $E_2/E_2/1$ queue, in which the interarrival times have the Erlang distribution with parameters (2, .5) and the service times also have the Erlang distribution with parameters (2, .4). The procedure is as follows:

1. Using the Erlang generator in §13.14, the random numbers R_1 and R_2 in columns 2 and 3 give the interarrival times t_{i-1} in column 4. From this, we can calculate the arrival time $T_i = T_{i-1} + t_{i-1}$ of the ith customer in column 5.
2. Using the same generator, the random numbers R_3 and R_4 in columns 7 and 8 give the service time s_i of the ith customer in column 9.
3. Column 6 now gives the waiting time q_i of the ith customer as

$$q_i = \max(q_{i-1} + s_{i-1} - t_{i-1}, 0)$$

Table 13.15.1 Simulating an $E_2/E_2/1$ Queue with $\mathcal{E}[t_1] = .5$ and $\mathcal{E}[s_1] = .4$

1 Customer # i	2 R_1	3 R_2	4 Interarrival Time t_{i-1}	5 Arrival Time T_i	6 Waiting Time q_i	7 R_3	8 R_4	9 Service Time s_i
1				0	0	0.557	0.744	0.176
2	0.478	0.292	0.493	0.493	0	0.517	0.054	0.714
3	0.081	0.829	0.676	1.169	0.038	0.534	0.716	0.192
4	0.971	0.324	0.289	1.458	0	0.206	0.325	0.541
5	0.525	0.535	0.317	1.776	0.224	0.923	0.105	0.467
6	0.215	0.662	0.487	2.263	0.204	0.704	0.266	0.335
7	0.680	0.488	0.276	2.539	0.263	0.399	0.982	0.188
8	0.912	0.117	0.560	3.099	0	0.514	0.200	0.455
9	0.852	0.588	0.173	3.272	0.282	0.677	0.841	0.113
10	0.576	0.749	0.210	3.482	0.184	0.712	0.121	0.490
11	0.361	0.155	0.721	4.204	0	0.332	0.329	0.443
12	0.026	0.565	1.052	5.256	0	0.926	0.576	0.126
13	0.741	0.541	0.229	5.485	0	0.178	0.383	0.537
14	0.238	0.642	0.470	5.954	0.067	0.933	0.208	0.328
15	0.540	0.638	0.267	6.221	0.128	0.014	0.756	0.913
16	0.100	0.840	0.619	6.840	0.422	0.399	0.182	0.524
17	0.509	0.613	0.292	7.132	0.655	0.124	0.307	0.654
18	0.856	0.310	0.332	7.463	0.978	0.728	0.262	0.332
19	0.576	0.139	0.631	8.094	0.678	0.548	0.266	0.385
20	0.607	0.506	0.295	8.389	0.768	0.269	0.954	0.272

From this simulation of 20 customers, the expected waiting time can be estimated as the average of column 6, which is 0.245.

13.16 SIMULATION AND ANALYTICAL METHODS

If simulation is that easy and far more versatile than analytical methods, why don't we just forget what we have been studying in this book and simply resort to simulation? Have we been wasting our time making a long journey in this book only to end up at the same starting point, namely, empirical observations? The answer is an emphatic no. Here are some reasons.

First, simulation has its own drawbacks. It is very detail-oriented and requires highly trained users. It can take a long time to write and debug a simulation program. Also, running *what-if* experiments with simulation can require a prohibitive amount of computer time.

Many simulation languages are available to help make simulation more concise and transparent. Some provide subroutines to facilitate simulation written in a general-purpose language; some offer high-level programming languages. Unfortunately, the more they make the simulation simple, the more they tend to be less versatile and less flexible. Using them often means losing control and not knowing exactly what they do. They are only suitable for less critical applications or for implementing approximate and ephemeral models in the early stages of a project.

Second, simulation is only an *approximation* technique. Each approximation requires *validation* to determine its accuracy and reliability. One way to validate an approximation is to compare it with another approximation. Results obtained by simulation therefore must be compared with other model approximations or analytical approximations.

Queuing Models in Action: Simulation and Analytical Solutions. The Cape Breton Development Corporation operates the coal mines of Cape Breton Island in the province of Nova Scotia. According to Allen et al. (1993),

> [It] had not been achieving the throughput needed from its transport system for moving coal from its coal storage sites to the shipping piers for loading onto ocean going vessels.
>
> . . . We designed the simulation model to help us to understand train movements and how these are affected by the basic tasks of loading, transit, and dumping, by shift changes and other details.
>
> . . . The simulation model proved accurate in duplicating the performance of the system.
>
> . . . Testing all the combinations of parameters of interest explicitly would involve an enormous number of simulation runs.
>
> . . . We needed a simple way of screening out the interesting cases. The simulation model is quite detailed and it allows the user to track the movements of each train. Each model run, however, takes 30 to 60 seconds and approximately another 30 seconds to print the results. We wanted to examine more than 900 combinations of parameters to screen out the interesting ones. We developed a simple queuing model implemented as a macro in the EXCEL spreadsheet package to carry out this screening process.
>
> The model we used is that of a finite population of customers being served by a single server.
>
> . . . Obviously the queuing model is a crude approximation to the real system. It is based on the assumption that both service times and round-trip times are from exponential distributions. This is clearly not true here; we used deterministic times in the simulation. Also, the queuing model ignores the issues of crew scheduling on both the trains and the Pier. However, the throughput of the queuing model is very simple to calculate and its results are consistent with the simulation model.
>
> . . . A combination of modeling approaches proved to be very helpful in producing dramatic improvements in the throughput of the coal-handling system. Each modeling approach made an important contribution. The simulation model showed that we were dealing with the operating details of the system properly. The queuing models made it possible for us to evaluate an enormous number of system configurations so that we could be sure that we were considering good alternatives. ■

Third, in this book, we have presented not only the results but also the probabilistic arguments leading to them. Even if the final and desirable results are not available, those probabilistic arguments should provide valuable insight to the system and could make the simulation of it more efficient and valid.

For example, one of the major problems in simulating a system's limiting behavior is the inability of beginning the system in the steady state. This has inspired many simulators to discard a sufficient number of data before collecting them, hoping that the transience had passed. In most cases, the determination of the number of data to be discarded is quite arbitrary.

For the simulation of a $G/G/1$ queue, with the knowledge presented in this book, we can do better! We should not throw away any customer's waiting times because we know now that the system is regenerative, and the data collected within the first busy cycle are independent with, and as good as, those in other cycles. Statistical techniques have been well developed to analyze these independent blocks of data (Crane and Iglehart, 1974). In Table 13.15.1, the first block is the waiting time q_1 alone, the second block is from q_2 to q_3, and so on.

As another example, note that in heavy traffic (that is, when $\rho \to 1$), the limiting mean and variance of the waiting times in a $G/G/1$ queue increase very rapidly to ∞ (see also §12.8). In this case, the use of a standard simulation technique to estimate W_q would result in a very large confidence interval. This is a serious drawback because queues in heavy traffic are the major source of concern for applied researchers, and the widths of the confidence interval obtained could easily become too large for any practical use.

Using a formula derived in this book, I proposed a very efficient *variance reduction technique* that makes the width of the confidence interval for W_q decrease (instead of increase) as the utilization factor ρ increases. This technique is based on the fact that, as the traffic becomes heavier, the waiting times become larger, but the idle times smaller. In Equation (13.3.1), the term $\mathcal{E}[(i_1)^2]/2\mathcal{E}[i_1]$ degenerates to 0 as $\rho \to 1$. Ignoring this term yields the approximation (13.5.3). Minh and Sorli (1983) suggested that, instead of ignoring it, we should estimate it by simulation. In other words, *let simulation pick up what analytical approximation discards*. Thus, instead of keeping records of the waiting times, we should concentrate on the idle times. The confidence interval for W_q now is the sum of $\mathcal{E}[u^2]/(-2\mathcal{E}[u])$ (which can be calculated analytically) and the confidence interval obtained by simulation for $\mathcal{E}[i^2]/2\mathcal{E}[i]$ (the width of which decreases as $\rho \to 1$). In fact, for the $M/M/1$ queues, this technique can reduce the variance of the estimator by more than 5000 times when $\rho > .88$!

We should not treat simulation and analytical methods as independent processes, but must realize that they are complementary; one can be used to support the other.

13.17 SUMMARY

Following are a few options we can take when the analytical results are not available:

1. **Model approximation:** We introduce additional assumptions to make the model more tractable.
2. **Analytical approximation:** We find an approximation, or the upper/lower bounds of the desired results.
3. **Numerical method:** We derive suitable algorithmic methods to obtain numerical answers for the problem.

4. Simulation: We use computers to run an artificial and controlled experiment to simulate the system behavior. The insights provided by analytical methods should be incorporated into simulation as far as possible.

PROBLEMS

13.1 Consider a $G/G/1$ queuing system. Suppose a customer arrives at an empty system at time $t = 0$. Recall that we denote the first busy period and the first idle period by p_1 and i_1, respectively. Use the regenerative property to explain why

- **a.** $b = \mathcal{E}[p_1]/\mathcal{E}[c_1]$, where b is the long-term time-average of the number of busy servers (§12.5).
- **b.** $\lim_{t\to\infty} \Pr\{\ell(t) = 0\} = \mathcal{E}[i_1]/\mathcal{E}[c_1]$, where $\ell(t)$ is the number of customers in the system at time t.

13.2 Consider an $M/G/1$ queuing system having arrival rate $\lambda = 1/\mathcal{E}[t_1]$. Use Little's result (12.5.4) and the results in Problem 13.1 to show that

- **a.** $\mathcal{E}[p_1] = \mathcal{E}[s_1]/(1 - \lambda\mathcal{E}[s_1])$
- **b.** $\lim_{t\to\infty} \Pr\{\ell(t) = 0\} = 1 - \lambda\mathcal{E}[s_1]$

13.3 Consider an $M/G/1$ queuing system with N-policy in which the server waits until there are N customers in the system before serving (§12.12). Let $p_{N,i}$ be its ith busy period, which starts when the number of customers in the system reaches N. (Thus, $p_{1,i}$ is the busy period of the normal $M/G/1$ queuing system.)

- **a.** Explain why $\mathcal{E}[p_{N,i}] = N\mathcal{E}[p_{1,i}]$ for all $N > 0$ and $i > 0$.
- **b.** Hence, find $\lim_{t\to\infty} \Pr\{\ell(t) = 0\}$.

13.4 Consider an $M/G/1/1$ queuing system.

- **a.** Find the fraction of customers that are lost. *Hint:* Use Problem 8.48.
- **b.** Hence, use Little's results to find $\lim_{t\to\infty} \Pr\{\ell(t) = 0\}$.

13.5 Consider an $M/E_k/1/1$ queuing system. Here, the service time of each customer has k *phases*, and the duration of each is exponentially distributed with rate $k\mu$.

- **a.** Represent the system as a continuous-time Markov chain to find $\lim_{t\to\infty} \Pr\{\ell(t) = 0\}$. *Hint:* Use Problem 11.9.
- **b.** Compare with the result in Problem 13.4.

13.6 Consider an $M/H_2/1/1$ queuing system in which the service time of each customer has a hyperexponential distribution (8.8.5) with $k = 2$.

- **a.** Represent the system as a continuous-time Markov chain to find

$$\lim_{t\to\infty} \Pr\{\ell(t) = 0\}$$

- **b.** Compare with the result in Problem 13.4.

13.7 Consider an $E_k/M/1/1$ queuing system. Here, each interarrival time has k phases, and the duration of each is exponentially distributed with rate $k\lambda$.

- **a.** Represent the system as a continuous-time Markov chain to find

$$\gamma_0 \equiv \lim_{t\to\infty} \Pr\{\ell(t) = 0\}$$

- **b.** When $k = 2$, find γ_0, L, and the fraction of customers lost.

13.8 Consider an $M/E_k/1$ queuing system. Here, the service of each customer has k phases, and the duration of each is exponentially distributed with rate $k\lambda$.

a. Apply the Pollaczek-Khinchin formula (13.4.1) to find L, L_q, W, and W_q.

b. Let $k = 1$; hence, we have an $M/M/1$ queuing system. Find L, L_q, W, and W_q. Compare with Equation (12.8.4).

13.9 Apply the Pollaczek-Khinchin formula (13.4.1) to an $M/D/1$ queuing system (in which the service times are a constant $1/\mu$) to find L, L_q, W, and W_q.

13.10 In an $M/G/1$ queuing system, let d_n be the number of customers in the system immediately after the departure of the nth customer.

a. Find the transition matrix of $\{d_n\}_{n=0,1,2,\ldots}$. Show that it has the form of Problem 7.12a. Compare with Problem 12.4.

b. Use the results in Problems 7.15c and 8.48 to find $\lim_{n\to\infty} \mathcal{E}[d_n]$.

c. From the Pollaczek-Khinchin formula (13.4.1), find L. Show that

$$L = \lim_{n\to\infty} \mathcal{E}[d_n]$$

13.11 Consider an $M/E_k/1$ queuing system. Here, the service time of each customer has k phases, and the duration of each is exponentially distributed with rate $k\mu$. Let $p(t)$ be the number of phases yet to be done for a customer being served at time t, if there is any. Let $r(t)$ be the total number of phases required by all customers in the system at time t; that is, $r(t) = k\ell_q(t) + p(t)$.

a. Show that $\{r(t)\}_{t\in[0,\infty)}$ is a continuous-time Markov chain. Write its balance equations but do not try to solve them.

b. Show how the limiting distribution of $\{\ell(t)\}_{t\in[0,\infty)}$ can be obtained from that of $\{r(t)\}_{t\in[0,\infty)}$.

13.12 We mentioned in §13.8 that the use of $m = 2^k$ makes the *modulo* operation mod(m) more efficient in computers. As we are more familiar with the decimal representation, let us demonstrate this fact by using the linear congruential generator (13.8.1) with $m = 10^2$, $b = 23$, $c = 0$, and $X_0 = 47$. Generate the first three random numbers using this generator and show that the modulo operation is equivalent to saving the rightmost k decimal digits.

13.13 Consider the following generator

Step 1. Let $i = 0$, $c = p(1-p)$, $r = (1-p)^n$, and $F = r$.
Step 2. Generate a random number R in $[0, 1)$.
Step 3. If $R < F$, deliver i, stop.
Step 4. Set $r = c(n-i)r/(i+1)$, $F = F + r$, $i = i + 1$, and go to step 3.

a. What is the distribution of the random variates generated by this generator?

b. What is the method used?

c. What is the expected number of comparisons needed to generate one random variate?

13.14 Use the inverse transform technique to generate random variates having the following distribution:

a. $F(t) = x^n$ for all $0 \le x \le 1$.

b. The *Weibull* distribution with

$$F(x) = 1 - e^{-(x/\alpha)^\beta} \quad \text{for all } t > 0$$

c. The *Cauchy* distribution with

$$F(x) = \frac{1}{\pi} \tan^{-1}\left(\frac{x-\mu}{\sigma}\right) + \frac{1}{2} \quad \text{for all } -\infty < x < \infty$$

d. The *logistic* distribution with

$$F(x) = \frac{1}{1+e^{-x}} \quad \text{for all } -\infty < x < \infty$$

e. The *Pareto* distribution with

$$F(x) = 1 - \left(\frac{k}{x}\right)^a \quad \text{for all } a > 0,\ 0 < k \leq x < \infty$$

f. The *extreme-value* distribution with

$$F(x) = 1 - \exp\left\{-\exp\left\{\frac{\xi - x}{\theta}\right\}^a\right\} \quad \text{for all } -\infty < x < \infty$$

13.15 Let $X_1, X_2, \ldots, X_n$ be n identically distributed random variables with common distribution function $F(x)$. Assume the inverse of $F(x)$ is simple to calculate. Use the inverse transform technique to derive the generators for $U \equiv \max(X_1, X_2, \ldots, X_n)$ and $V \equiv \min(X_1, X_2, \ldots, X_n)$. *Hint:* Use Problem 1.52.

13.16 Consider the gamma distribution (13.13.1). It can be shown that $\Gamma(3/2) = \sqrt{\pi}/2$. Use the exponential distribution $g(x) = \lambda e^{-\lambda x}$ as the envelope to develop a generator for this distribution when $\alpha = 3/2$.

13.17 Consider a Poisson stream with rate λ. This stream has exactly i event occurences during $(0, 1]$ if $T_i \leq 1 < T_{i+1}$ (Figure 8.11.1).

a. Use this property to derive an algorithm to generate a Poisson variate.

b. On the average, how many random numbers are required to generate one Poisson variate by this generator?

13.18 (Marsaglia, 1964) Use $g(x) = x \exp\{-\left(x^2 - d^2\right)/2\}$ as an envelope to derive a generator for a variate X from the *tail* of the standard normal distribution; that is, X has the standard normal distribution and $X \geq d$. *Hint:* Use Problem 8.4.

13.19 Consider a queuing system in which the interarrival times of the customers have the Erlang distribution with parameters (3, 0.4 hour); their service times have the triangle distribution $f(x) = 2 - 2x$ for $0 \leq x \leq 1$. Find W_q and L_q

a. by using the Kingman heavy traffic approximation

b. by using the Marchal approximation

c. by using the M/G/1 model

d. by simulating the waiting times of 20 customers

APPENDIX A

Some Useful Formulas

$$\lim_{x\to\infty}\left(1+\frac{a}{x}\right)^x = e^a \quad \text{for all } -\infty < x < \infty \tag{A.1}$$

$$\sum_{i=0}^{\infty}\frac{x^i}{i!} = e^x \quad \text{for all } -\infty < x < \infty \tag{A.2}$$

$$\sum_{i=0}^{\infty}\frac{x^i}{(i+1)!} = \frac{e^x-1}{x} \quad \text{for all } -\infty < x < \infty \tag{A.3}$$

$$\sum_{i=1}^{\infty}\frac{x^{(2i-1)}}{(2i-1)!} = \frac{e^x-e^{-x}}{2} \quad \text{for all } -\infty < x < \infty \tag{A.4}$$

$$\sum_{i=1}^{\infty}\frac{x^i}{i} = \ln\left(\frac{1}{1-x}\right) \quad \text{for all } |x| < 1 \tag{A.5}$$

$$\sum_{i=0}^{k}x^i = \frac{1-x^{k+1}}{1-x} \quad \text{for all } x \neq 1 \tag{A.6}$$

$$\sum_{i=1}^{k}x^i = \frac{x-x^{k+1}}{1-x} \quad \text{for all } x \neq 1 \tag{A.7}$$

$$\sum_{i=0}^{\infty}x^i = \frac{1}{1-x} \quad \text{for all } |x| < 1 \tag{A.8}$$

$$\sum_{i=1}^{k}ix^i = x\frac{(1-x^k)-kx^k(1-x)}{(1-x)^2} \quad \text{for all } x \neq 1 \tag{A.9}$$

$$\sum_{i=1}^{\infty}ix^i = \frac{x}{(1-x)^2} \quad \text{for all } |x| < 1 \tag{A.10}$$

$$\sum_{i=1}^{\infty} i^2 x^i = \frac{x(1+x)}{(1-x)^3} \quad \text{for all } |x| < 1 \tag{A.11}$$

$$\sum_{i=1}^{\infty} i(i+1)x^i = \frac{2x}{(1-x)^3} \quad \text{for all } |x| < 1 \tag{A.12}$$

$$\sum_{i=0}^{\infty} \frac{(i+n)!}{i!} x^i = \frac{n!}{(1-x)^{n+1}} \quad \text{for all } |x| < 1 \text{ and } n \geq 0 \tag{A.13}$$

$$\sum_{i=0}^{\infty} \binom{N+i-1}{i} x^{-i} = \left(\frac{x}{x-1}\right)^N \quad \text{for all } |x| < 1 \tag{A.14}$$

$$\int t e^{-\lambda t}\, dt = -\frac{1+t\lambda}{\lambda^2} e^{-\lambda t} \tag{A.15}$$

$$\int t^2 e^{-\lambda t}\, dt = -\left[\frac{t^2}{\lambda} + \frac{2t}{\lambda^2} + \frac{2}{\lambda^3}\right] e^{-\lambda t} \tag{A.16}$$

$$\int_0^{\infty} t^n e^{\lambda t}\, dt = \frac{n!}{\lambda^{n+1}} \quad \text{for } n = 1, 2, 3, \ldots \tag{A.17}$$

$$\begin{pmatrix} a & b \\ c & d \end{pmatrix}^{-1} = \frac{1}{(ad-bc)} \begin{pmatrix} d & -b \\ -c & a \end{pmatrix} \tag{A.18}$$

$$\begin{pmatrix} 1 & -s_1 & 0 & 0 \\ 0 & 1 & -s_2 & 0 \\ 0 & 0 & 1 & -s_3 \\ 0 & 0 & 0 & 1 \end{pmatrix}^{-1} = \begin{pmatrix} 1 & s_1 & s_1 s_2 & s_1 s_2 s_3 \\ 0 & 1 & s_2 & s_2 s_3 \\ 0 & 0 & 1 & s_3 \\ 0 & 0 & 0 & 1 \end{pmatrix} \tag{A.19}$$

$$\begin{aligned} &\begin{pmatrix} s_0 & -s_0 & 0 & 0 \\ -1+s_1 & 1 & -s_1 & 0 \\ -1+s_2 & 0 & 1 & -s_2 \\ -1+s_3 & 0 & 0 & 1 \end{pmatrix}^{-1} \\ &= \frac{1}{s_0 s_1 s_2 s_3} \begin{pmatrix} 1 & s_0 & s_0 s_1 & s_0 s_1 s_2 \\ 1 - s_1 s_2 s_3 & s_0 & s_0 s_1 & s_0 s_1 s_2 \\ 1 - s_2 s_3 & (1 - s_2 s_3) s_0 & s_0 s_1 & s_0 s_1 s_2 \\ 1 - s_3 & (1 - s_3) s_0 & (1 - s_3) s_0 s_1 & s_0 s_1 s_2 \end{pmatrix} \end{aligned} \tag{A.20}$$

REFERENCES

Aguiar, R. J.; Collares-Pereira, M.; and Conde, J. P. (1988) "Simple procedure for generating sequences of daily radiation values using a library of Markov transition matrices," *Solar Energy*, 40, 269–279.

Allen, G.; Gunn, E.; and Rutherford, P. (1993) "Improving throughput of a coal transport system with the aid of three simple models," *Interfaces*, 23, 88–103.

Anderson, R. B. W. (1974) "A Markov chains model of medical specialty choice," *Journal of Mathematical Sociology*, 3, 259–274.

Azzalini, A. and Bowman, A. W. (1990) "A look at some data on the Old Faithful geyser," *Applied Statistics*, 39, 357–365.

Badger, G. J. and Vacek, P. M. (1987) "A Markov model for a clinical episode of recurrent genital herpes," *Biometrics*, 43, 399–408.

Bertram, B. C. R. (1980) "Vigilance and group size in ostriches," *Animal Behaviour*, 28, 278–286.

Beshers, J. M. and Laumann, E. O. (1967) "Social distance: A network approach," *American Sociological Review*, 32, 225–236.

Bessent, E. W. and Bessent, A. M. (1980) "Student flow in a university department: Results of a Markov analysis," *Interfaces*, 10, 52–59.

Bourne, L. S. (1976) "Monitoring change and evaluating the impact of planning policy on urban structure: A Markov chain experiment," *Plan Canada*, 16, 5–14.

Bowler, S. (1990) "When will a big quake hit eastern US?" *New Scientist*, 29 September, 26.

Burke, P. J. (1976) "Proof of a conjecture on the interarrival-time distribution in an $M/M/1$ queue with feedback," *IEEE Transactions on Communications*, COM-24, 575–576.

Butler, J. P.; Spratt, D. I.; O'Dea, L. St. L.; and Crowley, W. F., Jr. (1986) "Interpulse interval sequence of LH in normal men essentially constitutes a renewal process," *American Journal of Physiology*, 250 (Endocrinol. Metab. 13), E338–E340.

Chin, E. H. (1977) "Modeling daily precipitation occurrence process with Markov chain," *Water Resources Research*, 13, 949–956.

Cotton, M. M. and Wood, W. P. (1984) "On probability models and avoidance learning," *Psychological Reports*, 55, 387–400.

Crane, M. A. and Iglehart, D. L. (1974) "Simulating stable stochastic systems, I: General multiserver queues," *Journal of the Association for Computing Machinery*, 21, 103–113.

Cyert, R. M.; Davidson, H. J.; and Thompson, G. L. (1962) "Estimation of the allowance for doubtful accounts by Markov chains," *Management Science*, 8, 287–303.

Daykin, D. E.; Jeacocke, J. E.; and Neal, D. G. (1967) "Markov chains and Snakes and Ladders," *The Mathematical Gazette*, 51, 313–317.

Deming, W. E. and Glasser, G. J. (1988) "A Markovian analysis of the life of newspaper subscriptions," *Management Science*, 14, B283–B293.

Drachman, D. (1981) "A residential continuum for the chronically mentally ill. A Markov probability model," *Evaluation and the Health Professions*, 4, 93–104.

Dryden, M. M. (1969) "Share price movements: A Markovian approach," *The Journal of Finance*, 24, 49–60.

Eilon, S. (1969) "A simpler proof of $L = \lambda W$," *Operations Research*, 17, 915–917.

Figlio, R. M. (1981) "Delinquency careers as a simple Markov process," in *Models in Quantitative Criminology* (Ed.: Fox, J. A.), New York: Academic Press, 25–37.

Foster, F. G. (1952) "A Markov chain derivation of discrete distributions," *Annals of Mathematical Statistics*, 23, 624–627.

Freed, J. R.; Marcus, M.; and Forsythe, A. B. (1979) "A Markovian model for evaluating dental care programs," *Community Dentistry Oral Epidemiology*, 7, 25–29.

Fuguitt, G. V. (1965) "The growth and decline of small towns as a probability process," *American Sociological Review*, 30, 403–411.

Garg, S. K.; Marshall, G.; Chase, H. P.; Jackson, W. E.; Archer, P.; and Crews, M. J. (1990) "The use of the Markov process in describing the natural course of diabetic retinopathy," *Archives Ophthalmology*, 108, 1245–1247.

Girard, D. M. and Sager, D. B. (1987) "The use of Markov chains to detect subtle variation in reproductive cycling," *Biometrics*, 43, 225–234.

Goldenberg, D. H. (1988) "Trading frictions and futures price movements," *Journal of Financial and Quantitative Analysis*, 23, 465–481.

Harary, F. and Lipstein, B. (1962) "The dynamics of brand loyalty: A Markovian approach," *Operations Research*, 10, 19–40.

Holgate, P. (1967) "The size of elephant herds," *The Mathematical Gazette*, 51, 302–304.

Holling, C. S. (1966) "The functional response of invertebrate predators to prey density," *Memoirs of Entomological Society of Canada*, 32, 22–32.

Hopkins, A. (1985) "Mathematical models of patterns of seizures," *Archives of Neurology*, 42, 463–467.

Houweling, H. W. and Kuné, J. B. (1984) "Do outbreaks of war follow a Poisson-process?" *Journal of Conflict Resolution*, 28, 51–61.

Hutchinson, M. F. (1990) "A point rainfall model based on a three-state continuous Markov occurrence process," *Journal of Hydrology*, 114, 125–148.

Ioannou, P. G. (1987) "Geologic prediction model for tunneling," *Journal of Construction Engineering and Management*, 113, 569–590.

Jackson, J. R. (1957) "Networks of waiting lines," *Operations Research*, 5, 518–521.

Jaggi, B. and Lau, H. S. (1974) "Toward a model for human resource valuation," *The Accounting Review*, 49, 321–329.

Jain, S. (1986) "Markov chain model and its application," *Computers and Biomedical Research,* 19, 374–378.

Kao, E. P. C. (1974a) "Modeling the movement of coronary patients within a hospital by semi-Markov processes," *Operations Research*, 22, 683–699.

Kao, E. P. C. (1974b) "A note on the first two moments of times in transient states in a semi-Markov process," *Journal of Applied Probability*, 11, 193–198.

Kendall, D. G. (1951) "Some problems in the theory of queues," *Journal of the Royal Statistical Society*, B13, 151–185.

Kulatilake, P. H. S. W. (1987) "Modelling of cyclical stratigraphy using Markov chains," *International Journal of Mining and Geological Engineering,* 5, 121–130.

Larson, R. C. (1987) "Perspectives on queues: Social justice and the psychology of queueing," *Operations Research*, 35, 895–905.

Leviton, A.; Schulman, J.; Kammerman, L.; Porter, D.; Slack, W.; and Graham, J. R. (1980) "A probability model of headache recurrence," *Journal of Chronic Diseases*, 33, 407–412.

Lichtenberg, J. W. and Hummel, T. J. (1976) "Counseling as stochastic process: Fitting a Markov chain model to initial counseling interviews," *Journal of Counseling Psychology*, 23, 310–315.

Lieberson, S. and Fuguitt, G. V. (1967) "Negro–white occupational differences in the absence of discrimination," *American Journal of Sociology*, 73, 188–200.

Liu, H. (1993) "On Cowan and Mecke's Markov chain," *Journal of Applied Probability*, 56, 554–560.

MacKinnon, W. J. (1966) "Elements of the SPAN technique for making group decisions," *The Journal of Social Psychology*, 70, 149–164.

Madsen, H.; Spliid, H.; and Thyregod, P. (1985) "Markov models in discrete and continuous time for hourly observations of cloud cover," *Journal of Climate and Applied Meteorology*, 24, 629–639.

Marchal, W. G. (1976) "An approximate formula for waiting time in single server queues," *AIIE Transactions*, 8, 473–474.

Marsaglia, G. (1964) "Generating a variable from the tail of the normal distribution," *Technometrics*, 6, 101–102.

McConway, K. J. (1982) "Simpler calculations for the SPAN technique," *The Journal of General Psychology*, 107, 91–98.

Meredith, J. (1974) "Program evaluation in a hospital for mentally retarded persons," *American Journal of Mental Deficiency*, 78, 471–481.

Midlarsky, M. I. (1981) "Equilibria in the nineteenth-century balance-of-power system," *American Journal of Political Science*, 25, 270–296.

Midlarsky, M. I. (1983) "Absence of memory in the nineteenth-century alliance system: Perspectives from queuing theory and bivariate probability distributions," *American Journal of Political Science*, 27, 762–784.

Minh, D. L. (1977) "A discrete time, single server queue from a finite population," *Management Science*, 23, 756–767.

Minh, D. L. (1980) "Analysis of the exceptional queueing system by the use of regenerative processes and analytical methods," *Mathematics of Operations Research*, 5, 147–159.

Minh, D. L. (1988) "Generating gamma variates," *ACM Transactions on Mathematical Software*, 14, 261–266.

Minh, D. L. and Sorli, R. M. (1983) "Simulating the $GI/G/1$ queue in heavy traffic," *Operations Research*, 31, 966–971.

Moore, A. D. (1990) "The semi-Markov process: A useful tool in the analysis of vegetation dynamics for management," *Journal of Environmental Management*, 30, 111–130.

Nagamatsu, A.; Koyanagi, T.; Hirose, K.; Nakahara, H.; and Nakano, H. (1988) "An application of the Markov process for quantitative prediction of labor progress," *Journal of Perinatal Medicine*, 16, 333–337.

Neuts, M. F. (1981) *Matrix-Geometric Solutions in Stochastic Models*, Baltimore and London: Johns Hopkins University Press, pp. ix–x.

Park, S. K. and Miller, K. W. (1988) "Random number generators: Good ones are hard to find," *Communications of the ACM*, 31, 1192–1201.

Parker, K. C. H. (1988) "Speaking turns in small group interaction: A context-sensitive event sequence model," *Journal of Personality and Social Psychology*, 54, 965–971.

Pavlov, I. P. (1927) *Conditional Reflexes*, London: Oxford University Press.

Perrin, E. B. and Sheps, M. C. (1964) "Human reproduction: A stochastic process," *Biometrics*, 20, 28–45.

Quinn, P.; Andrews, B.; and Parsons, H. (1991) "Allocating telecommunications resources at L. L. Bean, Inc.," *Interfaces*, 21, 75–91.

Rao, C. R. and Kshirsagar, A. M. (1978) "A semi-Markovian model for predator–prey interactions," *Biometrics*, 34, 611–619.

Resnick, S. I. (1992a) *Adventures in Stochastic Processes*, Boston: Birkhauser, p. 138.

Resnick, S. I. (1992b) *Adventures in Stochastic Processes*, Boston: Birkhauser, p. 406.

Richardson, L. F. (1944) "The distribution of wars in time," *Journal of the Royal Statistical Society*, 107, 242–250.

Robbins, H. (1952) "Some aspects of the sequential design of experiments," *Bulletin of the American Mathematical Society*, 58, 527–535.

Robinson, V. B. (1980) "On the use of Markovian equilibrium distributions for land use policy evaluation," *Socio-Economic Planning Sciences*, 14, 85–89.

Roldan-Cañas, J.; Garcia-Guzman, A.; and Losada-Villasante, A. (1982) "A stochastic model for wind occurrence," *Journal of Applied Meteorology*, 21, 740–744.

Saffer, S. I.; Mize, C. E.; Bhat, U. N.; and Szygenda, S. A. (1976) "Use of non-linear programming and stochastic modeling in the medical evaluation of normal–abnormal liver function," *IEEE Transactions on Biomedical Engineering*, BME-23, 200–207.

Schach, E. and Schach, S. (1972) "A continuous time stochastic model for the utilization of health services," *Socio-Economic Planning Sciences*, 6, 263–272.

Segal, E. S. (1985) "Projections of internal migration in Malawi: Implications for development," *The Journal of Modern African Studies*, 23, 315–329.

Shanthikumar, J. G. and Buzacott, J. A. (1980) "On the approximations to the single server queue," *International Journal of Production Research*, 18, 761–773.

Sheps, M. C. and Menken, J. A. (1973) *Mathematical Models of Conception and Birth*, Chicago: The University of Chicago Press, p. 282.

Sheynin, O. B. (1988) "A. A. Markov's work on probability," *Archive for History of Exact Science*, 39, 337–377.

Singer, B. and Spilerman, S. (1976) "The representation of social processes by Markov models," *American Journal of Sociology*, 82, 1–54.

Smith, D. R. and Smith, W. R. (1984) "Patterns of delinquent careers: An assessment of three perspectives," *Social Science Research*, 13, 129–158.

Styan, G. P. H. and Smith, H., Jr. (1964) "Markov chains applied to marketing," *Journal of Marketing Research*, 1, 50–55.

Suri, R.; Diehl, G. W. W.; Treville, S. D.; and Tomsicek, M. J. (1995) "From CAN-Q to MPX: Evolution of queuing software for manufacturing," *Interfaces*, 25, 128–150.

Taylor, R. G., Jr. and Reid, W. M. (1987) "Forecasting the salaries of professional personnel: An application of Markov analysis to school finance," *Journal of Education Finance*, 12, 421–428.

Thompson, R. A. (1970) "Stochastics and structure: Cultural change and social mobility in a Yucatec town," *Southwestern Journal of Anthropology*, 26, 354–374.

Thompson, W. W., Jr. and McNeal, J. U. (1967) "Sales planning and control using absorbing Markov chains," *Journal of Marketing Research*, 4, 62–66.

Ulmer, S. S. (1982) "Supreme Court appointments as a Poisson distribution," *American Journal of Political Science*, 26, 113–116.

Weiss, G. H. (1994) *Aspects and Applications of the Random Walk*, Amsterdam, Netherlands: North Holland, p. 269.

Whitaker, D. (1978) "The derivation of a measure of brand loyalty using a Markov brand switching model," *Journal of Operational Research*, 29, 959–970.

Yang, M. C. K. and Hursch, C. J. (1973) "The use of a semi-Markov model for describing sleep patterns," *Biometrics*, 29, 667–676.

INDEX